Big Data Analytics Using Multiple Criteria Decision-Making Models

T0144467

The Operations Research Series

Series Editor: A. Ravi Ravindran
Professor, Department of Industrial and Manufacturing Engineering
The Pennsylvania State University, University Park, PA

Published Titles

Big Data Analytics Using Multiple Criteria Decision-Making Models

Edited by
Ramakrishnan Ramanathan,
Muthu Mathirajan, and A. Ravi Ravindran

CRC Press
Taylor & Francis Group
Boca Raton London New York

CRC Press is an imprint of the
Taylor & Francis Group, an **informa** business

CRC Press
Taylor & Francis Group
6000 Broken Sound Parkway NW, Suite 300
Boca Raton, FL 33487-2742

© 2017 by Taylor & Francis Group, LLC
CRC Press is an imprint of Taylor & Francis Group, an Informa business

International Standard Book Number-13: 978-1-4987-5355-5 (Hardback)
978-1-1387-4765-4 (Paperback)

Library of Congress Cataloging-in-Publication Data

Names: Ramanathan, R., 1966- editor. | Mathirajan, M., editor. | Ravindran, A., 1944- editor.
Title: Big data analytics using multiple criteria decision-making models / edited by Ramakrishnan Ramanathan, Muthu Mathirajan, A. Ravi Ravindran.
Description: Boca Raton : Taylor & Francis, CRC Press, 2017. | Series: The operations research series | Includes bibliographical references and index.
Identifiers: LCCN 2016056409| ISBN 9781498753555 (hardback : alk. paper) | ISBN 9781498753753 (ebook)
Subjects: LCSH: Big data. | Multiple criteria decision making. | Business logistics--Decision making.
Classification: LCC QA76.9.B45 B5535 2017 | DDC 005.7--dc23
LC record available at https://lccn.loc.gov/2016056409

Visit the Taylor & Francis Web site at
http://www.taylorandfrancis.com

and the CRC Press Web site at
http://www.crcpress.com

Contents

Preface

The idea to produce this book originated in an International Symposium (http://www.ravisymposium.org/) organized in honor of Professor Ravi Ravindran in Bangalore during 12th–13th March 2015. The symposium was organized to commemorate 70th birthday of Professor Ravindran by his former doctoral students. The aim of the symposium was to create a platform and facilitate knowledge sharing on the applications of operations research (OR) in multiple criteria decision making, analytics, healthcare delivery systems, supply chain engineering, and project management. At the valedictory session, various themes for a book were brainstormed and it was decided to pursue the first two themes of the symposium, namely multiple criteria decision-making and analytics, in developing a book that would fittingly honor the extraordinary achievements of Professor Ravindran for a long time. It was also decided that the book will include chapters from Professor Ravindran's legacy—from his PhD students who are very successful academics/industrialists—in various parts of the world.

Multicriteria decision making (MCDM) is a subfield of operational research and is one of the so-called decision-making tools. A decision-making problem is characterized by the need to choose one or a few from among a number of alternatives. A good decision-making process should not only improve the clarity of the problem to the decision maker, but it should also shed new light into the problem by generating newer alternatives. The field of MCDM has been succinctly defined in the literature as making decisions in the face of multiple conflicting objectives. Chapter 2 of this volume provides a detailed description of the various MCDM models and also a comparative perspective on these models.

The field of MCDM assumes special importance in this era of big data and business analytics (BA). In the modern digital world, a wealth of so-called big data is being generated every minute, every second, even every nanosecond. Thanks to the astounding technological revolution, everything around us is being captured in someway or the other, stored in some form, and it is believed that this has the potential to make better business decisions. BA involves an appropriate use of analytic tools on big data to provide new predictive/prescriptive/descriptive insights that will allow businesses perform better. Since big data and BA are relatively recent phenomena, studies on understanding the power of big data and BA are rare with a few studies being reported in the literature. Chapter 3 of this volume is dedicated to address the basics of big data and BA. BA involves both modeling-based tools and statistics-based tools. The modeling-based tools involve use of operational research models. In this volume, the focus is primarily on modeling-based tools for BA, with exclusive focus on the subfield of MCDM within the domain of operational research.

We believe that the two themes of the book, MCDM and big data, address a very valuable research gap. While there are several textbooks and research materials in the field of MCDM, there is no book that discusses MCDM in the context of emerging big data. Thus, the present volume addresses the knowledge gap on the paucity of MCDM models in the context of big data and BA.

There was an instant response from Professor Ravindran's students and colleagues for the call for contributions of the book. A total of 15 chapters were considered in the first round of review. Though all of them were of good quality, after careful review and evaluation for the fit for the theme of the book, it was decided to include 13 chapters in this volume. At least five of these chapters have been authored by students and close associates of Professor Ravindran. There are contributions from authors based in the United States (5 chapters), from the United Kingdom (2), and from India (6).

This volume starts with a fitting Festschrift in Honor of Professor Ravindran by Professor Adedeji B. Badiru. The rest of the volume is broadly divided into three sections. The first section, consisting of Chapters 2 and 3, is intended to provide the basics of MCDM and big data analytics. The next section, comprising Chapters 4 through 10, discusses applications of traditional MCDM methods. The last section, comprising the final three chapters, discusses the application of more sophisticated MCDM methods, namely, data envelopment analysis (DEA) and the analytic hierarchy process.

Due to the topical nature of the theme of big data, it has been a challenge to ensure that the contributions of this volume, from traditional MCDM researchers, had adequate treatment of big data. We believe that the chapters of this book illustrate how MCDM methods can be fruitfully employed in exploiting big data, and will kindle further research avenues in this exciting new field. We also believe that the book will serve as a reference for MCDM methods, big data, and linked applications.

Ramakrishnan Ramanathan
Muthu Mathirajan
A. Ravi Ravindran

Editors

Ramakrishnan Ramanathan is the Director of Business and Management Research Institute, in the Business School of the University of Bedfordshire, Luton, the United Kingdom. In the past, he has worked and taught in a number of countries, including the United Kingdom, Finland, the Netherlands, Oman, and India. He has taught basic and advanced courses on operations management, production systems management, supply chain management, optimization theory, DEA, management science, business statistics, simulation, energy and environment, energy and environmental economics, energy and transport economics, and others. His research interests include operations management, supply chains, environmental sustainability, economic and policy analysis of issues in the energy, environment, transport, and other infrastructure sectors. He works extensively on modeling using techniques such as optimization, decision analysis, DEA, and the analytic hierarchy process.

Ram has successfully completed a number of research projects across the world. He is on the editorial boards of several journals and in the technical/advisory committees of several international conferences in his field. He is an advisory board member of an innovative new online resource, *The Oxford Research Encyclopedia of Business and Management*. He is a member of ESRC Peer Review College in the United Kingdom. He has produced four books (including an introductory textbook on DEA), more than 119 research publications in journals, and more than 141 conference presentations. His research articles have appeared in many prestigious internationally refereed journals, including *Omega, Tourism Economics, International Journal of Production Economics, Supply Chain Management, International Journal of Operations & Production Management, European Journal of Operational Research, Transport Policy,* and *Transportation Research*.

More details about Professor Ramanathan are available at the following links:

Profile: http://www.beds.ac.uk/research-ref/bmri/centres/bisc/people/ram-ramanathan

http://www.beds.ac.uk/howtoapply/departments/business/staff/prof-ramakrishnan-ramanathan

LinkedIn: http://www.linkedin.com/pub/ramakrishnan-ramanathan/12/50a/204

ResearcherID: http://www.researcherid.com/rid/H-5206-2012

Google Scholar: http://scholar.google.co.uk/citations?user=1CBQZA8AAAAJ

Contributors

Camelia Al-Najjar
Case Western Reserve University
Cleveland, Ohio

Abdulaziz Altowijri
Case Western Reserve University
Cleveland, Ohio

Vivek Ananthakrishnan
Pennsylvania State University
University Park, Pennsylvania

P. Y. Yeshwanth Babu
Latentview Analytics
Chennai, Tamil Nadu, India

A. B. Badiru
Department of Systems Engineering
    and Management
Air Force Institute of Technology
Dayton, Ohio

Sankaralingam Ganesh
Latentview Analytics
Chennai, Tamil Nadu, India

Raghav Goyal
Pennsylvania State University
University Park, Pennsylvania

N. Srinivasa Gupta
Manufacturing Division
School of Mechanical Engineering
VIT University
Vellore, Tamil Nadu, India

Pulipaka Kiranmayi
Indian Institute of Science
Bengaluru, Karnataka, India

Mohammad Komaki
Department of Electrical Engineering
    and Computer Science
Case Western Reserve University
Cleveland, Ohio

U. Dinesh Kumar
Indian Institute of Management
    Bangalore
Bengaluru, Karnataka, India

Rainer Leisten
Department of Mechanical and
    Process Engineering
University of Duisburg-Essen
Duisburg, Germany

Sakthivel Madankumar
Department of Management Studies
Indian Institute of Technology
    Madras
Chennai, Tamil Nadu, India

Behnam Malakooti
Department of Electrical Engineering
    and Computer Science
Case Western Reserve University
Cleveland, Ohio

Muthu Mathirajan
Department of Management Studies
Indian Institute of Science
Bengaluru, Karnataka, India

Pusapati Navya
Department of Management Studies
Indian Institute of Technology
    Madras
Chennai, Tamil Nadu, India

and journal publications. He got us engaged in funded projects to expose us to the world of pursuing funded projects, writing proposals, and executing projects. He would come to our offices and inquired what we were working on and whether we were aware of some latest opportunity out there for a funded project or a journal publication. At that time, we did not see the value of his inquisitive ways. It was later that we realized how much his gentle prying would put on a solid platform of becoming successful tenured professors. In fact, we, the assistant professors, often joked among ourselves that if we see Ravi coming down the hallway, we would go the other way because each time you meet him, he would have a new idea of something new and worthwhile for us to be doing.

Ravi's altruistic disposition was evident in the fact that he offered me a job at all. At that time, I was still on a student visa and not many departments were eager to make academic appointments without an existing "green card" or an official work permit. Ravi was among the handful of department heads willing to go out on a limb to offer jobs to inexperienced foreign students. Ravi took a chance with my offer and I am greatly appreciative of that opportunity. Even today, whenever I strive to achieve an even loftier goal, I, subconsciously, credit the endeavor to a justification of the incipient opportunity that Ravi gave me so long ago in 1984.

1.2 The Tinker Projects

The multicriteria leadership of Professor Ravi Ravindran led to many exciting times in the Industrial Engineering Department at the University of Oklahoma. The most notable of these were the series of projects we did under contract for Tinker Air Force Base in Oklahoma City. Ravi brought me into his Tinker research team in 1985, my first year at the University. The project won us international acclaim and many awards and helped launch our academic careers on a positive trajectory. The primary publications that emanated from the projects are the six references cited in this contribution (Ravindran et al., 1988, 1989; Foote et al., 1988, 1992; Leemis et al., 1990; Badiru et al., 1993). Several other publications followed these six primary articles. It is noteworthy that Ravindran et al. (1989) was recognized as one of the 20 best papers of the decade published in *TIMS Interfaces* (1980–1990). The "Tinker Projects," as they were affectionately called, culminated in the team's international recognition with the 1988 Finalist Achievement Award for the Franz Edelman Management Science Award from The Institute of Management Sciences. Thus, the team's accomplishment is permanently enshrined in the annals of the award winners and is still recognized annually until today. Figure 1.1 shows an image of the Edelman Laureate Ribbon presented to me at the INFORMS conference in Philadelphia, Pennsylvania

FIGURE 1.1
Edelman Laureate Ribbon recognizing Ravi Ravindran's team accomplishment with the Tinker Air Force Base Project in 1988.

in November 2015. So, even after 30 years, the Tinker Projects continue to bear intellectual fruits.

1.3 Reprint of a Tinker Project Article*

This section contains a reprint of one of the seminal journal articles published on the Tinker Project.

Abstract

We developed a large simulation model to aid reconstruction efforts after a disastrous fire at Tinker Air Force Base (TAFB). The model, developed in SLAM, facilitated the analysis and efficient design of the modular repair center (cellular type) layout that replaced the precious machine-based layout in the engine overhaul facility. It has been used extensively to determining the appropriate number of machines to place within the repair center, the stacker capacity for in-process inventory, the location of elevators for sending parts to the conveyor, and the optimal design and routing scheme for the overhead conveyor system. The new layout, as predicted by the simulation model, has proven to be quite effective. The new design has decreased material handling by 50 to 80 percent, decreased flow times, allowed better management control of part transfers, saved $4.3 million from the elimination of excess machine capacity, and saved $1.8 million from higher direct labor efficiency.

Tinker Air Force Base (TAFB), located in Oklahoma City, Oklahoma, is one of the five overhaul bases in the Air Force Logistics Command. It overhauls and repairs six types of jet engines and various aircraft and engine

---

* Reprinted verbatim with Permission from: Ravindran, A.; Foote, B. L.; Badiru, Adedeji B.; Leemis, L. M.; and Williams, Larry (1989), "An Application of Simulation and Network Analysis to Capacity Planning and Material Handling Systems at Tinker Air Force Base," *TIMS Interfaces*, Vol. 19, No. 1, Jan.–Feb., 1989, pp. 102–115.

accessories, and it manages selected Air Force assets worldwide. This engine overhaul facility is responsible for logistical support for a series of Air Force engines. Engines are returned from service activities for periodic overhaul or to complete a modification or upgrade. The engine is disassembled, and each part is inspected for wear and possible repair. Individual parts are repaired or modified to a like-new condition or are condemned and replaced with a new part. The majority of the parts are overhauled and returned to service for a fraction of the cost of a new part. A major overhaul may cost less than five percent of the cost of a new engine in terms of labor, material, and replaced parts, Between November 11 and 14, 1984, a fire devastated Building 3001, which contained the Propulsion (Engine) Division in the Directorate of Maintenance. The division consists of over 2800 employees and produces over 10 million earned hours to support Department of Defense overhaul requirements each year. In February 1985, the Air Force published a statement of work requesting assistance from industry to model and develop a simulation of the engine overhaul process to assist in the redesign and layout of approximately 900,000 square feet of production floor space. Three commercial firms attended an onsite prebid conference to learn the scope of the project, the nature of the data the Air Force could provide, and the time frame in which a finished product had to be delivered. The model was expected to predict the number and type of machines, the personnel, the queuing space required, the material-handling distribution, and the volume between and within organizations. The facility engineers needed various management reports to help them to lay out the plant. The project was to be completed within 120 days. Of the three firms, one elected not to respond, the second bid $225,000 with the first report in nine months, and the third quoted $165,000 to study the problem with an expected projected cost of over $300,000. Each was a highly reputable organization with considerable expertise and success in the field. Since TAFB had budgeted only $80,000 and time was running out, TAFB contacted the University of Oklahoma. It had not been considered earlier because of conflicts with class schedules. As the month of May approached, the university became a potential vendor. A contract was let on May 1, 1985, and the first product was delivered by June 15, 1985.

Project Scope

The scope of the project was to take advantage of the disaster and forge a state-of-the-art facility for overhauling engines with the most efficient and cost-effective organizational structure and physical layout. The relocation team was charged with developing and implementing a total change in the philosophy of engine overhaul that would maximize flexibility while minimizing facility and plant equipment costs. Of equal importance was the task of developing a means to predict and forecast resource requirements as work load mixes changed.

The eight-member relocation team comprised midlevel managers from engineering and production with detailed knowledge of the inner working of the facility plus four faculty members and four graduate students from the University of Oklahoma School of Industrial Engineering. The Corps of Engineers was to construct the building based on specifications provided by the internal engineering department. Mechanical and industrial engineers designed shop layouts. The university provided the skill and knowledge to develop a capacity-planning and material-handling simulation model using data provided by TAFB. The university team simulated repair activities to a level of detail never attempted or realized before. Its responsibility was to analyze the data available from TAFB and determine what, if any, additional specific data elements were required, to assist in developing techniques to obtain that data from existing systems, to check the data for outliers, and to develop and implement interface programs to obtain data for the simulation model.

The baseline data base consisted of 117 fields with over 2500 records used to describe the requirements of the organization by individual type of part being repaired. The university used this raw data to create forecasts and net equipment requirements for each individual modular repair center (MRC) based on variable mixes of workloads and resources. The data base provided the following information:

A work control document (WCD), a unique identifier for each engine part;

The annual requirement of each end item (engine or subassembly) of which the part to be repaired is a component (the WCD attached to each part carried this information):

- The sequential routes of the part and resource requirements coded by the industrial process code with labor and machine/process time required at each resource;
- The size and weight of the part so that storage and queue space can be estimated, and
- The number of units per assemble (UPA) required of each part to make up the end item.

Prior to the fire, the division was organized along functional operational lines with each department responsible for a specific process, such as machining, welding, cleaning, or inspection. This organization structure was developed in 1074 when engine overhaul functions were consolidated into one organization. At that time, such functional shop layouts maximized equipment utilization and skill concentrations since a typical long-flow part would require 30 to 50 production operations and change organizations only seven to 10 times. Today, the same part requires over 120 production operations and changes organizations as many as 30 to 50 times. This increase has been caused by incremental introduction of technology and by improved repair procedures that offset wear of critical engine parts and reduce replacement costs. The additional repairs increased routing that overburdened the mechanized conveyor system. Since 1974, the only major change was an experiment three years prior to the first to consolidate one part-type family, combustion cans, into a partially self-contained work center.

The reconstruction period after the fire gave TAFB a unique opportunity to design a modern production system to replace the one destroyed. TAFB manufacturing system analysts changed the repair process from a process specialization type of operation to a family (group) type of operation. Staff from the University of Oklahoma helped to solve the problems associated with ling flow types, lack of clear responsibility for quality problems, and excessive material handling. The plan for reconstruction was based on the concept of a modular repair canter (MRC), a concept similar to the group technology cell (GTC) concept except that it is more interrelated with other centers than a GTC.

We created and defined the modular repair center concept as a single organization to inspect and repair a collection of parts with similar geometries and industrial processes so as to provide the most economical assignment of equipment and personnel to facilitate single point organizational responsibility and control. An example of such a center is the blade MRC, which repairs all turbine blades from all engine types. With the exception of initial chemical cleaning, disassembly, plating, paint, and high temperature heat treatment, all industrial equipment and processes were available for assignment to an MRC.

Since TAFB lost an entire overhead conveyor system in the fire, implementing the MRC concept required a new conveyor design in terms of routing, size, and location of up and down elevators. The new system needed a conveyor to move parts to their respective MRCs from the disassembly area and to special areas such as hear treatment, painting, or plating and back to engine reassembly. When an engine arrives for repair, its turbine blades are removed and routed via the overhead conveyor to the blade MRC, out to heat treatment, painting, and plating, back to the MRC, and finally returned to be assembled back into an engine. A stacker (mechanized inventory storage system) in each MRC handles excess in-process queues that are too large for the finite buffer storage at each machine. One of the functions of the simulation model was to compute the capacity of the buffers and stacker.

Data Analysis

Standard sources at TAFB provided the information for analysis. The first source, the work control documents (WCD), gives the operation sequences for all the parts. It tells which MRC a part goes to and the sequence of machines the part will visit within the MRC. There are 2,600 different WCDs, with as few as 11 assigned to combustion cans and as many as 700 assigned to the general shop.

TABLE 1.1

Material Handling Codes

Alphanumeric Code			Weight of Item (lbs.)	
A			0–1	
B			1–5	
C			5–10	
D			10–25	
E			25–50	
F			>50	

W	0–6	6–12	12–24	24–48	>48
L					
0–6	1	2	4	7	11
6–12	2	3	5	8	12
12–24	4	5	6	9	13
14–48	7	8	9	10	14
>48	11	12	13	14	15

The second data source, the engine repair plan, showed how many engines of each type were expected to be repaired each year. We used the fiscal '85 requirements and a projected annual work load of 2000 engine equivalents to determine how many units of each family type would enter the system. Table 1.1 presents the material handling codes. The top part of the Table gives the code for the six different weight categories while the lower part (the matrix) expresses the code for 15 different categories of length and width of the base of the part, which rests on the pallet. Each number represents a combination of length (L) and width (W), measured in inches. D9, for instance, means a part that weighs 10–25 pounds and has a base whose length is between 12 and 24 inches and whose width is between 24 and 48 inches.

The third source of data, the TAFB standard material handling (MH) coding of each part, was based on the size and weight of each part. Parts move on pallets at TAFB. We used the MH coding to estimate the number of parts per pallet (see Table 1.1). TAFB engineers had decided on the shop configuration and location of the MRCs but had not determining their physical dimensions prior to our analysis. The configuration was based on groupings of jet engine parts with similar geometries, metal types, and repair processes (for example, major cases, rotating components). The MRCs are N-nozzle, S-seal, B-bearing housing, GX-gear box, TR-turbine compressor rotor, K-combustion can, BR-blade, AB-after burner, C-case, CR-compressor rotor, ZH-general handwork, ZM-general machining, ZW-general welding. In addition, general purpose shops handle painting, plating, heat treatment, blasting and cleaning. Since several hundred units of each WCD are processed, the facility handles over one-half million units annually. Each WCD is assigned to one of the MRCs and goes through several processes, comprising 25 to 100 operations each. Each MRC handles from 11 to 700 WCDs and has between 19 and 83 processes assigned to it.

Material Handling Characteristics

To conserve space and energy, parts are stored on pallets; in some cases, two different WCDs are stored on the same pallet. Parts that are large in two dimensions but small in a third are stacked. The number of WCDs on a pallet is a random variable depending on how many parts arrive when. To convert a flow of WCDs to a flow of pallets, we used a simple formula to estimate the total number of pallets that would flow between operations given the number of each part type that would move between those two operations. Table 1.2 shows the parts per pallet material handling (MH) codes. The matrix in the Table expresses the number of parts that can be placed on a pallet, given the MH code from Table 1.1. Two parts per pallet can be carried if the part is coded D9. The small values represent fixed loads based on weight and size, while larger values are averages of actual usage, since a pallet can carry many small parts.

To obtain the pallet factor estimate OPF (see appendix), we used the TAFB material handling codes (Tables 1.1 and 1.2). If OPF = 0.2 and 50 units of all types flow from area i to area j per half hour, then $(0.2)(50) = 10$ pallets will move on the average per half hour. Other technical details can be found in Foote et al. (1988).

TABLE 1.2

Parts per Pallet Material Handling Codes

Weight Code Size Code	A	B	C	D–E	F
1	50	30	20	10	5
2	8	8	8	4	4
3	8	8	8	4	4
4	8	6	5	3	2
5	8	6	5	3	2
6	8	6	5	3	2
7	8	6	5	3	2
8	4	4	4	2	2
9	4	4	4	2	2
10	4	4	4	2	2
11	4	4	4	2	2
12	2	2	2	1	1
13	2	2	2	1	1
14	2	2	2	1	1
15	2	2	2	1	1

General Model for Conveyor Design

To establish a basis for building a minimum-size conveyor system to handle the work load, we constructed a network model of the material-handling system. A minimum conveyor system is one that has the least length with the most flexibility and that meets all production volume requirements without logjams. In the network representation, 56 nodes represent (1) different MRCs and their possible associated loading/unloading points, (2) the assembly areas, (3) the possible transfer points in the conveyor, and (4) general purpose shops. Arcs or links in the network represent the possible different sections of the conveyor linking nodes. The arrows on the arcs show the direction items flow (one way or two way). Using the conveyor system drawings, we calculated the distances between all pairs of nodes to find the linear feet of conveyor. The numbers associated with arcs represent these distances. We used Floyd's algorithm (Floyd, 1962), which analyzed in Dreyfus (1969), to calculate the shortest distance between all pairs of nodes. The algorithm also determines the shortest path, namely, the optimal sequence of arcs (conveyor sections) to travel in order to minimize the total travel time from any department to any other department. Ravindran et al. (1988) cover the details of the conveyor design, including how the random variation in pallet flow on the conveyor was handled and how conveyor bottlenecks were eliminated. Figure 1.2 shows the old and new layouts and the associated conveyor systems. The figure presents the conveyor system pre- and postfire. The top layout shows Building 3001 as it was functionally laid out before the fire. The bottom figure shows the new layout based on a cellular manufacturing organization with conveyor routes optimized by Floyd's Algorithm.

Computer Generated Data

From the processing sequence on the work control document and the numbers of engines that need to be maintained, we calculated the flow from each MRC to other MRCs. We wrote a computer program to scan the processing sequence and determine when a move out of the MRC would be made. For example, when the process code for heat treatment appeared, the item would move from its MRC to hear treatment and then back to the MRC. The number of items of each type moving was the number of engines times the number of parts of this type per engine. We then summed the movements between each pair of locations over all part types, and converted the movement in terms of parts to pallets moved per half hour.

The Simulation Model

We wrote the simulation model, called the Tinker Integrated Planning Simulation (TIPS), using the discrete event orientation in SLAM (Pritsker, 1986); it contains approximately 1750 lines of FORTRAN code. TIPS is designed to simulate a single MRC at a time. The entities in the model are

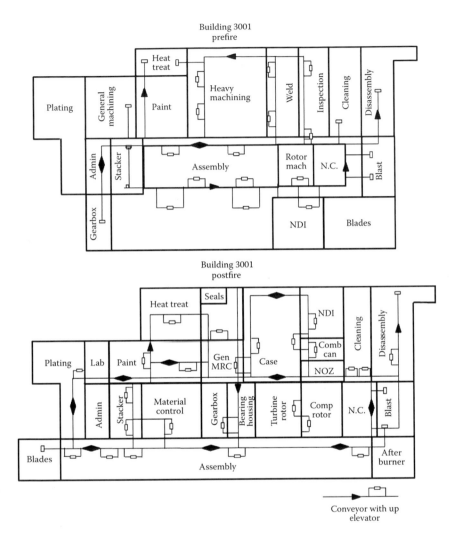

FIGURE 1.2
Conveyor system pre- and postfire.

the WCDs flowing through one particular MRC. Features of the TIPS model include three shifts, transfer to other MRC operations (that is, painting, plating, and heat treatment), and stackers to model WCD storage when machine queue lengths are exceeded. The simulation model is capable of storing 70,000 entities (concurrent WCDs) in an MRC. Despite this, three of the MRCs were so large that they had to be broken into smaller family groups.

Figure 1.3 illustrates the system concept of the MRC and how material flows inside and to external shops. This allows the stacker to be sized by the simulation; the maximum load will determine the size of the stacker installed. The part shown in Figure 1.3 has a 1-4-3-painting-5 machine sequence.

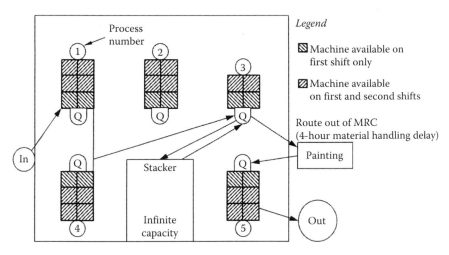

FIGURE 1.3
Sample five-machine MRC configuration.

The in-process queue area at the machine is limited. When this is full, the overflow goes to the infinite capacity stacker. The simulation computed the maximum stacker storage requirement needed. A route out after machine number three to painting and return after an 8-hour material handling delay is shown. The cross hatching on the machines in the diagram indicated the shifts when they are available. For example, there are six machines of type number one available during the day shift, and only four available during the second shift. If a WCD is on a machine when the shift change occurs, it is assumed that the machine completes processing the WCD prior to the changeover.

Tinker Air Force Base supplied the data used to determine the rate of flow of WCDs through each MRC. The data for each MRC came in two sets, the 1985 fiscal year data and the data for 2000 engine equivalents (when the facility would run at full capacity). Both data sets contained a list of the WCDs for the MRC, the operations sequence for each WCD, the corresponding machine process time for each WCD, the corresponding standard labor time for each WCD, the UPA (units per assembly) number for each WCD the data included, and a vector containing the relative frequencies of each WCD. In addition, the projected size of each MRC (for example, number of machines of each type) and information needed to calculate a From-To matrix (for inter- and intra-MRC transfers). We transformed all the data to a format that allowed SLAM to execute the discrete event model.

Two features of the TIPS simulation model make it unique. First, the model was so large that it used the SLAM language at its maximum configuration to run a single MRC. We had to consult with Pritsker and Associates to determine how to extend SLAM's storage limits in the source code. Second, the model integrated both physical (machines) and skill (labor) resources in a single model that supported a bottle-neck analysis, space analysis, and overhead-conveyor-routing analysis. We designed the model for managers and held two training sessions at Tinker AFB to allow managers to use TIPS for decision making.

A final feature of the system is its generality. Originally, 13 MRCs were to be modeled. This expanded to 17. We had three months to develop the model and had to meet the due date. We developed the program using a special format that allowed the model to be restructured for any MRC. Thus, the type of machines, their number, and their operations in an MRC were standard input. The process plan for each WCD was an input data set. With this structure, any MRC could be simulated. Some features of the model required special considerations as we constructed the model of the proposed shop configuration:

- *Downtime*: Machine breakdown affects the flow time and through-put for an MRC. We assumed that after each machine processes a part, a breakdown occurs with probability that depends on the machine. This assumption is based on the fact that uncompleted work can be finished on other machines and breakdowns are rare. This simplified the code, which helped meet the deadline. The time to repair a machine is exponentially distributed. We based the distributions and parameters used in the simulation on estimated by TAFB personnel.

- *Interarrival and service time distribution*: Since no data were available on the interarrival distribution of engine inputs, we used a deterministic interarrival time based on the annual volume of that particular WCD. This was reasonable in that repairs of engines are scheduled uniformly over the year. The service time was a truncated normal random variable with the range set at $\mu \pm 0.05\mu$.

- *Labor utilization*: The modeling of a WCD being processed on a machine had to incorporate the fact that both a machine and an operator are required to service the part. In addition, sick leave, training leave, and vacations for machine operators are modeled.

Calculating the Number of Machines Needed

A prime use of the simulation was to determine the number of machines of a particular type needed in each MRC. The stated objective was to have 95 percent availability for each machine type; that is, 95 percent of the time a machine will be available at a machine center when a component arrives. To determine the smallest number of machines needed to provide 95 percent availability, we first ran the simulation assuming ample machine capacity so that there was no queuing at the machine center. We then used the utilization statistics for the case of ample capacity to determine the minimum number of machines necessary for 95 percent availability.

We had to design the simulation for three shift operations because in emergencies all three shifts are used, and even on one-shift operations, some equipment, such as painting, plating, and equipment designed at TAFB, runs three shifts because it is not economical to duplicate the equipment. Thus, even though most of the equipment is manned for only one shift, the simulation must be run for three shifts per day. When considering a machine center that would operate for only one (or two) shift(s) per day, we had to rescale the utilization statistics from a three-shift day to reflect the shorter work day before calculating the number of machines needed for 95 percent availability.

Problems with Large MRCs

Some MRCs were too large to be handled by the simulation model both in terms of unacceptably long run times and memory requirements. As a result, the large MRCs had to be broken up into smaller family groups. For example, we broke the combustion can shop (MRC K) down into six families. (We could do this because the six families in the K shop shared only such entering processes as inspection, and it was easy to split inputs into six groups.) TIPS can handle approximately 70,000 entities (or WCDs) at one time. The run times varied depending on the size of the MRC. For example, the simulation for the estimated repair work load of 2000 engines for the gear box MRC took approximately one hour to run on an IBM 3081. The run times on a VAX 11/780 were generally eight times longer than the IBM run times. In one specific case, MRC CC2 (one of the smallest families in MRC K), the simulation took 1.5 minutes to run on IBM and 7.8 minutes on VAX. The CC2 shop contains a maximum of five WCDs and can handle up to 73 different processes. It has an annual work load of about 821 parts. By comparison, the gear box MRC handles about 329 WCD typed and up to 72 different processes. Its annual repair volume is over 100,000 parts. We used a warm-up period of 13 weeks (one quarter) for each simulation run, and collected statistics on MRCs starting with the 14th week. Simulation outputs were printed in 13-week time intervals to match regular production runs at TAFB. In the early testing, we compared the outputs from different warmup periods (13 and 26 weeks) using a *t*-test to determine if we needed an extra warm-up quarter. The differences in mean values were insignificant.

Verification

To determine whether the simulation model was working as intended, we took the following verification steps:

1. We developed the model incrementally. This made it easier to debug the programs.
2. We analyzed the outputs of each modular component of the overall model for reasonableness (does the output seem to represent real-world

expectations?), consistency (does the output remain about the same for similar inputs?), reasonable run time (does the program run longer than expected for the given MRC?), and output (a 10 percent increase in load should show more than a 10 percent increase in waiting time).

Validation

To validate the model, we made a diagnostic check of how closely the simulation model matched the actual system, taking the following steps:

1. We cross-checked the model assumptions. For example, is the assumption of normally distributed processing time correct?
2. We compared statistically analyzed results to actual historical data using a representative MRC simulation, checking both average output and range of output.

Output

The output from TIPS consisted of two documents: the standard SLAM summary report and a custom printout generated by a FORTRAN subroutine. The custom output presented the SLAM output in a format and at a level of detail suitable for prompt managerial decision making. The statistics in the output included:

1. Machine availability by shift for each process,
2. Maximum queue length in front of each process,
3. Average processing time,
4. Average waiting time for each process,
5. Number of units for each WCD type entering and leaving the system,
6. Part flow (in units) entering and leaving the stacker for each process,
7. Time spent in the stacker waiting for a specific process,
8. Utilization level for each process per shift, and
9. Total time in the system for each WCD, including waiting time, handling time, and processing time. Labor time is assumed to overlap with processing time.

We wrote supplementary FORTRAN programs to generate certain input data for the TIPS program. For example, we used a bottleneck program to set the initial number of machines available for each process. The TIPS program is, in effect, the nucleus of an integrated system of management decision aids, as shown in Figure 1.4. The simulation model is used at Tinker Air Force Base to perform the functions listed, such as capacity planning, production planning, and analysis of part flow, and to provide data for process design, management control, process capability analysis, and process

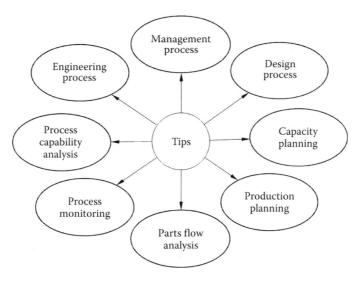

FIGURE 1.4
Central role of TIPS in shop management.

monitoring to determine if production goals are met, and to meet the needs of engineering in designing new overhaul procedures.

Design Constraints

Because the fire was so destructive and because slowing maintenance operations for a long time would seriously affect the national defense, TAFB set a time limit of three months for designing the analytic computer models. We developed quick approximations of such items as pallet flow so that we could test model validity quickly. The closeness of predicted need to actual need showed that these approximations were acceptable based on the criterion specified by TAFB (plus or minus five percent). The predicted requirements have ranged from 100 to 115 percent of actual need.

Input Data Verification

The data for the simulation were delivered on a magnetic tape and consisted of at least a quarter of a million individual elements. These data were visually scanned on a random basis and anomalies were noted. We reviewed these anomalies in joint meetings. We then developed rules for scanning the entire data set for data errors. These rules formed part of a rudimentary expert system to improve the quality of the data. The following are examples of ruled to check processing sequences and time standard data:

1. The heat treatment can never follow painting,
2. Labor or machine standard time could never be zero,

3. Machine time is always greater than or equal to labor time, and

4. Drilling times can vary between one and four minutes (these ranges were MRC specific).

The program checked for missing processes to see if other processes were required to follow or precede it based on technological constraints and then determining if the process plan met this constraint.

Errors found by use of the above rules and others were reported to TAFB personnel for analysis and correction, if necessary. TAFB printed out violations of the rules and corrected the errors. Some errors were transpositions and easily corrected, others necessitated quick time studies or verification of process sequence. Some data rejected by the tests were actually correct. This phase took two months, but it overlapped the design of the simulation and material-handling models. The simulation had to be developed in some detail to calculate the space needed for sequential queues as parts moved from machine to machine and to capture the material-handling sequence and routes as we played what-if games with resources and work-load assignments. We also needed this detail to document and solve bottleneck problems within the flow of a single part or for a combination of parts.

Project Summary

The Air Force started the project in January 1985, approved the organizational concept in February 1985, developed the industrial process code concept and started data collection for the data base in late January, and created rough-cut capacity plans and organizations in March–April 1985, and it required data to meet material and scheduling lead times by June 15, 1985. We had to complete all simulations by September 1985 to finalize the resource allocations and to allow for design lead times. The simulations were used to allocate personnel, machines, and floor space to the various organizations.

Our analysis to aid the transition to the MRC layout included formatting the TIPS simulation program, designing the overhead conveyor system, laying out the plant, and making a routing analysis of the inter-MRC transfers. The TIPS program proved valuable in aiding the transition to the new layout by estimating performance measures (for example, flow-time and queue statistics) that helped determine the number of machines of each type to place in an MRC. In one particular instance, the nozzle MRC, the Production and Engineering Department called for 24 work stations of a particular type. The TIPS analysis indicated that between 11 and 13 work stations were needed. Based on the TIPS results, only 12 work stations were installed and this has proved to be sufficient. At Tinker Air Force Base, simulation proved to be a valuable tool in assessing the effectiveness of the new plans and in determining the right parameters for each new MRC.

The fire at TAFB was a disaster that turned out to be a great opportunity. It would have been very hard to justify dislocating the entire facility for over a year and expending the sum of money required to redesign the facility. In the long run, the benefits of the new system may pay back the cost of the fire with interest.

The University of Oklahoma design team responded to the emergency with accomplishments we are very proud of. We developed and verified a general simulation model in three months, when national consulting firms estimated four times that long. The new system met design expectations, which is rare. Part flow times have been reduced by 35 to 50 percent depending on size. Labor savings have been $1.8 million in 1987, $2.1 million in 1988, and continue to rise. $4.3 million was saved in equipment purchases.

Space requirements were reduced by 30,000 square feet. The percent defective had dropped three percent in 1987 and five percent further in 1988. The conveyor system has had no jam-ups due to overloading. Finally, other Air Logistics Centers have adopted the TIPS concept to plan redesign of their facilities and the report (Ravindran et al., 1986) has been distributed for review by over 60 organizations at their request. We have proved that modern management science techniques can be applied quickly and with great impact.

Original 1989 Paper Acknowledgments

We express our appreciation to Margarita Beneke, K. Janikiraman, Wan-Seon Shin, Doug Stewart, and Murali Subramaniam for their assistance in developing and running the TIPS model, to Sandee Boyer for patiently typing several drafts of this paper, and to Hussein Saber for his help with Figure 1.2. In addition, Stephen Graves, Mary Haight, and two referees have improved the readability of this paper.

Appendix

The overall pallet factor is computed by

$$OPF = \frac{\left\{ \sum_{i=1}^{M} W_1 P_1 \right\}}{\left\{ \sum_{i=1}^{M} W_1 \right\}}$$

where
   P_1 = estimate of pallets/parts given its MH code. For example, if P_1 has code
      1A, $P_1 = 1/50 = /02$ (see Table 1.2), $i = 1,2, ..., M$,
   $W_1 = (N_1)(UPA_1)$, $i = 1,2, ..., M$,
   N_1 = number of WCD's of type i per year, $i = 1,2, ..., M$,

UPA_1 = number of units per WCD, $i = 1,2, ..., M$,
OPF = overall pallet factor, and
M = number of WCD's.

1.4 Conclusion

As can be seen in the case example of the Tinker Project presented in this festschrift, Ravi Ravindran has made several multicriteria leadership contributions to several people and several organizations. Under his unflinching leadership, several PhD and MSc students graduated on the basis of the Tinker Project. Those individuals continue to make professional and intellectual contributions internationally. The seminal publications that resulted from Ravi's leadership of the project continue to guide management science and operations research practitioners around the world. My personal exposure to Ravi's academic leadership actually started before I went to work for him at the University of Oklahoma. In 1982, I was offered admission to Purdue University for my PhD studies. Ravi Ravindran was at Purdue University at that time and he was assigned as my initial academic adviser, as was the practice for all new incoming students at that time. Although I opted to attend the UCF instead of Purdue University, for financial assistantship package reasons, the written communications with Ravi played a key role in mind as I continue to dedicate myself to the challenges of doctoral education. The Purdue admission letter, dated May 17, 1982 and signed by Professor James W. Barany, associate head, introduced me to Professor Ravi Ravindran, assistant head. Ravi engaged with me positively through a series of written communications to encourage me to attend Purdue. It was fortuitous for me that Ravi later left Purdue University to become the department head at the University of Oklahoma, where I ended up working for and with him for several years until he left to go to the Pennsylvania State University around 1997. Ravi's mentoring and nurturing ways continue to influence my professional activities even today. I am, thus, delighted to be able to contribute this festschrift in his honor.

References

Badiru, A. B.; B. L. Foote; L. Leemis; A. Ravindran; and L. Williams 1993, Recovering from a crisis at Tinker Air Force Base, *PM Network*, Vol. 7, No. 2, Feb. pp. 10–23.
Dreyfus, S. E. 1969, An appraisal of some shortest-path algorithms, *Operations Research*, Vol. 17, No. 3, pp. 395–412.

Foote, B.; A. Ravindran; A. B. Badiru; L. Leemis; and L. Williams 1988, Simulation and network analysis pay off in conveyor system design, *Industrial Engineering*, Vol. 20, No. 6, June, pp. 48–53.

Foote, B. L.; A. Ravindran; A. B. Badiru; L. Leemis; and L. Williams 1992, An application of simulation and network analysis to capacity planning and material handling systems at Tinker Air Force Base. In *Excellence in Management Science Practice—A Readings Book*, A. Assad, E. Wasil, and G. Lilien, eds., Prentice-Hall, Upper Saddle River, New Jersey, pp. 349–362.

Floyd, R. W. 1962, Algorithm 97: Shortest path, *Communications of the ACM*, Vol. 5, No. 6, p. 345.

Leemis, L.; A. B. Badiru; B. L. Foote; and A. Ravindran 1990, Job shop configuration optimization at Tinker Air Force Base, *Simulation*, Vol. 54, No. 6, June, pp. 287–290.

Pritsker, A and Alan B. 1986, *Introduction to Simulation and SLAM II*, third edition, Halsted Press, John Wiley & Sons, New York.

Ravindran, A.; B. L. Foote; A. B. Badiru; and L. Leemis 1988, Mechanized material handling systems design & routing, *Computers & Industrial Engineering*, Vol. 14, No. 3, pp. 251–270.

Ravindran, A.; B. L. Foote; A. B. Badiru; L. Leemis; and L. Williams 1986, *Job shop configuration optimization at Tinker Air Force Base, Final Technical Report*, The University of Oklahoma, Norman, Oklahoma.

Ravindran, A.; Foote, B. L.; Badiru, A. B.; Leemis, L. M.; and L. Williams 1989, An application of simulation and network analysis to capacity planning and material handling systems at Tinker Air Force Base, *TIMS Interfaces*, Vol. 19, No. 1, Jan.–Feb., pp. 102–115. Reprinted with permission.

2

Multi-Criteria Decision Making: An Overview and a Comparative Discussion

Ramakrishnan Ramanathan, A. Ravi Ravindran, and Muthu Mathirajan

CONTENTS

2.1 Introduction

Multi-criteria decision making (MCDM) is a subfield of operations research. It is a special case of the so-called decision-making problems. A decision-making problem is characterized by the need to choose one or a few from among a number of alternatives. The person who is to choose the alternatives is normally called the decision maker (DM). His preferences will have to be considered in choosing the right alternative(s). In MCDM, the DM chooses his most preferred alternative(s) on the basis of two or more criteria or attributes (Dyer et al., 1992). The terms criteria, attributes, and objectives are closely related and will be discussed in more detail later in this chapter.

The field of MCDM has been succinctly defined as making decisions in the face of multiple conflicting objectives (Zionts, 1992, 2000). According to Korhonen (1992), the ultimate purpose of MCDM is "to help a decision maker to find the 'most preferred' solution for his/her decision problem." Several perspectives are available in the literature to characterize a good decision-making process. Stewart (1992) suggests that "the aim of any MCDM technique is to provide help and guidance to decision maker in discovering his or her most desired solution to the problem" (in the sense of the course of action which best achieves the DM's long-term goals). According to French (1984), a good decision aid should help the DM explore not just the problem but also himself. Keeney (1992), in his famous book on value-focused thinking, says that we should spend more of our decision-making time concentrating on what is important, and that we should evaluate more carefully the desirability of the alternatives. He also mentions that we should articulate and understand our values and using these values we should select meaningful decisions to ponder and to create better alternatives. Howard (1992) describes decision analysis as a "quality conversation about a decision designed to lead to clarity of action." Finally, Henig and Buchanan (1996) say that a good decision process will force the DM to understand his or her preferences and allow the set of alternatives to be expanded. Thus, a good decision-making process should not only improve the clarity of the problem to the DM, but it should also shed new light into the problem by generating newer alternatives.

We are normally concerned with a single DM. If more number of DMs are involved, it is important to aggregate all their preferences, leading to a group decision-making (GDM) situation.

Starr and Zeleny (1977) provide a brief historical sketch of the early developments in MCDM. Some special issues of journals have been devoted to the field of MCDM, including *Management Science* (Vol. 30, No. 1, 1984), *Interfaces* (Vol. 22, No. 6, 1992) (devoted to decision and risk analysis), and *Computers and Operations Research* (Vol. 19, No. 7, 1994). The *Journal of Multi-Criteria Decision Analysis*, starting from the year 1992, publishes articles entirely devoted to MCDM. Special issues of the *Journal of the Operational Research Society* (April, 1982) and *Interfaces* (November–December, 1991) provide a range of applications. Issue No. 2, Volume 133 (January 2001) of *European Journal of Operational Research* is a special issue on goal programming and contains a collection of some papers presented at the Third International Conference on Multi-Objective Programming and Goal Programming Theories and Applications. Given the growing popularity of MCDM approaches in specific disciplines, a number of recent articles have reviewed applications of MCDM in supplier evaluation and selection (Govindan et al., 2015; Ho et al., 2010), energy planning (Pohekar and Ramachandran, 2004; Scott et al., 2012; Wang et al., 2009), infrastructure management (Kabir et al., 2014), construction (Jato-Espino et al., 2014), and municipal solid waste management (Soltani et al., 2015).

In this chapter, attempt is made to review the basic concepts of MCDM, present salient features of some important MCDM methods, and provide a comparative discussion. It is not our intention here to provide a detailed review of all MCDM methods presented in the literature.

2.2 MCDM Terminologies

- Several terminologies are normally used when dealing with a decision problem that has multiple criteria. The terms goals, objectives, criteria, and attributes are commonly found in the MCDM literature and can be used with interchangeable (and confusing) ease (Henig and Buchanan, 1996). The general meaning of these words is similar in most cases. Useful definitions of different terminologies used in MCDM literature are available in Roy (1999).

- Alternatives form the most fundamental entities in an MCDM model. They represent one of several things or courses of action to be chosen by the DM. They are also called solutions, especially when dealing with continuous variables, in the mathematical programming context.

Alternatives are normally compared with each other in terms of the so-called criteria. Identification of the criteria for a particular problem is subjective, that is, varies for each problem. Criteria are normally developed in a hierarchical fashion, starting from the broadest sense (usually called the goal of the problem) and refined into more and more precise sub- and sub-sub goals. There is no unique definition for the term "criterion," but a useful general definition is from Bouyssou (1990), who has defined criterion as a tool allowing comparison of alternatives according to a particular significance axis or point of view. Edwards (1977) calls criteria as the *relevant dimensions of value for evaluation of alternatives.* Henig and Buchanan (1996) consider criteria to be the *raison d'être* of the DM. The term "Criterion" is defined by Roy (1999) as a tool constructed for evaluating and comparing potential actions according to a well-defined point of view.

In general, some rules should be followed in identifying criteria for any decision problem (Keeney and Raiffa, 1976; Saaty, 1980; von Winterfeldt and Edwards, 1986). They have to be mutually exclusive or independent, collectively exhaustive, and should have operational clarity of definition.

Criteria of a decision problem are usually very general, abstract, and often ambiguous and it can be impossible to directly associate criteria with alternatives. Each criterion can be normally represented by a surrogate measure of performance, represented by some measurable unit, called the *attributes*, of the consequences arising from implementation of any particular decision alternative. Thus while *warmth* is a criterion, *temperature* measured in a suitable (say Celsius or Fahrenheit) scale is an attribute. Attributes are objective and measurable features of the alternatives. Thus, the choice of attributes reflects both the objectively measurable components of the alternatives and the DM's subjective criteria. Attributes of alternatives can be measured independently from DM's desires and expressed as mathematical functions of the decision variables.

Objectives, used in mathematical programming problems, represent directions of improvement of the attributes. A maximizing objective refers to the case where "more is better," while a minimizing objective refers to the case where "less is better." For example, profit is an attribute, while maximizing profit is an objective. The term "criterion" is a general term comprising the concepts of attributes and objectives. It can represent either attribute or objective depending on the nature of the problem. Perhaps, that is why MCDM is considered to encompass two distinct fields, namely, multi-attribute decision making (MADM) and multi-objective decision making (MODM) (e.g., Triantaphyllou, 2013). These fields are discussed in the next section.

2.3 Classification of Different MCDM Approaches

A wide variety of MCDM methods have been reported in the literature in the last few decades (Stewart, 1992). Some methods use rigorous mathematical programming approaches. Examples include the goal programming (GP) (Schniederjans, 1995), compromise programming (Zeleny, 1982), multi-objective linear programming (MOLP) (Zeleny, 1982), and data envelopment analysis (DEA) (Charnes et al., 1978). Some other methods are not programming-based. Examples include the multi-attribute utility theory (MAUT) (Keeney and Raiffa, 1976), the simple multi-attribute rating technique (SMART) (Edwards, 1977), the analytic hierarchy process (AHP) (Saaty, 1980), and the ELECTRE (Elimination Et Coix Traduisant la Realite or elimination and choice translating algorithm) methods (Roy, 1996).

Classifications of MCDM methods in terms of different criteria are discussed by Hanne (1999). According to him, MCDM methods can be categorized in terms of suitability of problem type (MADM for problems with a finite set of alternatives, and MODM for problems with a continuous and infinite set of alternatives) and in terms of solution concepts (aspiration levels, pairwise comparisons, interactivity or non-interactivity, outranking, etc.).

Stewart (1992) has classified MCDM methods as:

1. Value- or utility-based approaches (comprising value theory, utility theory, the AHP, and some interactive methods using value functions)
2. Goals or reference point approaches (including GP, reference point methods, the step method [STEM], etc.)
3. Methods using outranking concepts (including the ELECTRE approaches, PROMETHEE [preference ranking organization method for enrichment evaluations], etc.)
4. Fuzzy set theory
5. Descriptive methods (including factor analysis, correspondence analysis, principal components analysis, multi-dimensional scaling, etc.)

Figure 2.1 illustrates a possible classification scheme for MCDM methods. Note that this classification need not be exhaustive as some methods can be classified in more than one category. For example, though MAUT can be classified as part of MADM in the figure, when the utility function is used as an objective function of a programming problem, it can also be classified as part of MODM. Similarly, the aspiration-level interactive method (AIM) can be classified under both the categories as it draws on many methods developed under both the categories. Several references (Evans, 1984; Stewart, 1992) were used in arriving at the structure shown in Figure 2.1.

The broad area of MCDM can be divided into two general categories, MADM and MODM. MADM involves cases in which the set of alternatives is defined explicitly by a finite list (Stewart, 1992) from which one or a few alternatives should be chosen that reflect DM's preference structure. MODM involves cases in which the set of alternatives is defined implicitly by a mathematical programming structure with objective functions. Such alternatives are usually defined in terms of continuous variables, which results in an infinite number of alternatives. MODM is also referred to in the literature as "multi-objective mathematical programming" (MMP) or "multi-objective optimization" (MOO) or vector optimization or simply "multi-objective programming" (MOP).

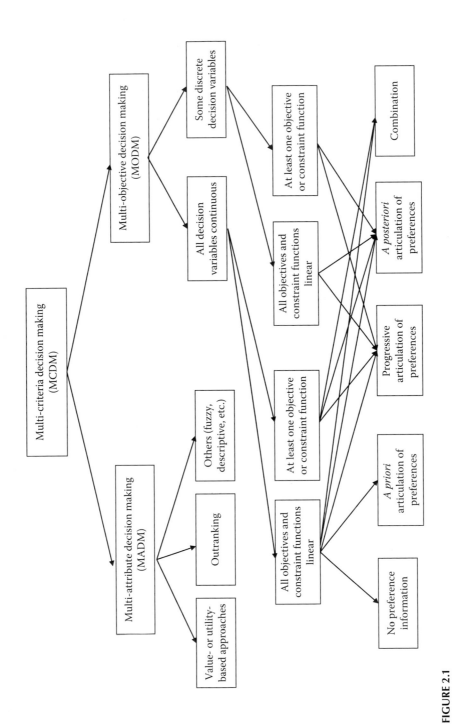

FIGURE 2.1
Classification of MCDM methods.

MADM methods can be further classified on the basis of the solution approaches. Value- or utility-based approaches attempt to capture the utility function of the DM in order to choose the alternative that maximizes the DM's overall utility. MAUT, multi-attribute value theory (MAVT), and SMART belong to this group. Sometimes, the AHP is also said to belong to this group though its proponents have argued against this classification. Outranking methods use the concept of outranking. They are used in ELECTRE methods and the PROMETHEE methods. Other MADM approaches include those based on fuzzy set theory, some descriptive and statistical approaches (such as factor analysis, principal components analysis, multi-dimensional scaling, etc.).

MODM methods normally use an optimization scheme to identify the non-dominated solutions (see the next section for a discussion of non-dominated solutions). There can be more than one, typically many, non-dominated solutions to an MODM problem. The DM often needs to select one or a few from these non-dominated solutions, that best satisfies his preferences.

Sometimes, it is easier for the DM to screen the large number of non-dominated solutions using some screening methods so as to obtain smaller number of solutions that can be managed easily by the DM. Graves et al. (1992) provide some statistics-based screening methods, including clustering and filtering.

Identification of these best compromise solution(s) from the set of all non-dominated solution requires that at least some additional information about the DM's preference structure is obtained. Hence, in MODM methods, in addition to an optimization scheme, some procedure to obtain this preference information is required. Hwang and Masud (1979) have classified solution techniques for MODM methods according to the timing of the requirement for preference information versus the optimization. They have classified the methods according to three pure approaches of articulation of the DM's preference structure:

- Prior to the optimization (*a priori* articulation of preferences)
- During, or in sequence with, the optimization (progressive articulation of preferences, and sometimes also termed as interactive articulation of preferences)
- After the optimization (*a posteriori* articulation of preferences)

In addition, some more categories (no preference information) are added in Figure 2.1. MOLP and DEA are examples of MODM methods that do not require any preference information. Of course, both these methods can only identify the non-dominated alternatives, and cannot choose the best one from these non-dominated alternatives that is in conformity with DM's preferences.

Evans (1984) has provided references for mathematical programming articles under most of the MODM categories listed in Figure 2.1. Many methods of MODM have been described in detail in Steuer (1986).

2.4 Multi-Objective Decision Making: Some Basic Concepts

In this section, a comparison of an MODM problem with a single-objective decision-making problem is used to introduce certain key concepts frequently used in MCDM literature. Note that many of the concepts introduced in this section are also applicable to MADM problems.

A single-objective decision-making (SODM) problem is generally written in the form:

$$\max f_1(x)$$
$$\text{subject to: } g_j(x) \le 0, \quad \text{for } j = 1, \dots, m, \tag{2.1}$$

where x is an n-vector of decision variables and $f_1(x)$ is the single-decision criterion or objective function to be optimized.

Let $S = \{x/g_j(x) \le 0, \text{ for all } j\}$.
$Y = \{y/f_1(x) = y \text{ for some } x \in S\}$.

S is called the decision space and Y is called the criteria or objective space. Methods for solving single-objective mathematical programming problems have been studied extensively for the past 40 years. However, almost every important real-world problem involves more than one objective. A general MODM problem has the following form:

$$\text{Max } F(x) = \{f_1(x), f_2(x), \dots, f_k(x)\}$$
$$\text{Subject to, } \quad g_j(x) \le 0 \quad \text{for } j = 1, \dots, m, \tag{2.2}$$

where x is an n-vector of *decision variables* and $f_i(x)$, $i = 1, \dots, k$ are the k *criteria/objective functions*.

Let $S = \{x/g_j(x) \le 0, \text{ for all } j\}$.
$Y = \{y/F(x) = y \text{ for some } x \in S\}$.

S is called the *decision space* and Y is called the *criteria or objective space* of the MODM problem. Without loss of generality, we can assume all the objective functions to be maximizing. Thus, the MODM problem is similar to an

TABLE 2.1

Hypothetical Options

	Cost ($)	Time (h)	Emissions (kg of Aggregated Pollutants)
Option A	50	16	10
Option B	40	20	12
Option C	60	16	12

SODM problem except that it has a stack of objective functions instead of only one.

Let us consider an example involving choice of transport modes. Let the objectives be as follows:

Minimize $f_1(x) =$ Cost

Minimize $f_2(x) =$ Time

Minimize $f_3(x) =$ Emissions

Subject to some constraints

Let us assume that we have three options (Table 2.1) that satisfy the constraints, that is, the feasible options, identified somehow.

Note that option C fares, on all objectives, worse than A and hence it should not be considered anymore. Options A and B are incomparable, as none of them is at least as good as the other in terms of all the objectives. While option A results in lesser time and emissions compared with option B, it is more expensive. Hence, options A and B are said to be non-dominated options while option C is called a dominated option.

In any MODM exercise, we are first interested in identifying the non-dominated options. Note that there may be more than one non-dominated option for any MODM problem.

2.4.1 Efficient, Non-Dominated, or Pareto Optimal Solution

A solution $x^o \in S$ to MODM problem is said to be *efficient* if $f_k(x) > f_k(x^o)$ for some $x \in S$ implies that $f_j(x) < f_j(x^o)$ for at least one other index j. More simply stated, an efficient solution has the property that an improvement in any one objective is possible only at the expense of at least one other objective.

A *dominated solution* is a feasible solution that is not efficient.

Efficient Set: Set of all efficient solutions is called the *efficient set* or *efficient frontier.*

Note: Even though the solution of MODM problem reduces to finding the efficient set, it is not practical because there could be an infinite number of efficient solutions.

EXAMPLE 2.1

Consider the following bicriteria linear program (BCLP):

Max $Z_1 = 5x_1 + x_2$

Max $Z_2 = x_1 + 4x_2$

Subject to: $x_1 \leq 5$

$x_2 \leq 3$

$x_1 + x_2 \leq 6$

$x_1, x_2 \geq 0$

Solution

The *decision space* and the *objective space* are given in Figures 2.2 and 2.3, respectively. Corner points C and D are efficient solutions, while corner points A, B, and E are dominated. The set of all efficient solutions is given by the line segment CD in both figures.

Ideal solution is the vector of individual optima obtained by optimizing each objective function separately ignoring all other objectives. In Example 2.1, the maximum value of Z_1, ignoring Z_2, is 26 and occurs at point D. Similarly, maximum Z_2 of 15 is obtained at point C. Thus, the ideal solution is (26, 15) but is *not* feasible or achievable.

Note: One of the popular approaches to solving MODM problems is to find an efficient solution that comes "as close as possible" to the ideal solution. We will discuss this approach later in Section 2.6.2.4.

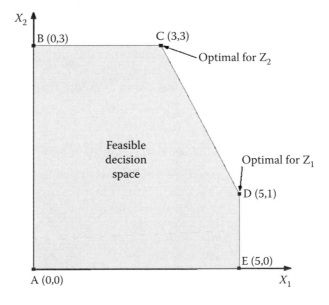

FIGURE 2.2
Decision space (Example 2.1).

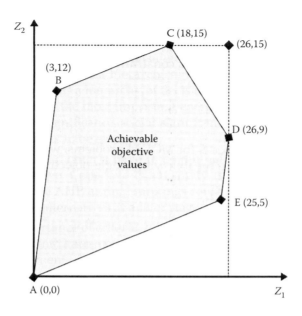

FIGURE 2.3
Objective space (Example 2.1).

2.4.2 Determining an Efficient Solution (Geoffrion, 1968)

For the MODM problem (2.2), consider the following single-objective optimization problem, called the P_λ *problem*. The P_λ problem is also known as the *weighted objective problem*.

$$\text{Max } Z = \sum_{i=1}^{k} \lambda_i f_i(x)$$

$$\text{Subject to} : x \in S \tag{2.3}$$

$$\sum_{i=1}^{k} \lambda_i = 1 \quad \text{and} \quad \lambda_i \geq 0.$$

Theorem 2.1 (Sufficiency)

Let $\lambda_i > 0$ for all i be specified. If $\mathbf{x}^o$ is an optimal solution for the P_λ problem (Equation 2.3), then $\mathbf{x}^o$ is an efficient solution to the MODM problem.

In Example 2.1, if we set $\lambda_1 = \lambda_2 = 0.5$ and solve the P_λ problem, the optimal solution will be at D, which is an efficient solution.

Warning: Theorem 2.1 is only a sufficient condition and is *not* necessary. For example, there could be efficient solutions to MODM problem which could

aggregation procedures (MCAP) and a comparative discussion of some MCDM methods are available in Guitouni and Martel (1998). In the next few pages, a brief description of some selected MCDM methods will be provided. For ease of exposition, we have divided the methods in terms of MADM methods and MODM methods, though some methods, such as AIM, can be classified in either category. For each method, an attempt is made here to outline the basic idea, specify important introductory and review articles, and outline strengths and criticisms, and some important applications. Reader interested in knowing more about a method should refer to the articles referred in the appropriate subsection.

2.6.1 MADM Methods

2.6.1.1 Multi-Attribute Utility Theory

Utility measures the subjective value "worth" of an outcome, even one that is not a money value. Often the utility function is not linear. Traditionally, utility functions are defined for stochastic problems that involve uncertainty. In the case of deterministic problems, the term value functions is more commonly used. The utility or value functions may be thought of as evaluative mechanisms that can be used to measure the value of a particular solution (Zionts, 1992).

As we noted above, utility functions are defined in terms of uncertainty and thus tie in the DM's preferences under uncertainty, revealing his risk preference for an attribute. An uncertain situation that is faced by a DM can be considered similar to a lottery—he can earn $X with a probability p, and earn Y with probability $(1 - p)$. In these situations, a rational DM is expected to maximize his *expected* utility, given by $\$(pX + (1 - p)Y)$.

Utility functions are assessed by giving the DM a sequence of choice between alternatives or between alternative lotteries. The responses are used to generate functions.

MAUT consists of assessing and fitting utility functions and probabilities for each attribute, and then using the functions and probabilities to come up with rankings of alternatives. The utility function for each attribute is aggregated to get the overall utility function. At least two methods of aggregation are used in MAUT: additive and multiplicative. Certain conditions need to be satisfied in order that these aggregations are valid: *preferential independence and utility independence.*

The additive aggregation is given by the following:

$$U(A) = \sum_i w_i u_i(a_i) \quad \text{where} \quad 0 \le u_i(a_i) \le 1$$

$$\sum_i w_i = 1, \quad w_i \ge 0,$$

where $u_i(a_i)$ is the utility function describing preferences with respect to the attribute i, a_i represents the performance of the alternative A in terms of the attribute i, w_i are scaling factors which define acceptable trade-offs between different attributes, and $U(A)$ represents the overall utility function of the alternative A when all the attributes are considered together. This form of additive aggregation is valid if and only if (sometimes referred to as "iff" in this book) the DM's preferences satisfy the *mutual preferential independence*. Suppose that there are a set of attributes, X. Let Y be a subset of X, and let Z be its complement, that is, $Z = X - Y$. The subset Y is said to be preferentially independent of Z, if preferences relating to the attributes contained in Y do not depend on the level of attributes in Z.

The condition *utility independence* is a stronger assumption. More details can be obtained from Keeney and Raiffa (1976). A brief discussion of MAUT is available in Kirkwood (1992).

The utility functions may also be used as objective functions for solving mathematical programming problems.

- Utility theory has been criticized because of its "strict" assumptions which are usually not empirically valid (Henig and Buchanan, 1996). Because of the strict assumptions, practical applications of MAUT are relatively difficult, though there are several practical successful applications of MAUT. This has led to some simplifications of the MAUT concepts. For example, MAVT is a simplification of MAUT where uncertainty and risk are not assumed. SMART is another simplification that makes weaker assumptions while eliciting utilities. They are described in the next few subsections.

2.6.1.2 Multi-Attribute Value Theory

MAVT is a simplification of MAUT: MAVT does not seek to model the DM's attitude to risk, while MAUT considers risk and uncertainty. As a result, MAVT is easier to implement compared with MAUT (Belton, 1999).

Value theory assumes that it is possible to represent a DM's preferences in a defined context by a value function, $V(\cdot)$, such that if alternative A is preferred to alternative B, then $V(A) > V(B)$. For this representation to be possible, the DM's preferences should satisfy two properties: the transitivity property (if A is preferred to B, B is preferred to C, then A should be preferred to C), and comparability (given to alternatives A and B, the DM must be able to indicate whether A is preferred to B, or B is preferred to A, or is indifferent between the two).

Note that the value function is an ordinal function, that is, it can be used to only rank alternative. In contrast, utility function is cardinal, that is, it can be used to measure the strength of preference among alternatives.

MAVT explicitly recognizes that the DM will use many attributes (criteria) while evaluating a set of alternatives. For each attribute i, a partial value function $v_i(a_i)$ describing preferences with respect to the attribute i is assessed from the DM, where a_i represents the performance of the alternative A in terms of the attribute i. Then, the overall value function $V(A)$ of the alternative when all the attributes are considered together is normally obtained using the additive form: $V(A) = \Sigma_i v_i(a_i)$. This is more generally expressed as follows:

$$V(A) = \sum_i w_i v_i(a_i) \quad \text{where} \quad 0 \leq v_i(a_i) \leq 1$$

$$\sum_i w_i = 1, \quad w_i \geq 0.$$

As mentioned above, w_i are scaling factors which define acceptable trade-offs between different attributes. Again, this additive value function is appropriate if and only if the DM's preferences satisfy the so-called *mutual preferential independence* discussed earlier.

More detailed information on the practical implementation of MAVT is available in Belton (1999).

2.6.1.3 *Simple Multi-Attribute Rating Technique*

SMART (Edwards, 1977; von Winterfeldt and Edwards, 1986) follows the steps described in the previous section for modeling a decision problem. It uses the simple weighting technique for the assessment of importance of criteria, and for the assessment of alternatives with respect to criteria.

To rate (i.e., assess the importance of) criteria, one will start by assigning the least important criterion an importance of 10. Then, he has to consider the next least important criterion, and ask as to how much more important (if at all) is it than the least important criterion, and assign a number that reflects that ratio. This procedure is continued till all the criteria are assessed, checking each set of implied ratios as each new judgment is made. The experts will be given the opportunity to revise previous judgments to make them consistent. Once the numbers are assigned, the relative importance of criteria is obtained by summing the importance weights, and dividing each by the sum. Thus, the relative importance of the criterion j (w_j) is the ratio of importance weight of this criterion to the sum. Note that $\Sigma_j w_j = 1$ by definition.

Alternatives are rated with respect to each criterion in a similar fashion. While MAUT requires the development of complex utility functions for each criterion, SMART prefers to produce the rating using a more straightforward approach: the expert is asked to estimate the position of the alternative on a criterion on a 0–100 scale, where 0 is defined as the minimum plausible value, while 100 is defined as the maximum plausible value.

Once the above two measures are available, the overall performance of an alternative i can be aggregated using the simple weighted average,

$$U_i = \sum_j w_i u_{ij},$$

where U_i is the overall performance rating of alternative i, w_j is the relative importance of criterion j, and u_{ij} is the rating of the alternative i with respect to the criterion j. The alternative that has the maximum U_i is the most preferred alternative to achieve the goal of the decision problem. The values of U_i can be used to provide the overall rankings of the alternatives.

MAUT, or its simplified versions MAVT or SMART, has been used for several practical applications. For example, MAUT or its variants have been planning a government research program (Edwards, 1977). Jones et al. (1990) have applied MAVT for the study of UK energy policy. Keeney and McDaniels (1999) have used this technique for identifying and structuring values for integrated resource planning, while Keeney (1999) has used the technique to create and organize a complete set of objectives for a large software organization. Duarte (2001) has used MAUT to identify appropriate technological alternatives to implement to treat industrial solid residuals. Some more MAUT applications are discussed by Bose et al. (1997).

2.6.1.4 Analytic Hierarchy Process

The AHP (Saaty, 1980) is one of the most popular and widely employed multi-criteria methods (Golden et al., 1989; Shim, 1989; Vargas, 1990). In this technique, the process of rating alternatives and aggregating to find the most relevant alternatives are integrated. The technique is employed for ranking a set of alternatives or for the selection of the best in a set of alternatives. The ranking/selection is done with respect to an overall goal, which is broken down into a set of criteria.

The application of the methodology consists of establishing the importance weights to be associated to the criteria in defining the overall goal. This is done by comparing the criteria pairwise. Let us consider two criteria C_j and C_k. The expert is asked to express his graded comparative judgment about the pair in terms of the relative importance of C_j over C_k with respect to the goal. The comparative judgment is captured on a semantic scale (equally important/moderately more important/strongly important, and so on) and is converted into a numerical integer value a_{jk}. The relative importance of C_k over C_j is defined as its reciprocal, that is, $a_{kj} = 1/a_{jk}$. A reciprocal pairwise comparison matrix A is then formed using a_{jk}, for all j and k. Note that $a_{jj} = 1$. It has been generally agreed (Saaty, 1980) that the weights of criteria can be estimated by finding the principal eigenvector w of the matrix A. That is,

$$Aw = \lambda_{\max} w.$$

When the vector w is normalized, it becomes the vector of priorities of the criteria with respect to the goal. λ_{max} is the largest eigenvalue of the matrix A and the corresponding eigenvector w contains only positive entries. The methodology also incorporates established procedures for checking the consistency of the judgments provided by the DM.

Using similar procedures, the weights of alternatives with respect to each criterion are computed. Then, the overall weights of alternatives are computed using the weighted summation,

$$\begin{pmatrix} \text{Overall weight} \\ \text{of alternative } i \end{pmatrix} = \sum_j \begin{pmatrix} \text{Weight of alternative } i \text{ with respect to } C_j \times \\ \text{Weight of } C_j \text{ with respect to the goal} \end{pmatrix}.$$

The popularity of AHP stems from its simplicity, flexibility, intuitive appeal, and its ability to mix quantitative and qualitative criteria in the same decision framework. Despite its popularity, several shortcomings of AHP have been reported in the literature, which have limited its applicability. However, several modifications have been suggested to the original AHP, such as the multiplicative AHP (MAHP) (Lootsma, 1999; Ramanathan, 1997), to overcome these limitations.

Some of the prominent limitations of AHP include the following:

- *Rank reversal* (Belton and Gear, 1983; Dyer, 1990): The ranking of alternatives determined by the original AHP may be altered by the addition of another alternative for consideration. For example, when AHP is used for a technology selection problem, it is possible that the rankings of the technologies get reversed when a new technology is added to the list of technologies. One way to overcome this problem is to include all possible technologies and criteria at the beginning of the AHP exercise, and not to add or remove technologies while or after completing the exercise. However, MAHP, the multiplicative variant of AHP, does not suffer from this type of rank reversal (Lootsma, 1999).

- *Number of comparisons*: AHP uses redundant judgments for checking consistency, and this can exponentially increase the number of judgments to be elicited from DMs. For example, to compare eight alternatives on the basis of one criterion, a total of 28 judgments are needed. If there are n criteria, then the total number of judgments for comparing alternatives on the basis of all these criteria will be $28n$. This is often a tiresome and exerting exercise for the DM. Some methods have been developed to reduce the number of judgments needed (Millet and Harker, 1990). Also, some modifications, such as MAHP, can compute weights even when all the judgments are not available.

AHP has been applied to a variety of decision problems in the litera-ture. Several detailed annotated bibliographies of AHP applications are available (Golden et al., 1989; Shim, 1989; Vargas, 1990). Some of the more recent applications of AHP include solar energy utilization (Elkarni and Mustafa, 1993), integrated resource planning (Koundinya et al., 1995), cli-mate change negotiations (Ramanathan, 1998), greenhouse gas mitigation (Ramanathan, 1999), and environmental impact assessment (Ramanathan, 2001a).

2.6.1.5 ELECTRE Methods

ELECTRE methods (Bouyssou and Vincke, 1997; Roy, 1996; Vincke, 1999) belong to the so-called outranking approaches. PROMETHEE method that will be discussed in the next section also belongs to this group of outrank-ing approaches. Outranking methods are especially popular in France and Belgium (Hanne, 1999). Sometimes, the outranking approaches are referred to as the French or European approaches (Roy and Vanderpooten, 1996; Vincke, 1999) (in contrast with MAUT or AHP, which are called American approaches).

There are several versions of ELECTRE methods, including ELECTRE I (historically the first of the versions), ELECTRE IS, ELECTRE II, ELECTRE III, ELECTRE IV, ELECCALC, and ELECTRE TRI. We shall briefly describe only ELECTRE I in this book, using the material presented in Vincke (1999). The reader is referred to other references (Bose et al., 1997; Roy, 1996) for more detailed descriptions of these methods.

It is important to understand some preliminary definitions for appreciat-ing the ELECTRE methods. The set of alternatives is denoted by A. A binary relation R on A is a subset of $A \times A$. It is said to be

- Symmetric iff $aRb \Rightarrow bRa$, $\forall a,b \in A$
- Asymmetric iff $aRb \Rightarrow b\overline{R}a$, $\forall a,b \in A$
- Complete iff $a\overline{R}b \Rightarrow bRa$, $\forall a,b \in A$
- Transitive iff $aRb, bRc \Rightarrow aRc$, $\forall a,b,c \in A$
- A *complete preorder* iff it is complete and transitive
- A *partial preorder* iff it is transitive and not complete.

A *criterion g* is defined as a real valued function on A in the sense that,

$$\begin{cases} g(a) > g(b) \text{ iff } a \text{ is preferred to } b, \\ g(a) = g(b) \text{ iff } a \text{ is indifferent to } b. \end{cases}$$

The preferences can be represented by a criterion iff the relation R defined by aRb iff a is preferred or indifferent to b is a complete preorder.

In ELECTRE I, like any other outranking approach, the so-called *outranking concept* is schematized as follows: an alternative a outranks b if, given the information about the preferences of the DM, there are sufficient arguments to affirm that a is at least as good as b and there is no really important reason to refuse this assertion. This outranking concept is operationalized for choosing alternatives, sorting them into categories or ranking them from the best to worst.

The basic information is a set of n criteria $\{g_1, g_2, ..., g_n\}$ on A and, for each of them:

- A weight w_j expressing the relative importance of criterion g_j,
- A veto threshold $v_j(g_j) > 0$.

For each ordered pair (a,b), a *concordance index* $c(a,b)$ is calculated as,

$$c(a,b) = \frac{1}{W} \sum_{j:g_j(a)>g_j(b)} w_j,$$

where

$$W = \sum_{j=1}^{n} w_j.$$

This index varies from 0 to 1 and can be considered as a measure of the arguments in favor of the assertion "*a* outranks *b*." A *concordance level s* is chosen to help declare that a outranks b, denoted by aSb iff:

$c(a,b) \geq s$,

$\forall j$ such that $g_j(a) < g_j(b)$, the interval $(g_j(a), g_j(b))$ is smaller than $v_j(g_j(a))$.

One can multiply all the weights by the same number, and if the new weights are integers, the building of the outranking relations in ELECTRE I can be interpreted as a voting procedure with a special majority rule (characterized by the concordance level).

ELECTRE methods have received many practical applications. Roy et al. (1986) have used ELECTRE for determining which Paris metro stations should be renovated. An application of the ELECTRE TRI method for business failure prediction is presented by Zoupounidis and Dimitras (1998). Hokkanen and Salminen (1997) have applied ELECTRE III and IV to environmental problems. Salminen et al. (1998) have applied ELECTRE III, PROMETHEE I, II, and SMART in the context of four different real applications to environmental problems in Finland.

According to Hanne (1999), outranking methods have been criticized on some theoretical grounds. Despite their appeal, the lack of an axiomatic basis makes their underlying logic unsound, which often leads to paradoxical results (Ballestero and Romero, 1998). Alley (1983) (see also Gershon and Duckstein, 1983) has pointed out the possibility of obtaining dominated solutions with the ELECTRE approach and considers the ranking process to be a "mystery to the DM." Stewart (1992) has supposed some outranking procedures to be "difficult to verify empirically as models of human preferences."

2.6.1.6 PROMETHEE Methods

The PROMETHEE methods also belong to the class of outranking approaches, and were proposed by Brans et al. (1984). They are also briefly described in Bouyssou and Vincke (1997) and Vincke (1999). The following brief discussion is based on Vincke (1999).

The basic information used by the PROMETHEE methods is a set of n so-called generalized criteria (g_j, F_j) on A, and a weight w_j expressing the relative importance of criterion g_j for each of them.

A valued strict preference relation is defined by calculation, for each ordered pair (a,b), the quantity

$$\pi(a,b) = \frac{1}{W} \sum_{j=1}^{n} w_j F_j(a,b),$$

where

$$W = \sum_{j=1}^{n} w_j$$

and $F_j(a,b)$ is the degree of preference of a over b for criterion j.

$F_j(a,b)$ is a number between 0 and 1. The possible variations of this number in the interval $g_j(a) - g_j(b)$ are given in Figure 2.4. For each criterion g_j, based on discussions with the DM, a particular function F_j is chosen and the corresponding parameters $(q_j, p_j, \text{or } \sigma_j)$ are fixed.

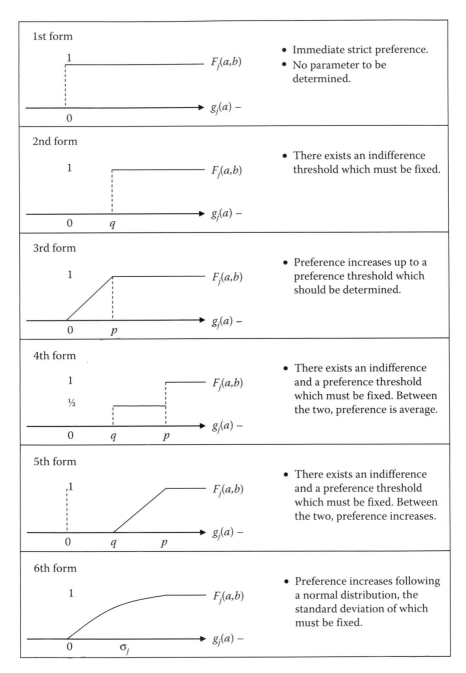

FIGURE 2.4
Preference functions used in PROMETHEE methods. (From Vincke, P. 1999.)

Just like the ELECTRE methods, two complete preorders are built: one consists of ranking the actions following the decreasing order of the numbers ($\phi^+(a)$),

$$\phi^+(a) = \sum_{b \in A} \pi(a,b) \quad \text{(outgoing flow)};$$

and the order following the increasing order of numbers ($\phi^-(a)$),

$$\phi^-(a) = \sum_{b \in A} \pi(b,a) \quad \text{(incoming flow)}.$$

Their intersection yields the partial preorder of the PROMETHEE I method. The PROMETHEE II method consists in ranking the actions following the decreasing order of the numbers $\phi(a)$ such that $\phi(a) = \phi^+(a) - \phi^-(a)$ and yields a unique preorder.

2.6.1.7 Fuzzy Set Theory

Fundamentally speaking, MCDM deals with imprecision in human preferences: Human beings are not able to specify their preferences unambiguously. It is not possible to exactly specify whether one alternative is preferred to another in terms of some criteria (and by how much). Fuzzy set theory has been proposed to deal with imprecision (Zadeh, 1965). The usefulness of fuzzy sets in management science has long been recognized (Bellman and Zadeh, 1970). One of the nice reviews of fuzzy set theory in operations research context is provided by Zimmermann (1983).

Fuzzy set theory assigns the so-called *membership function* (a measure of the degree of membership) to each alternative in the fuzzy set of good (or satisfactory, or the like) in terms of each criterion. The overall membership of the alternative (in terms of membership of all the criteria taken together) is normally defined as the intersection of all these single-criterion fuzzy sets.

Fuzzy sets have received several applications in MCDM literature. Smithson (1987) has described several applications of fuzzy set theory to behavioral and social sciences. Ammar and Wright (2000) have described three applications of fuzzy set theory to multi-criteria evaluation. Fuzzy sets have also been used in conjunction with other methods to incorporate the concept of fuzziness. For example, Bisdorff (2000) has described the application of fuzzy ELECTRE methods. Özelkan and Duckstein (2000) have used fuzzy sets with regression for carrying multi-criteria evaluation of rainfall–runoff relationships pertaining to hydrologic system models. Parra et al. (2001) have used fuzzy GP for portfolio selection.

Use of fuzzy sets has certain practical problems, including the assessment of membership functions, and the operational definition of fuzzy intersection. See Stewart (1992) for a brief discussion of these problems.

2.6.2 MODM Methods[*]

For most of the MODM problems, we can start with the following mathematical programming problem (presented earlier in Section 2.4).

$$\text{Max } F(x) = \{f_1(x),\ f_2(x),\dots, f_k(x)\}$$
$$\text{Subject to,} \quad g_j(x) \leq 0 \quad \text{for } j = 1,\dots, m,$$

(2.2)

where $\mathbf{x}$ is an *n*-vector of *decision variables* and $f_i(\mathbf{x})$, $i = 1, \dots, k$ are the *k criteria/ objective functions*.

Let $S = \{\mathbf{x}/g_j(\mathbf{x}) \leq 0, \text{ for all } j\}$,
$Y = \{\mathbf{y}/F(\mathbf{x}) = \mathbf{y} \text{ for some } \mathbf{x} \in S\}$.

S is called the *decision space* and Y is called the *criteria or objective space* of the MODM problem.

In MODM problems, there are often an infinite number of efficient solutions and they are not comparable without the input from the DM. Hence, it is generally assumed that the DM has a real-valued *preference function* defined on the values of the objectives, but it is not known explicitly. With this assumption, the primary objective of the MODM solution methods is to find the *best compromise solution*, which is an efficient solution that maximizes the DM's preference function.

In the last three decades, most MCDM researches have been concerned with developing solution methods based on different assumptions and approaches to measure or derive the DM's preference function. Thus, the MODM methods can be categorized by the basic assumptions made with respect to the DM's preference function as follows:

1. When *complete* information about the preference function is available from the DM.
2. When *no* information is available.
3. Where *partial* information is obtainable progressively from the DM.

In the following sections, we first talk about a simpler extension of SODM called MOLP. We will then discuss three more MODM methods—GP, *compromise programming*, and *interactive methods*, as examples of categories 1, 2, and 3 type approaches, respectively. More methods such as DEA are discussed later in this section.

---

[*] This section is based on Ravindran (2016).

2.6.2.1 Multi-Objective Linear Programming

Note that if there were only one objective function, the above problem can be easily solved by the traditional simplex method. It is possible to conceive a variation of the simplex method to deal with the case of multiple objectives. It is usually called the multi-criteria simplex method. This method has been described in detail in Zeleny (1982). The procedure is very similar to that of the single-criterion simplex method, except that the so-called minimum ratio rule for identifying the outgoing basic variable in any iteration is modified to take into the presence of many objective functions, and additional checks are incorporated for identifying whether the basic solution of any iteration is dominated or not. Note that the multi-criteria simplex method can identify only the non-dominated solutions of an MODM problem. As it incorporates no preference information from the DM, it cannot reduce the set of non-dominated solutions to one or a few that is acceptable to the DM.

MOLP has been applied to MODM problems. Zeleny (1982) discusses some applications. Leung et al. (2001) provide a recent application of MOLP to fish resource utilization.

2.6.2.2 Goal Programming (Ravindran et al., 2006)

GP, likely the oldest school of MCDM approaches, has been developed in the 50s as an extension of linear programming (Charnes and Cooper, 1961; Charnes et al., 1955). One way to treat multiple criteria is to select one criterion as primary and the other criteria as secondary. The primary criterion is then used as the optimization objective function, while the secondary criteria are assigned acceptable minimum and maximum values and are treated as problem constraints. However, if careful considerations were not given while selecting the acceptable levels, a feasible design that satisfies all the constraints may not exist. This problem is overcome by GP, which has become a practical method for handling multiple criteria. GP falls under the class of methods that use completely prespecified preferences of the DM in solving the MODM problem.

In GP, all the objectives are assigned target levels for achievement and relative priority on achieving these levels. GP treats these targets as *goals to aspire for* and not as absolute constraints. It then attempts to find an optimal solution that comes as "close as possible" to the targets in the order of specified priorities.

Before we discuss the formulation of GP models, we should discuss the difference between the terms *real constraints* and *goal constraints* (or simply *goals*) as used in GP models. The real constraints are absolute restrictions on the decision variables, while the goals are conditions one would like to achieve but are not mandatory. For instance a real constraint given by

$$x_1 + x_2 = 3$$

requires all possible values of $x_1 + x_2$ to always equal 3. As opposed to this, a goal requiring $x_1 + x_2 = 3$ is not mandatory, and we can choose values of $x_1 + x_2 \geq 3$ as well as $x_1 + x_2 \leq 3$. In a goal constraint, positive and negative deviational variables are introduced as follows:

$$x_1 + x_2 + d_1^- - d_1^+ = 3 \quad d_1^+, d_1^- \geq 0.$$

Note that, if $d_1^- > 0$, then $x_1 + x_2 < 3$, and if $d_1^+ > 0$, then $x_1 + x_2 > 3$.

By assigning suitable weights w_1^- and w_1^+ on d_1^- and d_1^+ in the objective function, the model will try to achieve the sum $x_1 + x_2$ as close as possible to 3. If the goal were to satisfy $x_1 + x_2 \geq 3$, then only d_1^- is assigned a positive weight in the objective, while the weight on d_1^+ is set to zero.

2.6.2.2.1 Goal Programming Formulation

Consider the general MODM problem given by Equation 2.2. The assumption that there exists an optimal solution to the MODM problem involving multiple criteria implies the existence of some preference ordering of the criteria by the DM. The GP formulation of the MODM problem requires the DM to specify an acceptable level of achievement (b_i) for each criterion f_i and specify a weight w_i (ordinal or cardinal) to be associated with the deviation between f_i and b_i. Thus, the GP model of an MODM problem becomes:

$$\text{Minimize } Z = \sum_{i=1}^{k} \left(w_i^+ d_i^+ + w_i^- d_i^- \right) \tag{2.4}$$

$$\text{Subject to}: f_i(x) + d_i^- - d_i^+ = b_i \quad \text{for } i = 1,\dots,k \tag{2.5}$$

$$g_j(x) \leq 0 \quad \text{for } j = 1,\dots,m \tag{2.6}$$

$$x_j, d_i^-, d_i^+ \geq 0 \quad \text{for all } i \text{ and } j. \tag{2.7}$$

Equation 2.4 represents the objective function of the GP model, which minimizes the weighted sum of the deviational variables. The system of equations (Equation 2.5) represents the *goal constraints* relating the multiple criteria to the goals/targets for those criteria. The variables, d_i^- and d_i^+, in Equation 2.5 are called *deviational variables*, representing *the under achievement* and *over achievement* of the *i*th goal. The set of weights (w_i^+ and w_i^-) may take two forms:

1. Prespecified weights (cardinal)
2. Preemptive priorities (ordinal)

Under prespecified (cardinal) weights, specific values in a relative scale are assigned to w_i^+ and w_i^- representing the DM's "trade-off" among the goals. Once w_i^+ and w_i^- are specified, the goal program represented by Equations 2.4 through 2.7 reduces to a single-objective optimization problem. The cardinal weights could be obtained from the DM using any of the methods discussed earlier, such as rating method and AHP. However, in order for this method to work, the criteria values have to be scaled properly.

In reality, goals are usually *incompatible* (i.e., incommensurable) and some goals can be achieved only at the expense of some other goals. Hence, *preemptive GP*, which is more common in practice, uses *ordinal ranking* or *preemptive priorities* to the goals by assigning incommensurable goals to different priority levels and weights to goals at the same priority level. In this case, the objective function of the GP model (Equation 2.4) takes the form

$$\text{Minimize } Z = \sum_p P_p \sum_i (w_{ip}^+ \, d_i^+ + w_{ip}^- \, d_i^-), \qquad (2.8)$$

where P_p represents priority p with the assumption that P_p is much larger than P_{p+1} and w_{ip}^+ and w_{ip}^- are the weights assigned to the ith deviational variables at priority p. In this manner, lower priority goals are considered only after attaining the higher priority goals. Thus, *preemptive GP* is essentially a sequence of single-objective optimization problems, in which successive optimizations are carried out on the alternate optimal solutions of the previously optimized goals at higher priority.

In both preemptive and non-preemptive GP models, the DM has to specify the targets or goals for each objective. In addition, in the preemptive GP models, the DM specifies a preemptive priority ranking on the goal achievements. In the non-preemptive case, the DM has to specify relative weights for goal achievements.

To illustrate, consider the following BCLP:

EXAMPLE 2.2 (BCLP)

$\text{Max } f_1 = x_1 + x_2$

$\text{Max } f_2 = x_1$

Subject to: $4x_1 + 3x_2 \leq 12$

$\qquad\qquad x_1, x_2, \geq 0$

Maximum f_1 occurs at $\mathbf{x} = (0, 4)$ with $(f_1, f_2) = (4, 0)$. Maximum f_2 occurs at $\mathbf{x} = (3, 0)$ with $(f_1, f_2) = (3, 3)$. Thus, the ideal values of f_1 and f_2 are 4 and 3, respectively, and the bounds on (f_1, f_2) on the efficient set will be:

$$3 \leq f_1 \leq 4$$
$$0 \leq f_2 \leq 3.$$

Let the DM set the goals for f_1 and f_2 as 3.5 and 2, respectively. Then the GP model becomes:

$$x_1 + x_2 + d_1^- - d_1^+ = 3.5 \tag{2.9}$$

$$x_1 + d_2^- - d_2^+ = 2 \tag{2.10}$$

$$4x_1 + 3x_2 \leq 12 \tag{2.11}$$

$$x_1, x_2, d_1^-, d_1^+, d_2^-, d_2^+ \geq 0. \tag{2.12}$$

Under the preemptive GP model, if the DM indicates that f_1 is much more important than f_2, then the objective function will be

$$\text{Min } Z = P_1 d_1^- + P_2 d_2^-$$

subject to the constraints (2.9) through (2.12), where P_1 is assumed to be much larger than P_2.

Under the non-preemptive GP model, the DM specifies relative weights on the goal achievements, say w_1 and w_2. Then the objective function becomes:

$$\text{Min } Z = w_1 d_1^- + w_2 d_2^-$$

subject to the same constraints (2.9) through (2.12).

2.6.2.2.2 *Partitioning Algorithm for Preemptive Goal Programs*

Linear Goal Programs: Linear GP problems can be solved efficiently by the *partitioning algorithm* developed by Arthur and Ravindran (1978, 1980a). It is based on the fact that the definition of preemptive priorities implies that higher-order goals must be optimized before lower-order goals are even considered. Their procedure consists of solving a series of linear programming subproblems by using the solution of the higher priority problem as the starting solution for the lower priority problem. Care is taken that higher priority achievements are not destroyed while improving lower priority goals.

Integer Goal Programs: Arthur and Ravindran (1980b) show how the partitioning algorithm for linear GP problems can be extended with a modified branch and bound strategy to solve both pure and mixed integer GP problems. They demonstrate the applicability of the branch and bound algorithm by solving a multiple-objective nurse scheduling problem (Arthur and Ravindran, 1981).

Non-Linear Goal Programs: Saber and Ravindran (1996) present an efficient and reliable method called the partitioning gradient-based (PGB) algorithm for solving non-linear GP problems. The PGB algorithm uses the partitioning technique developed for linear GP problems and the generalized reduced gradient

(GRG) method to solve single-objective non-linear programming problems. The authors also present numerical results by comparing the PGB algorithm against a modified pattern search method for solving several non-linear GP problems. The PGB algorithm found the optimal solution for all test problems proving its robustness and reliability, while the pattern search method failed in more than half the test problems by converging to a non-optimal point.

Kuriger and Ravindran (2005) have developed three intelligent search methods to solve non-linear GP problems by adapting and extending the simplex search, complex search, and pattern search methods to account for multiple criteria. These modifications were largely accomplished by using partitioning concepts of GP. The paper also includes computational results with several test problems.

Some of the major GP formulations, including MINSUM GP, least absolute value regression, MINMAX GP, preemptive GP, fractional GP, and non-linear GP, have been reviewed by Lee and Olson (1999). Aouni and Kettani (2001) have sketched the developments in the nearly 40 years of the history of GP.

Review articles on GP have been published in leading journals on a regular basis (Romero, 1986; Tamiz et al., 1995). Recently, Lee and Olson (1999) have provided a detailed overview of the applications of GP, in a variety of fields including engineering, operations management, business, agriculture, and public policy.

GP has been criticized by several authors (Min and Storbeck, 1991; Stewart, 1992; Zeleny, 1980). For example, it can, in some circumstances, choose dominated solution (Ballestero and Romero, 1998). The criticisms could be overcome by careful applications of the method. Some of the related issues are reviewed in Lee and Olson (1999).

2.6.2.3 *Method of Global Criterion and Compromise Programming*

Method of global criterion (Hwang and Masud, 1979) and *compromise programming* (Zeleny, 1982) fall under the class of MODM methods that do not require any preference information from the DM.

Consider the MODM problem given by Equation 2.2. Let

$$S = \{x/g_j(x) \le 0, \text{ for all } j\}.$$

Let the ideal values of the objectives $f_1, f_2, \ldots, f_k$ be $f_1^*, f_2^*, \ldots, f_k^*$. The method of global criterion finds an efficient solution that is "closest" to the ideal solution in terms of the L_p distance metric. It also uses the ideal values to normalize the objective functions. Thus the MODM problem reduces to:

$$\text{Minimize } Z = \sum_{i=1}^{k} \left(\frac{f_i^* - f_i}{f_i^*} \right)^p$$

subject to : $x \in S$.

The values of f_i^* are obtained by maximizing each objective f_i subject to the constraints $\mathbf{x} \in S$, *but* ignoring the other objectives. The value of p can be 1, 2, 3, ..., etc. Note that $p = 1$ implies equal importance to all deviations from the ideal. As p increases, larger deviations have more weight.

2.6.2.4 Compromise Programming

Compromise programming is similar in concept to the method of global criterion. It finds an efficient solution by minimizing the weighted L_p distance metric from the ideal point as given below.

$$\text{Min } L_p = \left[\sum_{i=1}^{k} \lambda_i^p \left(f_i^* - f_i \right)^p \right]^{1/p} \tag{2.13}$$

$$\text{subject to } \mathbf{x} \in S \text{ and } p = 1, 2, \ldots, \infty,$$

where λ_i's are weights that have to be specified or assessed subjectively. Note that λ_i could be set to $1/(f_i^*)$.

Theorem 2.3

Any point x^* that minimizes L_p (Equation 2.13) for $\lambda_i > 0$ for all i, $\Sigma \lambda_i = 1$ and $1 \le p < \infty$ is called a *compromise solution*. Zeleny (1982) has proved that these compromise solutions are non-dominated. As $p \to \infty$, Equation 2.13 becomes

$$\text{Min } L_\infty = \text{Min } \max_i \left[\lambda_i \left(f_i^* - f_i \right) \right]$$

and is known as the Tchebycheff metric.

Compromise programming has received some applications in the MCDM literature. For example, Romero (1996) has used this technique in environmental economics applications. Lee et al. (2001) have used compromise programming for helping goal setting process in rural telecommunications establishment. More than one MCDM method has been used in this article as the weights of different objective functions have been estimated using AHP.

2.6.2.5 Interactive Methods

Interactive methods for MODM problems rely on the progressive articulation of preferences by the DM. These approaches can be characterized by the following procedure:

Step 1: Find a solution, preferably feasible and efficient.

Step 2: Interact with the DM to obtain his/her reaction or response to the obtained solution.

Step 3: Repeat steps 1 and 2 until satisfaction is achieved or until some other termination criterion is met.

When interactive algorithms are applied to real-world problems, the most critical factor is the functional restrictions placed on the objective functions, constraints, and the *unknown* preference function. Another important factor is *preference assessment styles* (hereafter, called *interaction styles*). According to Shin and Ravindran (1991), the typical interaction styles are:

1. *Binary pairwise comparison*: The DM must compare a pair of two dimensional vectors at each interaction.
2. *Pairwise comparison*: The DM must compare a pair of p-dimensional vectors and specify a preference.
3. *Vector comparison*: The DM must compare a set of p-dimensional vectors and specify the best, the worst or the order of preference (note that this can be done by a series of pairwise comparisons).
4. *Precise local trade-off ratio*: The DM must specify precise values of local trade-off ratios at a given point. It is the *marginal rate of substitution* between objectives f_i and f_j: in other words, trade-off ratio is how much the DM is willing to give up in objective j for a unit increase in objective i at a given efficient solution.
5. *Interval trade-off ratio*: The DM must specify an interval for each local trade-off ratio.
6. *Comparative trade-off ratio*: The DM must specify his preference for a given trade-off ratio.
7. *Index specification and value trade-off*: The DM must list the indices of objectives to be improved or sacrificed, and specify the amount.
8. *Aspiration levels* (or reference point): The DM must specify or adjust the values of the objectives which indicate his/her optimistic wish concerning the outcomes of the objectives.

Shin and Ravindran (1991) also provide a detailed survey of MODM interactive methods. Their survey includes the following:

- A classification scheme for all interactive methods.
- A review of methods in each category based on functional assumptions, interaction style, progression of research papers from the first publication to all its extensions, solution approach, and published applications.
- A rating of each category of methods in terms of the DM's cognitive burden, ease of use, effectiveness, and handling inconsistency.

Lofti et al. (1992) have compared the performance of interactive methods with other methods such as AHP (implemented on the computer using the software *Expert Choice*), and according to them, interactive methods outperformed the latter on numerous measures, but for no measure was the reverse true.

2.6.2.6 Data Envelopment Analysis

We have seen that SMART or AHP simply aggregates the ratings of alternatives using criteria weights. However, if experts find it difficult to provide importance weights of the criteria, it is not possible to use these methods. An alternative in such cases is to identify *non-dominated* alternative, that is, the alternatives that have been rated better than, and not rated below, others. We have seen that the MOLP can help identify the non-dominated alternatives. The DEA is a relatively recent technique that can be effectively used to identify the non-dominated solutions in an MODM problem.

The methodology of DEA is relatively more complex compared to SMART or AHP as it uses linear programming for identifying the non-dominated alternatives. The method has been first proposed by Charnes et al. (1978) for assessing the relative performance of a set of firms that use a variety of identical inputs to produce a variety of identical outputs. Firms that produce maximum possible outputs using a given set of inputs or that consume minimum possible inputs for a given set of outputs are considered more efficient than others. To identify these efficient firms, a mathematical programming problem, that maximizes the ratio of weighted sum of outputs to weighted sum of inputs subject to the condition that similar ratios for all the firms are less than one, is used. Obviously, efficient firms will have the maximized value of their ratio to be unity or 100%.

DEA was originally developed as a tool for performance measurement (Ramanathan, 2003), and is a relatively late entrant to the field of MCDM. However, over the last few years, the linkages between the fields of DEA and MCDM have been explored, and DEA is now widely accepted as a tool for MCDM. The most fundamental use of DEA in MCDM is the identification of *non-dominated* alternatives. Note that identification of non-dominated alternatives does not require any preference information, and is purely based on available data. A non-dominated alternative is identified as the most efficient alternative in DEA with 100% efficiency.

Doyle and Green (1993) have highlighted that the absence of DEA in the set of MCDM approaches presented in the critical survey of MCDM techniques presented by Professor Stewart (1992) was an important omission, which was later accepted by Stewart (1994). Belton and Stewart (1999) further explored the relationship between DEA and MCDM from a decision theoretic perspective. Joro et al. (1998) provide in detail the comparisons between DEA and MOLP. Recent reviews of multi-criteria methods (Malczewski and Jackson, 2000; Urli and Nadeau, 1999) have included DEA as an MCDM tool in their discussion. The book on MCDM by Shi and Zeleny (2000) has included a separate section on DEA. One of the eight parts and three of the 21 papers included in the book are on DEA. Incorporating preference information in a DEA analysis provides a natural extension of DEA toward MCDM. This issue has been discussed for a long time in DEA literature (Halme et al., 1999; Thanassoulis and Dyson, 1992).

Since DEA was proposed more than two decades ago, the methodology has received numerous traditional as well as novel applications. Seiford (1996) has presented one of the recent bibliographies on DEA. A very comprehensive DEA bibliography is maintained at the University of Warwick by Emrouznejad (1995–2001). This bibliography is available on the Internet at the site, http://www.deazone.com/. Some prominent applications of DEA include the education sector (Ganley and Cubbin, 1992; Ramanathan, 2001b), banks (Yeh, 1996), comparative risk assessment (Ramanathan, 2001c), health sector (Bates et al., 1996), transport (Ramanathan, 2000), energy (Lv et al., 2015), environment (Bi et al., 2015; Fare et al., 1996; Zhao et al., 2016), and international comparisons on carbon emissions (Ramanathan, 2002).

2.6.3 Other MCDM Methods

Several MCDM methods, including those discussed in this chapter, have been reviewed by Jacquet-Lagrèze and Siskos (2001). They have listed several methods, mainly based on outranking concepts, provided a list of software support available to these methods, and reviewed their application areas. Miettinen (1999) provides an overview of several methods (including some described in this chapter) for non-linear MOO. As mentioned earlier, Guitouni and Martel (1998) have provided a comprehensive list of various MCDM methods, references to their application literature, and a comparative study of various methods, in their effort to provide tentative guidelines for choosing an appropriate method for a given application. Similarly, Zanakis et al. (1998) have compared performance of eight different MADM methods (ELECTRE, TOPSIS [the Technique for Order of Preference by Similarity to Ideal Solution], multiplicative exponential weighting, simple additive weighting, and four versions of AHP). A variety of applications of MCDM methods are available in Karwan et al. (1997).

The reference point approach developed at the International Institute of Applied Systems Analysis (IIASA) is another popular MODM method. It is closely related to the aspiration-level concept of GP. This approach is described in Wierzbicki (1998, 1999). The PRIME method or preference ratio through intervals in multi-attribute evaluation (Salo and Hamalainen, 2001) is a relatively recent one that deals with incomplete information about the preferences of the DMs. The MACBETH (measuring attractiveness by a categorical based evaluation technique) approach by Bana e Costa and Vansnick (1995) has been developed in Europe in the 1990s. It is an interactive approach for cardinal measurement of judgments about the degrees of attractiveness in decision processes.

Other methods based on outranking approach include QUALIFLEX method (Paelnick, 1978), ORESTE method (Roubens, 1982), and the TACTIC method (Vansnick, 1986). Jacquet-Lagrèze and Siskos (1982) have proposed the regression based UTA (utilitiés additives) method. An application of UTA method for business failure prediction is presented by Zoupounidis and Dimitras (1998). Another MCDM methodology is the ZAPROS method (Larichev, 1999, 2000; Larichev and Moshkovich, 1995). NIMBUS (Miettinen and Mäkelä, 2000) is an MOO method capable of solving non-differentiable and non-convex problems.

2.7 A General Comparative Discussion of MCDM Methodologies

In this section, various MCDM methods are compared in general terms with reference to some vital parameters when used in practical applications:

- *Uncertainty*: In terms of their purpose, it is possible to make comparisons between MCDM methods. Some methods, such as MAUT and PRIME, have been designed to deal with uncertainty. Initially, SMART and the AHP were developed for the analysis of decision problems where one of the available alternatives is to be selected in a situation where several incommensurate criteria need to be accounted for. Thus, they assume that the objectives and the alternatives are both known, whereby the uncertainties that pertain, for example, to the performance of the alternatives with regard to the criteria are not explicitly modelled. However, these methods, and the AHP, in particular, have found applications beyond this relatively narrow problem context, as they have supported the construction of R&D project portfolios, among others. Other methods are generally not designed to handle uncertainty.

- *Incomplete information*: Practical applicability of the methods is often constrained by the unavailability of complete information. For example, in MADM methods, DM may not want to comment on some criteria, or may not want to rate some alternatives on the basis of some criteria. This may arise when DM feels that he does not have the expertise to comment on the specific comparison. In MODM problems, the DM may not be able to compare some criteria. Note that, in many MODM methods, the preference information on criteria is often required as external input. For example, GP or compromise programming requires that the weights of criteria are available before the actual implementation of the methods, and normally do not specify how this information should be obtained.

- When only incomplete information is available on the preference information of the DM, it is important that the methodologies provide rankings based on the available information. Some MCDM methods can provide rankings based on incomplete information, for example, PRIME and MAHP.

- *Number of questions*: All the MADM methods require the elicitation of DMs' opinions for rating criteria and alternatives. MODM methods also require this elicitation for obtaining the information on importance of criteria. This may be done using written questionnaires, direct interviews, and/or by the use of suitable software. More often, the number of questions to be answered by the expert tends to rise with the number of elements (criteria or alternatives) that are to be compared. The problem is the worst in methods that are based on pairwise comparison of decision elements (criteria or alternatives), such as the AHP or the outranking methods. For example, for comparing n elements using AHP, the number of questions to be answered by an expert will be $n(n-1)/2$. This is because AHP elicits many redundant judgments from experts so that the consistency of their judgments can be verified. Of course, some modifications of AHP do not need so many comparisons, but they also do not verify consistency of judgments.

- *Checking the consistency of judgments*: Because of the redundant judgments, AHP is able to provide a check on the consistency of judgments given by the experts. This is done using the so-called the consistency index. More details are available in the last chapter. If the expert's judgments are not found to be consistent, then it is not advisable to use the judgments and the expert may be asked to provide his judgments once again. Note that many modifications of AHP do not provide this consistency check. Similarly, most of the other MCDM methods also do not provide a mathematically verifiable consistency check.

- *Group decisions*: Most practical uses of MCDM methods require elicitation of judgments from many DMs. It is then important to appropriately aggregate the judgments of all the DMs. This group decision problem is integral to all the MCDM methods. As indicated earlier, the best way to obtain the overall group judgments is to encourage unanimous decisions by the group through discussions. However, sometimes it may not be possible to arrive at a single group opinion by consensus, and it may be necessary to mathematically aggregate the opinions expressed by individual experts. In most of the MCDM methods, mathematical aggregation methods, usually in the form of weighted additive or multiplicative aggregation, where the weights are the relative importance measures associated with each DM, are used for the group preference aggregation. However, it

Bouyssou, D. 1990. Building criteria: A prerequisite for MCDA, In: C. A. Bana e Costa (Ed.), 58–80. *Readings in Multiple Criteria Decision Aid*. Springer, Berlin.

Bouyssou, D. and Ph. Vincke. 1997. Ranking alternatives on the basis of preference relations: A progress report with special emphasis on outranking relations. *Journal of Multi-Criteria Decision Analysis*. 6: 77–85.

Brans, J. P., B. Mareschal, and Ph. Vincke. 1984. Promethee: A new family of outranking methods in multicriteria analysis, In: J. P. Brans (Ed.), 408–421. *Operational Research '84*. Elsevier Science Publishers B.V, North-Holland.

Buede, D. M. 1992. Software review: Overview of the MCDA software market. *Journal of Multi-Criteria Decision Analysis*. 1: 59–61.

Charnes, A. and W. W. Cooper. 1961. *Management Models and Industrial Applications of Linear Programming*. Wiley, New York.

Charnes, A. W. and W. Cooper, and R. O. Ferguson. 1955. Optimal estimation of executive compensation by linear programming. *Management Science*. 1(2): 138–151.

Charnes, A., W. W. Cooper, and E. Rhodes. 1978. Measuring the efficiency of decision making units. *European Journal of Operational Research*. 2: 429–444.

Chen, H., R. H. Chiang, and V. C. Storey. 2012. Business intelligence and analytics: From big data to big impact. *MIS Quarterly*. 36(4): 1165–1188.

Doyle, J. and R. Green. 1993. Data envelopment analysis and multiple criteria decision making. *Omega*. 21(6): 713–715.

Duarte, B. P. M. 2001. The expected utility theory applied to an industrial decision problem—What technological alternative to implement to treat industrial solid residuals. *Computers & Operations Research*. 28: 357–380.

Dyer, J. S. 1990. Remarks on the analytic hierarchy process. *Management Science*. 36/3: 249–258.

Dyer, J. S., P. C. Fishburn, R. E. Steuer, J. Wallenius, and S. Zionts. 1992. Multiple criteria decision making, multiattribute utility theory: The next ten years. *Management Science*. 38(5): 645–654.

Edwards, W. 1977. How to use multiattribute utility measurement for social decision making. *IEEE Transactions on Systems, Man, and Cybernetics*. 7/5: 326–340.

Elkarni, F. and I. Mustafa. 1993. Increasing the utilization of solar energy technologies (SET) in Jordan: Analytic hierarchy process. *Energy Policy*. 21: 978–984.

Emrouznejad, A. 1995–2001. *Ali Emrouznejad's DEA HomePage*. Warwick Business School, Coventry CV4 7AL, UK.

Evans, G. W. 1984. An overview of techniques for solving multiobjective mathematical programs. *Management Science*. 30(11): 1268–1282.

Fare, R., S. Grosskopf, and D. Tyteca. 1996. An activity analysis model of the environmental performance of firms—application to fossil-fuel-fired electric utilities. *Ecological Economics*. 18: 161–175.

French, S. 1984. Interactive multi-objective programming: Its aims, applications and demands. *Journal of the Operational Research Society*. 35: 827–834.

French, S., R. Hartley, L. C. Thomas, and D. J. White. 1983. *Multi-Objective Decision Making*. Academic Press, London.

Ganley, J. A. and J. S. Cubbin. 1992. *Public Sector Efficiency Measurement: Applications of Data Envelopment Analysis*. North-Holland, Amsterdam.

Geoffrion, A. 1968. Proper efficiency and theory of vector maximum. *Journal of Mathematical Analysis and Applications*. 22: 618–630.

Gershon, M. and L. Duckstein. 1983. Reply. *Water Resources Research*. 19(1): 295–296.

Golden, B. L., E. A. Wasil, and D. E. Levy. 1989. Applications of the analytic hierarchy process: A categorized, annotated bibliography, In: B. L. Golden, E. A. Wasil, and P. T. Harker (Eds.), 37–58. *The Analytic Hierarchy Process: Applications and Studies*. Springer-Verlag, Berlin.

Govindan, K., S. Rajendran, J. Sarkis, and P. Murugesan. 2015. Multi criteria decision making approaches for green supplier evaluation and selection: A literature review. *Journal of Cleaner Production*. 98: 66–83.

Graves, S. B., J. L. Ringuest, and J. F. Bard. 1992. Recent developments in screening methods for nondominated solutions in multiobjective optimization. *Computers & Operations Research*. 19(7): 683–694.

Guitouni, A. and J.-M. Martel. 1998. Tentative guidelines to help choosing an appropriate MCDA method. *European Journal of Operational Research*. 109: 501–521.

Halme, M., T. Joro, P. Korhonen, S. Salo, and J. Wallenius. 1999. A value efficiency approach to incorporating preference information in data envelopment analysis. *Management Science*. 45: 103–115.

Hanne, T. 1999. Meta decision problems in multiple criteria decision making, In: T. Gal, T. J. Stewart, and T. Hanne (Eds.), *Multicriteria Decision Making: Advances in MCDM Models, Algorithms, Theory, and Applications*, Kluwer Academic Publishers, Boston, pp. 6.1–6.25.

Henig, M. I. and J. T. Buchanan. 1996. Solving MCDM problems: Process concepts. *Journal of Multi-Criteria Decision Analysis*. 5: 3–21.

Ho, W., X. Xu, and P. K. Dey. 2010. Multi-criteria decision making approaches for supplier evaluation and selection: A literature review. *European Journal of Operational Research*. 202(1): 16–24.

Hokkanen, J. and P. Salminen. 1997. ELECTRE III and IV decision aids in an environmental problem. *Journal of Multi-Criteria Decision Analysis*. 6: 215–226.

Howard, R. 1992. Heathens, heretics and cults: The religious spectrum of decision aiding. *Interfaces*. 22(6): 15–27.

Hwang, C. L. and A. Masud. 1979. *Multiple Objective Decision Making—Methods and Applications: A State of the Art Survey*. Lecture notes in economics and mathematical systems. 164, Springer-Verlag, Berlin.

Jacquet-Lagrèze, E. and J. Siskos. 1982. Assessing a set of additive utility functions for multicriteria decision making: The UTA method. *European Journal of Operational Research*. 10: 151–164.

Jacquet-Lagrèze, E. and Y. Siskos. 2001. Preference disaggregation: 20 years of MCDA experience. *European Journal of Operational Research*. 130: 233–245.

Jato-Espino, D., E. Castillo-Lopez, J. Rodriguez-Hernandez, and J. C. Canteras-Jordana. 2014. A review of application of multi-criteria decision making methods in construction. *Automation in Construction*. 45: 151–162.

Jones, M., C. Hope, and R. Hughes. 1990. A multi-attribute value model for the study of UK energy policy. *Journal of the Operational Research Society*. 41(10): 919–929.

Joro, T., P. Korhonen, and J. Wallenius. 1998. Structural comparison of data envelopment analysis and multiple objective linear programming. *Management Science*. 40: 962–970.

Kabir, G., R. Sadiq, and S. Tesfamariam. 2014. A review of multi-criteria decision-making methods for infrastructure management. *Structure and Infrastructure Engineering*. 10(9): 1176–1210.

Karwan, M. H., J. Spronk, and J. Wallenius. (Eds.). 1997. *Essays in Decision Making: A Volume in Honour of Stanley Zionts.* Springer-Verlag, Berlin.

Keeney, R. L. 1992. *Value Focused Thinking: A Path to Creative Decision-making.* Harvard University Press, Cambridge, MA.

Keeney, R. L. 1999. Developing a foundation for strategy at Seagate software. *Interfaces.* 29/6: 4–15.

Keeney, R. L. and T. L. McDaniels. 1999. Identifying and structuring values to guide integrated resource planning at bc gas. *Operations Research.* 47/5: 651–662.

Keeney, R. L. and H. Raiffa. 1976. *Decisions with Multiple Objectives: Preferences and Value Tradeoffs.* Wiley, New York. Another edition available in 1993 from Cambridge University Press.

Kirkwood, C. W. 1992. An overview of methods for applied decision analysis. *Interfaces.* 22(6): 28–39.

Korhonen, P. 1992. Multiple criteria decision support: The state of research and future directions. *Computers & Operations Research.* 19(7): 549–551.

Koundinya, S., D. Chattopadhyay, and R. Ramanathan. 1995. Combining qualitative objectives in integrated resource planning: A combined AHP—Compromise programming model. *Energy Sources.* 17(5): 565–581.

Kuriger, G. and A. Ravindran. 2005. Intelligent search methods for nonlinear goal programs. *Information Systems and Operational Research.* 43: 79–92.

Larichev, O. I. 1999. Normative and descriptive aspects of decision making, In: T. Gal, T. J. Stewart, and T. Hanne (Eds.), *Multicriteria Decision Making: Advances in MCDM Models, Algorithms, Theory, and Applications,* Kluwer Academic Publishers, Boston, pp. 5.1–5.24.

Larichev, O. I. 2000. Qualitative comparison of multicriteria alternatives, In: Yong, S. and M. Zeleny (Eds.), *New Frontiers of Decision Making for the Information Technology Era,* World Scientific Publishing Co. Pte. Ltd., Singapore, pp. 207–224.

Larichev, O. I. and H. Moshkovich. 1995. ZAPROS-LM—A method and system for ordering multiattribute alternatives. *European Journal of Operational Research.* 82(3): 503–521.

Lee, H., Y. Shi, S. M. Nazem, S. Y. Kang, T. H. Park, and M. H. Sohn. 2001. Multicriteria hub decision making for rural area telecommunication networks. *European Journal of Operational Research.* 133: 483–495.

Lee, S. M. and D. L. Olson. 1999. Goal programming, In: T. Gal, T. J. Stewart, and T. Hanne (Eds.), *Multicriteria Decision Making: Advances in MCDM Models, Algorithms, Theory, and Applications,* Kluwer Academic Publishers, Boston, pp. 8.1–8.33.

Leung, P. S., K. Heen, and H. Bardarson. 2001. Regional economic impacts of fish resources utilization from the Barents Sea: Trade-offs between economic rent, employment, and income. *European Journal of Operational Research.* 133: 432–446.

Lofti, V., T. J. Stewart, and S. Zionts. 1992. An aspiration-level interactive model for multiple criteria decision making. *Computers & Operations Research.* 19(7): 671–681.

Lootsma, F. A. 1999. *Multi-Criteria Decision Analysis Via Ratio and Difference Judgement.* Dordrecht: Kluwer Academic Publishers.

Lv, W., X. Hong, and K. Fang. 2015. Chinese regional energy efficiency change and its determinants analysis: Malmquist index and Tobit model. *Annals of Operations Research.* 228(1): 9–22.

Malczewski, J. and M. Jackson. 2000. Multicriteria spatial allocation of educational resources: An overview. *Socio-Economic Planning Sciences.* 34: 219–235.

Mareschal, B. 1988. Weight stability intervals in multicriteria decision aid. *European Journal of Operational Research.* 33: 54–64.

Mareschal, B. and J. P. Brans. 1988. Geometric representations for MCDA. *European Journal of Operational Research.* 34: 69–77.

Maxwell, D. T. 2000. Decision analysis: Aiding insight V—Fifth biennial survey. *ORMS Today.* October: 28–35. Also available on the internet at the URL: http://www.lionhrtpub.com/orms/surveys/das/das.html

Miettinen, K. 1999. *Nonlinear Multiobjective Optimization.* Kluwer Academic Publishers, Boston.

Miettinen, K. and M. M. Mäkelä. 2000. Interactive multiobjective optimization system WWW-NIMBUS on the Internet. *Computers & Operations Research.* 27: 709–723.

Millet, I. and P. T. Harker. 1990. Globally effective questioning in the analytic hierarchy process. *European Journal of Operational Research.* 48: 88–97.

Min, H. and J. Storbeck. 1991. On the origin and persistence of misconceptions in goal programming. *Journal of the Operational Research Society.* 42(4): 301–312.

Özelkan, E. C. and L. Duckstein. 2000. Multi-objective fuzzy regression: A general framework. *Computers & Operations Research.* 27: 635–652.

Paelnick, J. 1978. Qualiflex, a flexible multiple criteria method. *Economics Letters.* 3: 193–197.

Parra, M. A., A. B. Terol, and M. V. Rodríquez Uría. 2001. A fuzzy goal programming approach to portfolio selection. *European Journal of Operational Research.* 133: 287–297.

Pohekar, S. D. and M. Ramachandran. 2004. Application of multi-criteria decision making to sustainable energy planning—A review. *Renewable and Sustainable Energy Reviews.* 8(4): 365–381.

Ramanathan, R. 1997. A note on the use of goal programming for Multiplicative AHP. *The Journal of Multi Criteria Decision Analysis.* 6: 296–307.

Ramanathan, R. 1998. A multi-criteria methodology to the global negotiations on climate change. *IEEE Transactions on Systems, Man and Cybernetics—Part C: Applications and Reviews.* 28(4): 541–548.

Ramanathan, R. 1999. Selection of appropriate greenhouse gas mitigation options. *Global Environmental Change: Human and Policy Dimensions.* 9(3): 203–210.

Ramanathan, R. 2000. A holistic approach to compare energy efficiencies of different transport modes. *Energy Policy.* 28(11): 743–747.

Ramanathan, R. 2001a. A note on the use of analytic hierarchy process for environmental impact assessment. *Journal of Environmental Management.* 63(1): 27–35.

Ramanathan, R. 2001b. A data envelopment analysis of comparative performance of schools in the Netherlands. *OPSEARCH—The Indian Journal of Operational Research.* 38(2): 160–182.

Ramanathan, R. 2001c. Comparative risk assessment of energy supply technologies: A data envelopment analysis approach. *Energy—The International Journal.* 26: 197–203.

Ramanathan, R. 2002. Combining indicators of energy consumption and CO_2 emissions: A cross-country comparison. *International Journal of Global Energy Issues.* 17(3): 214–227.

Ramanathan, R. 2003. *An Introduction to Data Envelopment Analysis: A Tool for Performance Measurement.* SAGE Publications, New Delhi.

Yeh, Q. 1996. The application of data envelopment analysis in conjunction with financial ratios for bank performance evaluation. *Journal of the Operational Research Society.* 47: 980–988.

Zadeh, L. 1965. Fuzzy sets. *Information and Control.* 8: 338–353.

Zanakis, S. H., A. Solomon, N. Wishart, and S. Dublish. 1998. Multi-attribute decision making: A simulation comparison of select methods. *European Journal of Operational Research.* 107: 507–529.

Zeleny, M. 1980. The pros and cons of goal programming. *Computers & Operations Research.* 8: 357–359.

Zeleny, M. 1982. *Multiple Criteria Decision Making (McGraw-Hill Series In Quantitative Methods For Management).* McGraw-Hill Book Co., New York.

Zhao, L., Y. Zha, K. Wei, and L. Liang. 2016. A target-based method for energy saving and carbon emissions reduction in China based on environmental data envelopment analysis. *Annals of Operations Research.* (in press).

Zimmermann, H. J. 1983. Using fuzzy sets in operational research. *European Journal of Operational Research.* 13: 201–216.

Zionts, S. 1992. Some thoughts on research in multiple criteria decision making. *Computers & Operations Research.* 19(7): 567–570.

Zionts, S. 2000. Some thoughts about multiple criteria decision making for ordinary decisions, In: Yong, S. and M. Zeleny (Eds.), *New Frontiers of Decision Making for the Information Technology Era*, World Scientific Publishing Co. Pte. Ltd., Singapore, pp. 17–28.

Zoupounidis, C. and A. I. Dimitras. 1998. *Multicriteria Decision Aid Methods for the Prediction of Business Failure.* Kluwer Academic Publishers, Boston.

3

Basics of Analytics and Big Data

U. Dinesh Kumar, Manaranjan Pradhan, and
Ramakrishnan Ramanathan

CONTENTS

3.1 Introduction

Business analytics (BA) and big data have become essential components that every organization should possess to compete effectively in the market. Hopkins et al. (2010) claimed that analytics sophistication is one of the primary differentiators between high-performing and low-performing organizations. BA is a set of statistical, mathematical and machine-learning management tools, and processes used for analyzing the past data, to understand hidden trends, which can assist in problem solving and/or drive fact-based decision making in an organization. In the 1980s, many organizations did not collect data or the data was not in an appropriate form to derive insights. Organizations at that point in time found decision making and/or problem solving arduous due to the nonavailability of data; with the advent of enterprise resource planning

(ERP) systems, most of the organizations have ensured the availability of data, which could be called upon whenever needed. However, for effective and efficient problem solving and decision making thereby, the data stored within the ERP systems needed to be analyzed, and this gave birth to the use of analytics.

Analytics can be grouped into three categories: descriptive analytics, predictive analytics, and prescriptive analytics. Descriptive analytics deals with describing past data using descriptive statistics and data visualization; useful insights may be derived using descriptive analytics. Predictive analytics aims to predict future events such as demand for a product/service, customer churn, and loan default. Prescriptive analytics on the other hand provides an optimal solution to a given problem or offers the best alternative among several alternatives. In other words, descriptive analytics captures what happened, predictive analytics predicts what is likely to happen, and prescriptive analytics provides the best alternative to solve a problem. Although all three components of analytics are important, the value-add and the usage of different analytics components are shown in Figure 3.1. For all the hype around analytics, vast majority of organizations use descriptive analytics in the form of business intelligence (BI). Significantly, a smaller group of organizations use predictive analytics, mainly for forecasting; the number of organizations using prescriptive analytics is minimal at this point in time in comparison with descriptive and predictive analytics. However, it is interesting to note that the value-add to a company increases many fold if organizations were to use predictive and prescriptive analytics conjointly as compared to descriptive analytics alone.

Today, with the ever-growing use of the Internet, social media platforms, smartphones, and Internet of things (IOTs), the amount of data that gets

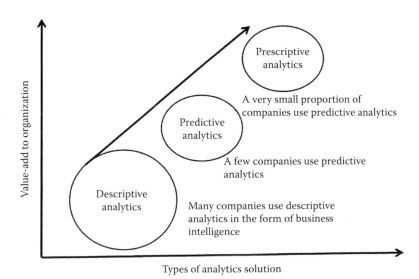

FIGURE 3.1
Types of analytics solution and the value-add.

generated everyday has increased several thousand fold over the past few years. An estimate by www.vcloudnews.com claims that 2.5 Exabytes of data gets generated every day and it will increase exponentially in the future; and all these data provide greater advantage to enterprises or organizations around the world that look to leverage these data to get diversified insights. Enterprises/organizations today can better understand their dynamic business environments, their customer's behavior and preferences, predict accurate market trends, weather forecasts, and thereby optimize resources at granular levels to increase efficiency to an extent they never believed was possible earlier.

One such wonderful case study was Google's flu trends, which predicts flu trends in real time even before public organizations like Centre for Disease Control (CDC) know it. Google built a system that extracted search terms, their search frequencies by regions and time from several billions of historical search requests it received over few years, and correlated with the actual incident reported by CDC; and after correlating millions of search terms with reported incidents, it found about 45 search terms which are highly correlated. Since 2009, Google has been using those very search term frequencies in real time to predict flu trends.

Unlike in the past, enterprises today do not only use transactional data for getting insights into business and customers but also use other sources of data like weblogs or clickstreams, social feeds, emails to get deeper understanding of customers. For example, e-commerce companies (which have mushroomed in recent times) do not wait until a new customer makes some purchase to understand his or her preferences, but know that from browsing patterns (i.e., by checking on the different links that the customer may have visited or spent time on). Thus, by this example one can possibly gage the data size that would be generated every day for these e-commerce sites. To illustrate this further and to put numbers, let us consider Amazon, which has about 188 million visitors (Anon, 2016) to its site every month, and it stores information about every single link they click, their wish lists, and purchases. With this humongous amount of data, it is a real challenge for e-commerce companies first to capture, store, and finally analyze the data for them to gain insights into their customers, understand their preferences, and thereby make purchase recommendations.

In this chapter, we will be discussing in detail various aspects of analytics and big data with few examples of real-life applications. Finally, we end with an example of how analytics is used in multicriteria decision making.

3.2 Analytics

The primary objective of analytics is enabling to take informed decisions as well as solve business problems. Organizations would like to understand the association between the key performance indicators (KPIs) and factors that have significant impact on the KPIs for effective management. Knowledge of relationship between KPIs and factors would then provide the decision maker with appropriate actionable items. Analytics thus is a knowledge repository consisting of statistical and mathematical tools, machine-learning algorithms, data management processes such as data extraction, transformation, and loading (ETL) and computing technologies such as Hadoop that create value by developing actionable items from data. Devonport and Harris (2009) reported that there was a high correlation between the use of analytics and business performance. They reported that a majority of high performers (measured in terms of profit, shareholder return, revenue, etc.) strategically apply analytics in their daily operations as compared to low performers.

The *theory of bounded rationality* proposed by Herbert Simon (1972) is becoming very evident in the current context of managing organizations and competing in the market. The increasing complexity of business problems, existence of several alternative solutions, and the paucity of time available for decision making demand a highly structured decision-making process using past data for the effective management of organizations. There are several reasons for the existence of bounded rationality such as uncertainty, incomplete information about alternatives, and lack of knowledge about cause and effect relationships between parameters of importance. Although decisions are occasionally made using the "highest paid person's opinion" (HiPPO) algorithm especially in a group decision-making scenario and Flipism (all decisions are made by flipping a coin), there is a significant change in the form of "data-driven decision-making" among several companies. Many companies use analytics as competitive strategy and many more are likely to use this in the near future, and here is why; a typical data-driven decision-making process uses the steps as shown in Figure 3.2.

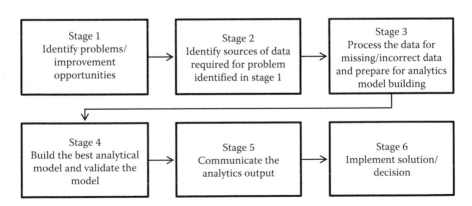

FIGURE 3.2
Data-driven decision-making flow diagram.

1. Identify the problem or opportunity for value creation.
2. Identify sources of data (primary as well as secondary data).
3. Preprocess the data for missing and incorrect data. Prepare the data for analytics model building, if necessary transform the data.
4. Build the analytical models and identify the best model using model validation.
5. Communicate the data analysis output and decisions effectively.
6. Implement solution/decision.

BA is a set of statistical, mathematical management tools and processes used for analyzing the past data that can assist in problem solving and/or drive fact-based decision making in an organization. Increasing complexities associated with businesses demand an unparalleled understanding of the customer expectations in order to serve better. According to the theory of bounded rationality proposed by Herbert Simon (1972), the human mind lacks the ability to choose and make the right decisions due to the complexity of problems that organizations face and the limited time available for decision making. In the 1980s, the culture of data collection was poor, several organizations did not collect data or the data was not in a form, which could be used for deriving insights, and this in turn resulted in organizations finding it difficult especially for prompt decision making. With the introduction of ERP systems, organizations today are ensured of the data availability that can be called upon whenever needed. However, the data sitting in the ERP systems need to be analyzed for problem solving and decision making; this need is now met using analytics. BA helps organizations to derive BI that helps organizations to manage the data-to-decision cycle. BA can be grouped into three types: descriptive analytics, predictive analytics, and prescriptive analytics. In the following sections, we discuss the three types of analytics in detail.

3.2.1 Descriptive Analytics

Descriptive analytics is the simplest form of analytics that mainly uses simple descriptive statistics, data visualization techniques, and business-related queries to understand the past data. As a utility tool, it captures the past and present data and is not able to predict the future, which is generally done by predictive analytics. Descriptive analytics is also known to be a power tool to communicate hidden facts and trends in data. The following are a few examples of descriptive analytics:

1. Most shoppers turn toward their right when they enter a retail store.
2. Men who kiss their wife before going to work earn more and live longer than those who do not.

3. Many marriages/relationships break up in January (known as relationship freeze).
4. Conversion rate among female shoppers is higher than male shoppers in consumer durable shops.

Simple descriptive statistics, charts (data visualization), or query can be used to extract such information, and at the same time it also lends very useful insights for businesses. For example, retailers keep products with higher profit on the right side of the store since most people turn right. There are many such strategies (or in the analytics nomenclature, "actionable items," used by decision makers).

Edward Tufte (2001), in his book titled *The Visual Display of Quantitative Information*, demonstrated how innovative visuals can be used to communicate data effectively. For businesses today, dashboards form the core of their BI and are an important element of analytics. Indian companies such as Gramener* have used innovative data visualization tools to communicate hidden facts in the data. Descriptive analytics may also be understood as the initial stages of creating the analytics capability; and descriptive statistics could help Small and Medium Enterprises (SME) uncover inefficiencies and thereby eliminate and/or minimize them.

Simple data analysis can lead to business practices that result in financial rewards. For instance, companies such as RadioShack and Best Buy found a high correlation between the success of individual stores and the number of female employees in the sales team (Underhill, 2009). Underhill (2009) also reported that the conversion rate (percentage of people who purchased something) in consumer durable shops was higher among female shoppers than among male shoppers. Sometimes, all you need is a simple query, which could even lead to fraud detection. Consider the following example: recently, China Eastern Airline found a man who had booked a first class ticket more than 300 times within a year and cancelled it just before its expiry for full refund so that he could eat free food at the airport's VIP lounge.† (It is surprising that the airline took so long to uncover this!)

3.2.2 Predictive Analytics

In the analytics capability maturity model (ACMM), predictive analytics comes after descriptive analytics and is the most important analytics capability that aims to predict the probability of occurrence of a future event such as customer churn, loan defaults, and stock market fluctuations. While descriptive analytics is used for finding what has happened in the past, predictive analytics is used for what is likely to happen in the future. The

---

* Source: https://gramener.com/
† Source: http://articles.timesofindia.indiatimes.com/2014-01-30/mad-mad-world/46827501_1_free-meals-single-ticket-issued-ticket

ability to predict a future event such as an economic slowdown, a sudden surge or decline in a commodity's price, which customer is likely to churn, what will be the total claims from auto insurance customer, how long a patient is likely to stay in the hospital, and so on will help organizations to plan their future course of action. Anecdotal evidence suggests that predictive analytics is the most frequently used type of analytics across several industries. The reason for this is that almost every organization would like to forecast demand for the products that they sell, the price of the material used by them, and so on. Irrespective of the type of business, organizations would like to forecast the demand for their products or services and understand the causes of demand fluctuations. For example, when Hurricane Charley struck the United States in 2004, Linda M. Dillman, Walmart's Chief Information Officer, wanted to understand the purchase behavior of its customers (Hays, 2004). Using data-mining techniques, Walmart found that the demand for strawberry pop-tarts went up over seven times during the hurricane compared to its normal sales rate; the prehurricane top-selling item was found to be beer. These insights were used by Walmart when the next hurricane—Hurricane Frances—hit the United States in August–September 2004; most of the items predicted by Walmart sold quickly. Although the high prehurricane demand for beer can be intuitively predicted, the demand for strawberry pop-tarts was a complete surprise. The use of analytics can reveal relationships that were previously unknown and are not intuitive.

The most popular example of the application of predictive analytics is Target's pregnancy prediction model. In 2002, Target hired statistician Andrew Pole; one of his assignments was to predict whether a customer is pregnant (Duhigg, 2012a). New parents are the holy grail of marketers since they are price-insensitive customers who would like to buy the best things for their new born. In 2010, it was reported that parents spent about USD 6800 on average on a child before his/her first birthday; the North American new baby market was worth USD 36.3 billion (Duhigg, 2012b). At the outset, the assignment put forward by the marketing department to Pole may look bizarre, but it made great business sense.

3.2.3 Prescriptive Analytics

Prescriptive analytics is the highest level of capability of analytics today, wherein firms decide what to do once they gain insights through descriptive and predictive analytics. Prescriptive analytics assists users in finding the optimal solution to a problem or in making the right choice/decision from among several alternatives. Unlike predictive analytics (which, in many cases, provides the probability of a future event), prescriptive analytics in most cases provides an optimal solution/decision to a problem. Operations research (OR) models form the core of prescriptive analytics. Ever since their introduction during World War II, OR models have been used in every sector and in every industry. The potent prescriptive analytics "tools" comprising of several applications have been widely used, and several companies across the world have benefitted from them.

Coca-Cola Enterprises (CCE) is the largest distributor of Coca-Cola products. In 2005, CCE distributed 2 billion physical cases containing 42 billion bottles and cans of Coca-Cola in the United States (Kant et al., 2008). CCE developed an OR model that would meet several objectives such as improved customer satisfaction and optimal asset utilization for its distribution network of Coca-Cola products from 430 distribution centers to 2.4 million retail outlets. The optimization model resulted in cost savings of USD 54 million and improved customer satisfaction. A similar distribution network problem (vehicle routing) was solved by the IIM Bangalore team for Akshaya Patra. The *Akshaya Patra Midday Meal Routing and Transportation Algorithm* (AMRUTA) was developed to solve the vehicle routing problem; this was implemented at Akshaya Patra's Vasanthapura campus, resulting in savings of USD 75,000 per annum (Mahadevan et al., 2013). A major challenge for any e-commerce company is to improve the conversion of visits to transactions and order sizes. Hewlett Packard (HP) established HPDirect.com in 2005 to build online sales. HP Global Analytics developed predictive and prescriptive analytics techniques to improve sales. The analytical solutions helped HP to increase conversion rates and order sizes (Rohit et al., 2013).

Inventory management is one of the teething problems that are most frequently addressed using prescriptive analytics. Samsung implemented a set of methodologies under the title "Short Life and low Inventory in Manufacturing" (SLIM) to manage all the manufacturing and supply chain problems. Between 1996 and 1999, Samsung implemented SLIM in all its manufacturing facilities, resulting in a reduction in the manufacturing cycle time of random access memory devices from more than 80 days to less than 30 days. SLIM enabled Samsung to capture additional markets worth USD 1 billion (Leachman et al., 2002). In the next section, we will discuss the more frequently used predictive and prescriptive analytics tools.

3.3 Big Data: Volume, Variety, Velocity, and Veracity

Big data has four main characteristics: volume, variety, velocity, and veracity. Volume specifies the need to deal with large amount of data; but how large is large enough to be called big data? One definition is the data is so large that it cannot be stored and processed in any of the traditional platforms that enterprises were using so far. Thus, it is related to the existing technology and its ability to store and process the data. There is a need for an alternate solution or platform that has the margin to scale in order to accommodate and process the exponentially increasing size of data.

Variety in big data refers to different types of data; all the data that are captured nowadays cannot be arranged into rows and columns, or in other words, not all data are structured. For example, data such as texts, images, and machine-generated data in its original form cannot be arranged in the traditional rows and columns. Such unstructured data can be in any form such as XML, Json, free text, images, audios, or videos, and so on and there is a need to analyze these data to get insights; for example, the sentiments expressed by customers on a product or service provided by a company are very important for improving the product and service overall.

Velocity is the rate of data growth. We have seen a little earlier that data is growing exponentially, and thus it is imperative to analyze data faster, almost in real time. As data starts to become backdated, their value diminishes hampering organizations to study and analyze data, which would have enabled them to improve service delivery and decision-making. Veracity refers to the quality and reliability of the data. Though we can leverage big data to get new insights, the real challenge lies in how to capture, store, and process these ever-increasing data size. Can the traditional platforms help or we need to take a fresh approach to deal with big data?

3.4 Limitations of Traditional Technologies for Big Data

Most of the traditional IT platforms were built to be deployed on single systems, and the amount of resources available on single systems is always limited. The default approach to add more compute and storage to a single system is called vertical scalability or scale-up approach. However, this approach is not highly scalable as beyond a certain point, no more resources can be added to a single system. The other approach is to keep adding more systems: this is an infinitely scalable approach and is called horizontal scalability or scale-out approach; but most of the software frameworks or platforms are not built to leverage this approach, mostly due to an increase in complexity at the software level. In a scale-out approach, the software typically needs to deal with more failure points due to the presence of multiple systems and complex coordination mechanisms thereof. Thus, a number of traditional softwares were developed to be a single system solution or scale-out in limited capacity, and therefore most of these systems cannot accommodate the sheer size of the data that need to be handled today.

3.5 Analytics Life Cycle

Typically in analytics, the data follows a life cycle that has multiple stages, such as:

- Data capture
- Data store
- Prepare
- Analyze
- Share

3.5.1 Data Capture, Store, Prepare, Analyze, and Share

Data Capture is the stage where data is received from multiple sources such as transactional databases, customer surveys, financial reports, weblogs, message queues, web services, or social feeds. It is to be noted here that at times the incoming data rate could be very high, which in effect could mean millions of messages in a second or minutes or billions of records per day. Let us look at an example: an energy analytics provider in North America called OPower receives about 200 readings from a single customer household every day and it has about 25 million household customers, which effectively translates to 5 billion records per day (25 million × 20)—the challenge quite understandably is to design a system that would capture those records within a single system. This is the big data challenge at the capture stage itself, as there are hardly any traditional platforms that can look to store several billion writes per day.

However, the primary problem is not just capturing the large volume of data in this case. Enterprises want the data to be stored for months and years altogether for future analysis. If the data received in a day is about 2 or 3 Tera bytes (TBs), by end of the year it becomes almost 1 Peta byte (PB). Therefore, in order to handle such huge data, enterprises need a platform that has elasticity and can scale as the data grows to store it for a longer period of time, thereby giving a vital leverage to build complex analytical systems. This is the big data problem at the *data store* stage.

Once the data is stored, the next herculean task is to navigate through these data sets and extract only the relevant information for analytics. In most cases, the captured data are not fit for analysis; lot of cleaning, aggregations, filtering, and data munging tasks need to be applied on the raw data in order to make it ready for analytics. Not to forget the series of algorithms, which has to scan 100 TBs and PBs of data, needing super computational capacity; this is the *data prepare* stage.

Analyze is a stage where the prepared data sets are used for applying algorithms like Structured Query Language (SQL) queries, statistical techniques, or machine-learning algorithms. In the final stage, that is, *share*, the analytics output is shared for visualization or fed to another system for usage. There could be a big data problem in any one or more of the above stages. We will discuss each stage in detail in later sections. The analytics life cycle is shown in Figure 3.3.

Data from multiple sources come through various channels such as files, web services, message queues, plain Transmission Control Protocol (TCP) sockets, or Rich Site Summary (RSS) feeds. Data does also flow at different rates, like a few dumps every hour or continuous streaming messages every second like social feeds or transactional records or messages from devices or online transactional processing (OLTP) systems. If the data inflow rate is very high, then the systems which capture these data are in all likelihood to run out of space and in the end may collapse; thus newer systems need to be designed and ready to be used as a queue or buffer between source systems and data capture systems. And finally, the data capture systems need to be scaled out to support high volume writes. The entire architecture of data capture in atypical big data environment is shown in Figure 3.4.

In summary, data capture poses three challenges:

- The ability to integrate multiple sources
- The ability to match the arrival rate of messages and capture rate
- The ability to capture/write data faster (millions of writes per second)

FIGURE 3.3
Analytics life cycle.

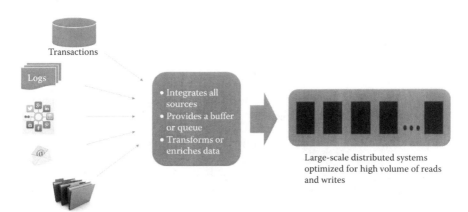

FIGURE 3.4
Data capture in big data environment.

There are open-source tools like *Apache Flume* which acts as an intermediate system listening to multiple sources and aggregates, transforms or enriches, and writes to systems capturing the data. Tools like *Apache Kafka* works as a queuing system between the source and destination with different rates of arrival and capture of data, and has the ability to scale out to a great extent to support millions of messages flowing through the system per second. Most of the Not only SQL (NoSql) Systems are designed to read and write at very high volume (millions of reads and writes per second) by scaling out systems.

3.5.2 NoSql Systems

Traditional relational database management systems (RDBMS) enforce schema very strictly. All data that are captured need to conform to the schema. Any minor changes to schema will end up altering all existing records and prove to be very expensive. In today's world, where data model keeps evolving and changing very rapidly, using this approach is very hard to manage. All we need is a system that would allow schema-less design and allow defining schema dynamically where data is inserted. This in turn will allow records to follow an overall schema, but is by and large free to add or remove specific attributes or elements to it.

Another challenge with RDBMS is that they are not very conducive for storing real-world object which are more hierarchical in nature; RDBMS tend to be more flat in structure and need an OR (Object to Relational) mapping for storing data. This approach is expensive as OR mapping divides the objects and stores into multiple tables. This makes all reads and writes of the objects very costly and slow and cannot perform high volume reads and writes.

Finally, most of the RDBMS are not highly scalable systems. If we have to store billions of records, we will need a system that can scale out to 100 s and 1000 s of servers. Traditional RDBMS systems do not support this architecture.

So, NoSql systems were developed to support these three features.

- Support schema-less design
- Support storing real-world object forms (deformalized forms)
- Highly scalable

If a table is very large and cannot be stored in a single system, NoSql system splits the table into multiple shards and distributes these shards across several servers. Now each server dealing with a set of shards manages all reads and writes for those shards. By splitting the tables appropriately into several hundred shards, we can load balance high volume incoming reads and writes to different servers and hence can support large-scale data ingestions.

There are four types of NoSQL databases:

- *Key-value store*: These databases are designed for storing data in a key-value fashion like a map. Each record within consists of an indexed key and a value. Examples: DyanmoDB, Reddis, Riak, BerkeleyDB.

- *Document database*: Store data in key-value fashion where values are stored as "documents," which are designed to store complex structure or objects. Each record is a document and assigned a unique key, which is used to retrieve the document. A document can be a JavaScript Object Notation (JSON, a lightweight data-interchange format) messages; the document elements can be indexed for advanced searches. Examples: MongoDB and CouchDB.
- *Column family store*: Store records with grouping-related columns as column families. Each record needs to have a set of column families, but can have different column attributes in their respective column families. These stores offer very high performance and a highly scalable architecture. Examples: HBase and HyperTable.
- *Graph database*: Based on graph theory, these databases are designed for data whose relations are well represented as a graph with edges and nodes.

3.5.3 Data Store: Distributed File Systems

Traditional file systems have failed to scale out sufficiently in order to store the large influx of data in recent times, and therefore new distributed file systems have been developed; some among them are:

- Google File System (GFS)
- GlusterFS
- Hadoop-Distributed File System (HDFS)
- OneFS
- XtremeFS

Among these, HDFS has been very popular as it is a part of Hadoop Distribution, which has become a default big data platform for many enterprises. We will explain the HDFS architecture at a high level in this section; other distributed file systems are pretty much similar in architecture (Figure 3.5).

HDFS is an abstract level file system which splits the files and stores across a cluster of machines called datanodes and maintains the metadata in a master machine called namenode. The metadata provides information about which file is split into how many chunks and stores in what all datanodes, and it facilitates read and write operations by redirecting clients to the appropriate datanodes. After splitting a file into multiple chunks, each chunk is replicated multiple times and stored in different datanodes so that it can withstand datanodes failures. As metadata is critical for all operations on the file systems, it is a stand by for the master node (namenode), in case the primary or active namenode fails.

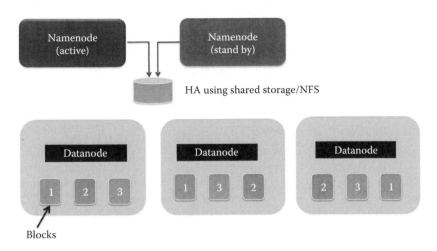

FIGURE 3.5
Hadoop-distributed system.

As data grows, more datanodes can be added to the cluster, thereby increasing its storage capacity. Theoretically it can grow to infinite scale. The largest deployed HDFS cluster in an enterprise today, at the time of this book writing, stores around 512 PBs of data.

3.6 Prepare and Analyze in Big Data

3.6.1 MapReduce Paradigm

After storing large volume of data, the next challenge is to process and analyze them. Conventional way of reading data from filesystem and presenting to a program or process cannot work anymore as the data is large and stored across multiple systems. So, now the program or process needs to be moved to the machines where data is stored. So, the programming paradigm of data to the process needs to change to process to data.

MapReduce programming paradigm as shown in Figure 3.6 provides the above feature of distributing the algorithms to process each split in datanodes and then provides a mechanism to distribute the intermediate outputs to different reduce nodes for final consolidation. The real challenge is to decompose every algorithm to map and reduce stages and a mechanism to distribute the data between maps and reduce stages.

3.6.2 Hadoop Ecosystem

Apache Hadoop is one of the most widely adopted platforms for storing and processing big data. The Hadoop framework has three core layers or

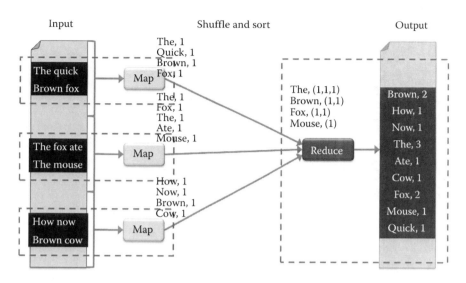

FIGURE 3.6
MapReduce approach to a simple word count task.

components; the bottom-most component is the distributed file system (as explained in the previous section), which is designed to scale out and store TBs and PBs of data. The next layer is Yet Another Resource Negotiator (YARN), which is responsible for keeping track of how many systems there are in a cluster, what resources are being currently consumed, what is remaining, and how to schedule analytical task on different systems as well as load balance the tasks. The final layer is MapReduce, which enables developing and executing tasks in parallel on the cluster nodes.

Hadoop is available as part of Apache Open Source license and being developed by a very large community of developers who work for companies such as Yahoo, Facebook, Twitter, LinkedIn, and so on. The software is free and designed to run on commodity hardware, which also brings down the cost of the implementation. One of the largest clusters is deployed by Yahoo and runs 40,000 nodes and stores, and processes around 512 PBs of data.

The Hadoop ecosystem as shown in Figure 3.7 also provides a few more vital components like Hive, Pig, and Mahout. As Hadoop is primarily developed in Java, the map-reducing algorithms need to be developed in Java, which in itself is a constraint as most data analysts are not familiar with programming languages. So, the Hadoop Ecosystem has developed a few abstractions like Hive, which supports SQL syntax or Pig, which supports a scripting interface for analyzing structure as well as unstructured data like log files or social feeds. Mahout is a dedicated library for machine-learning algorithms to be applied on big data. It supports algorithms such as Regression, Classification, Clustering, and Collaborative filtering.

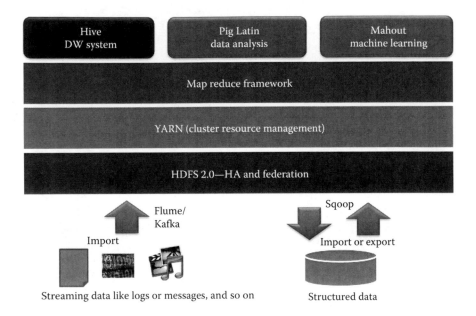

FIGURE 3.7
Hadoop and its ecosystem.

After data is analyzed and an insight is derived, these can be integrated with BI tools to visualize—like create graphs, charts, and so on. Or in some cases, big data platforms can be used to build analytical models, which can be then fed into real-time systems to make predictions in real time.

3.7 Big Data Analytics, Multicriteria Decision Making, and BI

One of the primary objectives of analytics and big data is to assist organizations with decision making. Organizations are engaged in collecting big data and analyzing the data using suitable analytics methods because they expect to generate useful business insights by doing so. As highlighted in Section 3.2, a number of descriptive, predictive, and prescriptive analytic tools are available in the literature. Mathematical modelling tools are a subset of these analytic tools.

MCDM methods presented in the previous chapter are a class of mathematical modelling tools when decisions have to be made in the presence of multiple criteria and sufficient data for developing such models can be generated. Many decision-making problems encountered by organizations have multiple criteria, and an optimal decision has to be arrived considering all criteria. One such example is performance-based contracts (PBC) that are becoming very popular among capital equipment users especially in the defense and the aerospace industry. In PBC, the customers demand performance as measured through multiple criteria such as reliability, availability, total cost of ownership, and logistic foot print.

Predicting reliability and availability involves collecting historical failure and maintenance data and finding the probability distribution of time to failure and time to maintain distribution function. Estimating cost of ownership will involve breaking down the total cost into different cost components and predicting the future costs (such as operation and maintenance costs) throughout the life of the systems. The original equipment manufacturer will have to use prescriptive analytics techniques optimize the various KPIs to provide equipment under PBC.

More examples are available in the remaining chapters of this book. The role of MCDM in big data is illustrated in Figure 3.8. Big data is important because the modern digital economy generates huge volumes of data every day and this data can be used to understand customers and ultimately help businesses. It has to be noted that data as such will not be sufficient for making business decisions. There are a number of analytic tools and MCDM models are one class of the whole range of analytics tools. However, big data and MCDM models (and other analytics tools) will not be sufficient; one

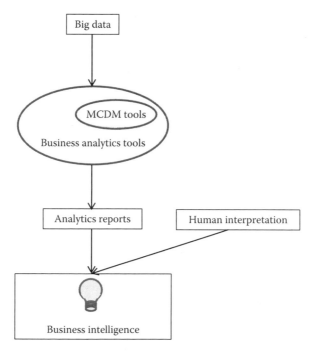

FIGURE 3.8
Big data analytics, MCDM, and business intelligence.

needs a shrewd business mind to make sense of the results from these analytics tools to generate BI insights (Ramanathan et al., 2012).

3.8 Conclusions

As enterprises try to understand their respective businesses more deeply to constantly deliver value to their customers, data is going to be the core focus in the time to come. As Angela Ahrendts, CEO of Burberry, says "whoever unlocks the reams of data and uses it strategically will win" (Andrew Gill, 2013). Hence, enterprises will have to deal with challenges of managing and analyzing big data. As technologies evolve over a period of time, enterprises need to adopt and learn quickly to benefit from it and remain competitive. Similarly, governments are also adopting digital platforms to administer and deliver services and quality life to its citizens. One such example is universal identification service (UID) implementation by Government of India. It has reached a billion users and promises to be the core platform to manage and deliver services under various social service programs.

Bibliography

Anon 2012. 2.5 Quintillion bytes created each day. *Storage Newsletter.com*, http://www.vcloudnews.com/every-day-big-data-statistics-2-5-quintillion-bytes-of-data-created-daily/, accessed on April 6, 2016.

Anon 2016. Most popular websites in the United States as of September 2015 ranked by visitors. http://www.statista.com/statistics/271450/monthly-unique-visitors-to-us-retail-websites/, accessed on April 6, 2016.

Brody, H., Rip, M.R., Johansen, P.V., Paneth, N., and Rachman, S. 2000. Map making and myth making in broad street: The London Cholera epidemic 1954. *The Lancet*, Vol. 356, pp. 64–69.

Bruhadeeswaran, R. 2012. Shubham Housing Development Finance raises $7.8M from Elevar, Helion, Others. *Vccircle*. www.vccircle.com/news/2012/11/12/shubham-housing-development-finance-raises-78m-elevar-helion-others.

Coles, P.A., Lakhani, K.R., and Mcafee, A.P. 2007. Prediction markets at Google. *Harvard Business School Case* (No 9-607-088).

Devonport, T.H. 2006. Competing on analytics. *Harvard Business Review*, January, pp. 1–10.

Devonport, T.H. and Harris, J.G. 2009. *Competing on Analytics—The New Science of Winning*, Harvard Business School Press, Boston, MA.

Devonport, T.H., Iansiti, M., and Serels, A. 2013. Managing with analytics at Proctor and Gamble. *Harvard Business School Case* (Case Number 9-613-045).

Devonport, T.H. and Patil, D.J. 2012. Data scientist: The sexiest job of 21st century. *Harvard Business Review*, 2–8.

Duhigg, C. 2012a. How companies learn your secret. *New York Times*, February 16, 2012.

Duhigg, C. 2012b. *The Power of Habit: Why We Do What We Do in Life and Business.* William Heinemann, London.

Grill, A. 2013. IBM CEO Ginni Rometty believes big data and social will change everything—How about other CEOs. *Leadership, Social Business and Social Media.* http://londoncalling.co/2013/03/ibm-ceo-ginni-rometty-believes-big-data-and-social-will-change-everything-how-about-other-ceos/

Hayes, B. 2013. First links in the Markov chain. *American Scientist*, Vol. 101, pp. 92–97.

Hays, C.L. 2004. What Wal-Mart knows about customers' habits. *New York Times*, November 14, 2004.

Hopkins, M.S., LaValle, S., Balboni, F., Kruschwitz, N., and Shockley, R. 2010. 10 insights: A first look at the new intelligence enterprise survey on winning with data. *MIT Sloan Management Review*, Vol. 52, pp. 21–31.

Howard, R. 2002. Comments on the origin and applications of Markov decision processes. *Operations Research*, Vol. 50, No. 1, pp. 100–102.

Kant, G., Jacks, M., and Aantjes, C. 2008. Coca-Cola enterprises optimizes vehicle routes for efficient product delivery. *Interfaces*, Vol. 38, pp. 40–50.

Leachman, R.C., Kang, J., and Lin, V. 2002. SLIM: Short cycle time and low inventory in manufacturing at Samsung Electronics. *Interfaces*, Vol. 32, pp. 61–77.

Lewis, M. 2003. *Moneyball: The Art of Winning an Unfair Game.* W W Norton & Company, London, UK.

Mahadevan, B., Sivakumar, S., Dinesh Kumar, D., and Ganeshram, K. 2013. Redesigning mid-day meal logistics for the Akshaya Patra foundation: OR at work in feeding hungry school children. *Interfaces*, Vol. 43, No. 6, pp. 530–546.

Mayer-Schonberger, V. and Cukier, K. 2013. *Big Data: A Revolution That Will Transform How We Live, Work and Think.* John Murray, New York.

Ramanathan, R., Duan, Y., Cao, G., and Philpott, E. 2012. Diffusion and impact of business analytics in UK retail: Theoretical underpinnings and a qualitative study. EurOMA Service Operations Management Forum, Cambridge, September 19–20, 2012.

Rohit, T., Chakraborty, A., Srinivasan, G., Shroff, M., Abdullah, A., Shamasundar, B., Sinha, R., Subramanian, S., Hill, D., and Dhore, P. 2013. Hewlett Packard: Delivering Profitable Growth for HPDirect.Com Using Operations Research. Interfaces, Vol. 43(1), pp. 41–61.

Savant, M.V. 1990. Ask Marilyn. *Parade Magazine*, p. 16, September 9, 1990.

Siegel, E. 2013. *Predictive Analytics: The Power to Predict Who Will Click, Buy, Lie or Die.* John Wiley and Sons, Hoboken, NJ.

Simon, H. 1972. Theories of bounded rationality. In *Decisions and Organizations.* McGuire, C.B., and Radner, R. (Eds.). North-Holland Publishing Company, New York, pp. 161–176.

Sirkin, H.L., Keenan, P., and Jackson, A. 2005. The hard side of change management. *Harvard Business Review*, Vol. 83, No. 10, pp. 109–118.

Snow, S.J. 1999. Death by Water: John Snow and cholera in the 19th century. *Liverpool Medical Institution.* http://www.lmi.org.uk/Data/10/Docs/11/11Snow.pdf

Stelzner, M.A. 2013. *2013 Social Media Marketing Industry Report—How Marketers Are Using Social Media to Grow Their Businesses.* Social Media Examiner Report 2013.

Suhruta, K., Makija, K., and Dinesh Kumar, U. 2013. 1920 Evil Returns—Bollywood and Social Media Marketing, IIMB Case Number IMB437.

Tandon, R., Chakraborty, A., Srinivasan, G., Shroff, M., Abdulla, A., Shamsundar, B., Sinha, R., Subramaniam, S., Hill, D., and Dhore, P. 2013. Hewlett Packard: Delivering profitable growth for HPDirect.Com using operations research. *Interfaces*, Vol. 43, No. 1, pp. 48–61.

Tufte, E. 2001. *Visual Display of Quantitative Information*, Graphics Press, Connecticut.

Underhill, P. 2009. *Why We Buy: The Science of Shopping*. Simon & Schuster (paperback), New York.

4

Linear Programming (LP)-Based Two-Phase Classifier for Solving a Classification Problem with Multiple Objectives

Sakthivel Madankumar, Pusapati Navya, Chandrasekharan Rajendran, N. Srinivasa Gupta, and B. Valarmathi

CONTENTS

4.1 Introduction

Classification models are used to predict the group/category, to which a new observation belongs, based on the training data set or set of observations for which the categories are known in advance. Linear programming (LP)-based classifiers for classification problem were discussed by Freed and Glover (1986), and the objective function of the classifier/LP model was either set to minimize the sum of errors, or set to minimize the maximum error for the training data set. Multiobjective LP model was used by Shi et al. (2001) for data mining in portfolio management. LP-based classifiers provide good results in terms of accuracy when the data set is linearly separable. To handle and to improve the accuracy of the classifier when the data set is not linearly separable, researchers and practitioners mostly consider logistic regression, support vector machines, and artificial neural networks (ANN) as classifiers. The problem of maximizing the accuracy of classification is computationally hard, and the methods such as LP-based classifier, logistic regression, support vector machines, and ANN work on their specific/respective objective functions with respect to each classifier, which indirectly improves the accuracy of the classification. In order to improve the accuracy of the conventional LP model when the data is not linearly separable, the concept of fuzzy measure was introduced and utilized by Yan et al. (2006).

The big data is generally characterized by volume, velocity, variety, and value of data. The technological advancements and the Internet of things (IoT) enable the massive collection of data with high velocity from various sources. The big data analytics is the process of understanding the hidden pattern in this large volume of data to get better insights about the data.

The multiple criteria decision making (MCDM) is a topic in Operations Research which deals with decision problems that involve multiple objectives. In order to solve the decision problem with multiple objectives, the researchers and practitioners consider goal programming or epsilon constraint-based models. The MCDM models can effectively be applied in the field of big data analytics to solve the underlying decision problems.

This chapter first proposes a mixed integer linear programming (MILP) model in order to maximize the accuracy of the classification. Since the problem is computationally hard, the MILP model can only solve the data set of small samples (e.g., in order of 100 samples) with reasonable execution time.

In big data analytics, we often encounter the classification problem which deals with large data sets. The study mainly concentrates on the development of computationally efficient LP-based classifiers that can handle the large volume of data, and also the ability to identify the non-dominated set of solutions with respect to multiple objectives.

In order to handle the large volume of data, batch processing technologies are used (e.g., Apache Hadoop) and to handle the high velocity of data, stream processing technologies are used (e.g., Apache Spark or Apache Storm). These technologies help in processing the big data to derive its meaning and also to convert the unstructured data from various sources into structured data. Once we have the structured data, we can apply the proposed LP-based classifiers to derive the hidden pattern/decisions in this large volume of data such that we can identify the non-dominated set of solutions with respect to multiple objectives.

So, in the next part of the chapter, we consider the conventional LP-based classifier with crisp boundary (with respect to the categories of classification) which can only solve the data set that is linearly separable. Subsequently, to further improve the accuracy, we propose an LP-based classifier with crisp boundary and with the consideration of interaction among the attributes and the contribution of attributes from their higher-order polynomial degrees. Our approach (unlike the conventional LP-based approach for classification) is able to capture the curvilinear/nonlinear boundary between the categories. In order to improve the accuracy, especially with respect to multiple objectives, and to produce non-dominated set of solutions with respect to multiple objectives, we propose a two-phase classifier for classification. Our LP-based two-phase classifier considers the interaction among the attributes and the contribution of attributes from their higher-order polynomial degrees to solve the data set that is not linearly separable. In the first phase, our proposed two-phase classifier runs the LP model by considering the bandwidth of boundary, and in the second phase, the classifier transforms the bandwidth of boundary into a crisp boundary, and in this process, the proposed classifier identifies a non-dominated set of solutions with respect to multiple objectives. Practitioners mostly consider the receiver operating characteristics (ROC) curve to determine the best threshold settings for the classifier, whereas the proposed LP-based two-phase classifier identifies the non-dominated set of solutions with respect to multiple objectives.

To evaluate the performance of the proposed LP-based classifiers, we consider two data sets that are already available in the literature (Yan et al., 2006; Maas et al., 2011). We also compare the accuracy of all the proposed LP-based classifiers with ANN, and the results indicate that the proposed LP-based two-phase classifier is able to give good results even when the data is not linearly separable, proving our claim that our LP-based two-phase classifier can handle the data set that is not linearly separable.

4.2 Research Problem Description and Assumptions

A set of observations for which the categories (two predetermined categories: category 0 and category 1) are known in advance is referred to as training data set, and a set of observations for which the categories need to be predicted is referred to as test data set or validation data set. Observations in both training data set as well as test data set have same number of features/attributes, and for the training data set, the value for the target variable (category) is known in advance and for the test and validation data set, the classifier predicts the value of the target variable on the basis of the training data set.

In Section 4.2.1, we propose a MILP model for solving the classification, and the same model can be extended to solve multiple objectives with the help of goal programming and epsilon-constraint method. However, this MILP model, when implemented, can handle only small-sized data set (a couple of hundred observations). In order to solve the classification problem that commonly arises in big data analytics and to overcome the MILP model's limitation to solve large data sets, in Section 4.2.2, we present the conventional LP-based classifier with crisp boundary (with respect to the classification of categories) that can solve the data set that is linearly separable. In order to further improve the accuracy, in Section 4.2.3, we propose an LP-based classifier (with crisp boundary) with the consideration of interaction among the attributes and the contribution of attributes from their higher-order polynomial degrees. This approach to data transformation is undertaken to capture the curvilinear/nonlinear boundary between the categories. Finally, in Section 4.3, we propose an LP-based two-phase classifier for classification which considers the interaction among the attributes and the contribution of attributes from their higher-order polynomial degrees to solve the data set that is not linearly separable and also to improve the accuracy with respect to multiple objectives. In the first phase, the classifier runs the LP model by considering the bandwidth of boundary, and in the second phase, the proposed classifier transforms the bandwidth of boundary into a crisp boundary, and in this process, the proposed classifier identifies a non-dominated set of solutions with respect to multiple objectives.

The salient contributions of the proposed LP-based two-phase classifier are in terms of treating the decision variables as unrestricted in sign, the contribution of attributes from their interaction effects, the contribution of attributes from their higher-order polynomial degrees, treating the classification threshold/cut-off as a decision variable, and converting the bandwidth of boundary of threshold to a crisp boundary, thereby behaving like a nonlinear classifier to an extent. We also consider multiple objectives to produce non-dominated set of solutions, when we determine the crisp boundary. The multiple objectives include: maximize the total accuracy with respect to both categories, maximize the accuracy with respect to category 1, and maximize the accuracy with respect to category 0. We consider such multiple objectives because in the application areas such as medical diagnosis, absence of an alarm (failing to predict the category 1) is more serious than a false alarm. The proposed LP-based two-phase classifier can find a non-dominated set of solutions (with respect to multiple objectives) for the enhanced decision support.

4.2.1 Proposed MILP Model

Parameters

A	Total number of attributes in each observation
N	Total number of observations in the training data set
j	Index for the attribute
i	Index for the observation
$a_{i,j}$	Value of attribute j for observation i
epsilon	A constant with a very small value 0.00001
normalize	Normalize is the function which converts *a priori* the value of the respective expression to a value between 0 and 1. For example, while applying the normalize function for the expression $a_{i,j}$ which corresponds to attribute j for observation i, it is expressed as follows:

$$normalize\ (a_{i,j}) = \frac{\left(a_{i,j} - \min_{i' \in N}\{a_{i',j}\}\right)}{\left(\max_{i' \in N}\{a_{i',j}\} - \min_{i' \in N}\{a_{i',j}\}\right)}.$$

Decision Variables

x''_j	An unrestricted variable to capture the coefficient for the term $a_{i,j}$
y	An unrestricted variable which acts as a surrogate constant to improve the accuracy
success$_i$	A binary variable to indicate whether the prediction is correct or not

The objective is to maximize the accuracy/success of the prediction:
Maximize

$$Z = \sum_{i=1}^{N} success_i \tag{4.1}$$

subject to the following constraints, for all i:
when the observation i is of category 0 (i.e., $\left(\sum_{j=1}^{A}(normalize(a_{i,j}) \times x''_j) + y\right)$ should be less than or equal to 0.5 in the case of category-0 observation), we have

$$\sum_{j=1}^{A} \left(normalize(a_{i,j}) \times x''_j\right) + y \leq 0.5 + (1 - success_i) \times M, \tag{4.2}$$

when the observation i is of category 1 (i.e., $\left(\sum_{j=1}^{A}(normalize(a_{i,j}) \times x''_j) + y\right)$ should be greater than or equal to $(0.5 + epsilon)$ in the case of category-1 observation), we have

$$\sum_{j=1}^{A} \left(normalize\ (a_{i,j}) \times x''_j\right) + y \geq 0.5 + epsilon - (1 - success_i) \times M. \tag{4.3}$$

In the above constraints (4.2) and (4.3), x_j'' and y are unrestricted in sign, and $success_i$ is a binary variable. The decision variable $success_i$ determines whether the observation i is predicted correctly with respect to the corresponding category, and objective function (4.1) maximizes the total accuracy of the prediction.

4.2.2 Proposed LP-Based Classifier (LP Model 1): Based on the Conventional Approach of Considering All the Attributes with Power Index Equal to 1

The set of parameters and decision variables (except $success_i$) given in Section 4.2.1 are also used in this section in addition to the following decision variable.

Decision Variable

$error_i$:	A real variable to capture the error value in the expression

The objective is to minimize the total errors:
Minimize

$$Z = \sum_{i=1}^{N} error_i \tag{4.4}$$

subject to the following constraints, for all i:
when the observation i is of category 0 (i.e., $\left(\sum_{j=1}^{A}(normalize(a_{i,j}) \times x_j'') + y\right)$ should be less than or equal to 0.5 in the case of category-0 observation), we have

$$error_i \geq \left(\sum_{j=1}^{A}\left(normalize\ (a_{i,j}) \times x_j''\right) + y\right) - 0.5, \tag{4.5}$$

when the observation i is of category 1 (i.e., $\left(\sum_{j=1}^{A}(normalize(a_{i,j}) \times x_j'') + y\right)$ should be greater than or equal to $(0.5 + epsilon)$ in the case of category-1 observation), we have

$$error_i \geq (0.5 + epsilon) - \left(\sum_{j=1}^{A}\left(normalize\ (a_{i,j}) \times x_j''\right) + y\right). \tag{4.6}$$

In the above constraints (4.5) and (4.6), x_j'' and y are unrestricted in sign, and $error_i$ is a real continuous variable. The decision variable $error_i$ captures the amount of deviation from their respective threshold for each observation (if present), and objective function (4.4) minimizes the total errors of all the observations in the training data set, so as to improve the accuracy of the prediction.

The proposed classifier (LP Model 1) splits the data set into training data set and test data set, and then uses the training data set to train the model, and uses the test data set to validate the model.

The proposed classifier (LP Model 1) runs the above LP model for the training data set to train the model. Then the proposed classifier (LP Model 1) makes use of the trained model to predict the category for the test data set. Note that the decision variable $error_i$ is used only in the above LP model (LP Model 1) for the training data set to train the model and then the value for the expression $\left(\sum_{j=1}^{A}\left(normalize(a_{i,j})\times x_j''\right)+y\right)$ is calculated for each observation i in the test data set to predict the category. If the value of the expression is less than or equal to 0.5, then the corresponding observation is predicted as category 0; and if the value of the expression is greater than or equal to $(0.5 + epsilon)$, then the corresponding observation is predicted as category 1.

Then the accuracy with respect to test data set is calculated for the following objectives:

maximize the total accuracy with respect to both categories,

$$\text{Objective 1} = \frac{(\text{number of samples predicted correctly})}{(\text{total number of samples})} \times 100; \quad (4.7)$$

maximize the accuracy with respect to category 1,

$$\text{Objective 2} = \frac{(\text{number of category} - 1 \text{ samples predicted correctly})}{(\text{total number of category} - 1 \text{ samples})} \times 100;$$

$$(4.8)$$

and maximize the accuracy with respect to category 0,

$$\text{Objective 3} = \frac{(\text{number of category} - 0 \text{ samples predicted correctly})}{(\text{total number of category} - 0 \text{ samples})} \times 100.$$

$$(4.9)$$

4.2.3 Proposed LP-Based Classifier with a Crisp Boundary (LP Model 2)

We consider the power index for each attribute in the range [1, 5] (range being chosen in this study), and the incorporation of interaction effects of attributes in the proposed LP-based classifier with a crisp boundary.

Parameters

A	Total number of attributes in each observation
N	Total number of observations in the training data set
j, j', j''	Indices for the attributes
i	Index for the observation
P_j	A set of power terms for attribute j, used with respect to the value of attribute's higher-order polynomial degrees
Q_j	A set of power terms for attribute j, used with respect to the interaction effects of attributes
p_j	An element from set P_j/* Each attribute can have its own set of power terms: For example, $P_1 = \{1, 2, 3, 4,\}$; $P_2 = \{1, 2, 3\}$ and we have $p_1 \in P_1$; $p_2 \in P_2$ */
q_j	An element from set Q_j /* Each attribute can have its own set of power terms with respect to the interaction effects of attributes: For example, $Q_1 = \{1, 2, 3\}$; $Q_2 = \{1, 2, 3\}$ and we have $q_1 \in Q_1$; $q_2 \in Q_2$ */
$a_{i,j}$	Value of attribute j for observation i
epsilon	A constant with a very small value 0.00001
normalize	*Normalize* is the function which converts *a priori* the value of the respective expression to a value between 0 and 1. For example, while applying the *normalize* function for the term $a_{i,j}^{p_j}$ which corresponds to attribute j for observation i, it is expressed as follows:

$$normalize\left(a_{i,j}^{p_j}\right) = \frac{\left(a_{i,j}^{p_j} - \min_{i' \in N}\left\{a_{i',j}^{p_j}\right\}\right)}{\left(\max_{i' \in N}\left\{a_{i',j}^{p_j}\right\} - \min_{i' \in N}\left\{a_{i',j}^{p_j}\right\}\right)};$$

and

$$normalize\left(a_{i,j'}^{q_{j'}} \times a_{i,j''}^{q_{j''}}\right) = \frac{\left(a_{i,j'}^{q_{j'}} \times a_{i,j''}^{q_{j''}} - \min_{i' \in N}\left\{a_{i',j'}^{q_{j'}} \times a_{i',j''}^{q_{j''}}\right\}\right)}{\left(\max_{i' \in N}\left\{a_{i',j'}^{q_{j'}} \times a_{i',j''}^{q_{j''}}\right\} - \min_{i' \in N}\left\{a_{i',j'}^{q_{j'}} \times a_{i',j''}^{q_{j''}}\right\}\right)}.$$

Decision Variables

$x_{j,pj}$	An unrestricted variable to capture the coefficient for the term/expression $a_{i,j}^{p_j}$
$x'_{j',j'',qj',qj''}$	An unrestricted variable to capture the coefficient for the interaction term with respect to $\left(a_{i,j'}^{q_{j'}} \times a_{i,j''}^{q_{j''}}\right)$
y	An unrestricted variable which acts as a surrogate constant to improve the accuracy
$error_i$	A real variable to capture the error value in the expression

The objective is to minimize the total errors:
Minimize

$$Z = \sum_{i=1}^{N} error_i \qquad (4.10)$$

subject to the following constraints, for all i:
when the observation i is of category 0 (i.e.,

$$\left(\sum_{j=1}^{A}\sum_{p_j\in P_j}\left(normalize\left(a_{i,j}^{p_j}\right)\times x_{j,p_j}\right)\right.$$

$$\left.+\sum_{j'=1}^{A-1}\sum_{j''=j'+1}^{A}\sum_{q_{j'}\in Q_{j'}}\sum_{q_{j''}\in Q_{j''}}\left(normalize\left(a_{i,j'}^{q_{j'}}\times a_{i,j''}^{q_{j''}}\right)\times x'_{j',j'',q_{j'},q_{j''}}\right)+y\right)$$

should be less than or equal to 0.5 in the case of category-0 observation), we have

$$error_i\geq\left(\sum_{j=1}^{A}\sum_{p_j\in P_j}\left(normalize\left(a_{i,j}^{p_j}\right)\times x_{j,p_j}\right)\right.$$

$$\left.+\sum_{j'=1}^{A-1}\sum_{j''=j'+1}^{A}\sum_{q_{j'}\in Q_{j'}}\sum_{q_{j''}\in Q_{j''}}\left(normalize\left(a_{i,j'}^{q_{j'}}\times a_{i,j''}^{q_{j''}}\right)\times x'_{j',j'',q_{j'},q_{j''}}\right)+y-0.5\right). \quad (4.11)$$

When the observation i is of category 1 (i.e.,

$$\left(\sum_{j=1}^{A}\sum_{p_j\in P_j}\left(normalize\left(a_{i,j}^{p_j}\right)\times x_{j,p_j}\right)\right.$$

$$\left.+\sum_{j'=1}^{A-1}\sum_{j''=j'+1}^{A}\sum_{q_{j'}\in Q_{j'}}\sum_{q_{j''}\in Q_{j''}}\left(normalize\left(a_{i,j'}^{q_{j'}}\times a_{i,j''}^{q_{j''}}\right)\times x'_{j',j'',q_{j'},q_{j''}}\right)+y\right)$$

should be greater than or equal to $(0.5+epsilon)$ in the case of category-1 observation), we have

$$error_i\geq(0.5+epsilon)-\left(\sum_{j=1}^{A}\sum_{p_j\in P_j}\left(normalize\left(a_{i,j}^{p_j}\right)\times x_{j,p_j}\right)\right.$$

$$\left.+\sum_{j'=1}^{A-1}\sum_{j''=j'+1}^{A}\sum_{q_{j'}\in Q_{j'}}\sum_{q_{j''}\in Q_{j''}}\left(normalize\left(a_{i,j'}^{q_{j'}}\times a_{i,j''}^{q_{j''}}\right)\times x'_{j',j'',q_{j'},q_{j''}}\right)+y\right). \quad (4.12)$$

In the above constraints (4.11) and (4.12), x_{j,p_j}, $x'_{j',j'',q_{j'},q_{j''}}$, and y are unrestricted in sign, and $error_i$ is a real variable. Constraints (4.11) and (4.12) capture the contribution of attributes from their higher-order polynomial degrees, and also capture the interaction effects among the attributes. The decision variable $error_i$ captures the amount of deviation from their respective threshold for each observation (if present). Objective function (4.10) minimizes the total errors of all the observations in the training data set, so as to improve the accuracy of the prediction.

The proposed classifier (LP Model 2) splits the data set into training data set and test data set, and then uses the training data set to train the model, and uses the test data set to validate the model.

The proposed classifier (LP Model 2) runs the above LP model for the training data set to train the model. Then this classifier makes use of the trained model to predict the category for the test data set. Note that the decision variable $error_i$ is used only in the above LP model (LP Model 2) for the training data set to train the model, and then the value for the expression

$$\left(\sum_{j=1}^{A} \sum_{p_j \in P_j} \left(normalize\left(a_{i,j}^{p_j}\right) \times x_{j,p_j} \right) \right. $$

$$\left. + \sum_{j'=1}^{A-1} \sum_{j''=j'+1}^{A} \sum_{q_{j'} \in Q_{j'}} \sum_{q_{j''} \in Q_{j''}} \left(normalize\left(a_{i,j'}^{q_{j'}} \times a_{i,j''}^{q_{j''}}\right) \times x'_{j',j'',q_{j'},q_{j''}} \right) + y \right)$$

is calculated for each observation i in the test data set to predict the category. If the value of the expression is less than or equal to 0.5, then the corresponding observation is predicted as category 0; and if the value of the expression is greater than or equal to (0.5 + *epsilon*), then the corresponding observation is predicted as category 1. Then the accuracy with respect to test data set is calculated for the objectives, Objective 1, Objective 2, and Objective 3.

4.2.4 Numerical Illustration for the Constraints in Proposed LP-Based Classifier (LP Model 1)

In this section, we present the numerical example for the constraints in proposed LP-based classifier (LP Model 1), and for the purpose of numerical illustration, let the number of attributes be 2, and values of attributes be in the range [0, 9]. Table 4.1 represents the samples with respect to each category for the numerical illustration.

As the values of attributes are in the range [0, 9], for attribute 2 of sample 1, the *normalize* function is expressed as follows:

$$normalize\,(5) = \frac{(5-0)}{(9-0)} = 0.56.$$

For the samples in Table 4.1, the constraints of the proposed LP model are expressed as follows:

With respect to sample 1, LP model Constraint (4.5) appears as follows:

$$error_i \geq \left(0.11 \times x_1'' + 0.56 \times x_2'' + y\right) - 0.5. \tag{4.13}$$

TABLE 4.1

Samples for the Numerical Illustration

Sample	Attribute 1	Attribute 2	Category
1	1	5	0
2	4	9	1

With respect to sample 2, LP model Constraint (4.6) appears as follows:

$$error_i \geq (0.5 + epsilon) - \left(0.44 \times x_1'' + 1.00 \times x_2'' + y\right). \tag{4.14}$$

4.2.5 Numerical Illustration for the Constraints in Proposed LP-Based Classifier (LP Model 2)

In this section, we present the numerical example for the constraints in proposed LP-based classifier (LP Model 2), and for the purpose of numerical illustration, let the number of attributes be 2, and values of attributes be in the range [0, 9]. Table 4.1 represents the samples with respect to each category for the numerical illustration. Sets of power terms with respect to higher-order polynomial degrees of attributes and with respect to interaction effects of attributes are as follows:

$$P_1 = \{1, 2, 3\}; \tag{4.15}$$

$$P_2 = \{1, 2, 3\}; \tag{4.16}$$

$$Q_1 = \{1, 2\}; \text{ and} \tag{4.17}$$

$$Q_2 = \{1, 2\}. \tag{4.18}$$

As the values of attributes are in the range [0, 9], for attribute 2 of sample 1 with power term 2, the *normalize* function is expressed as follows:

$$normalize \ (5^2) = \frac{((5 \times 5) - 0)}{((9 \times 9) - 0)} = 0.31.$$

For the samples in Table 4.1, the constraints of the proposed LP model are expressed as follows:

With respect to sample 1, LP model Constraint (4.11) appears as follows:

$$error_i \geq (0.11 \times x_{1,1} + 0.01 \times x_{1,2} + 0.00 \times x_{1,3} + 0.56 \times x_{2,1} + 0.31 \times x_{2,2} + 0.17$$
$$\times x_{2,3} + 0.06 \times x_{1,2,1,1}' + 0.03 \times x_{1,2,1,2}' + 0.01 \times x_{1,2,2,1}' + 0.00 \times x_{1,2,2,2}' + y) - 0.5.$$

$$\tag{4.19}$$

Note: The coefficients are presented in this entire chapter in two decimal points precision, and hence the expression $0.00 \times x_{1,3}$ in Constraint (4.19) represents actually a small coefficient for the variable $x_{1,3}$.

With respect to sample 2, LP model Constraint (4.12) appears as follows:

$$
\begin{aligned}
error_i \geq\ & (0.5 + epsilon) - (0.44 \times x_{1,1} + 0.20 \times x_{1,2} + 0.09 \times x_{1,3} + 1.00 \times x_{2,1} \\
& + 1.00 \times x_{2,2} + 1.00 \times x_{2,3} + 0.44 \times x'_{1,2,1,1} + 0.44 \times x'_{1,2,1,2} + 0.20 \times x'_{1,2,2,1} \\
& + 0.20 \times x'_{1,2,2,2} + y).
\end{aligned}
$$

$$(4.20)$$

4.3 Proposed LP-Based Two-Phase Classifier

In this section, we propose an LP-based two-phase classifier for classification which considers the interaction among the attributes and the contribution of attributes from their higher-order polynomial degrees to solve the data set that is not linearly separable. In the first phase, the classifier runs the LP model by considering the bandwidth of boundary (without invoking the crisp boundary), and this bandwidth is treated as a decision variable, and in the second phase, the proposed classifier transforms the bandwidth of boundary into a crisp boundary. During this process, the proposed classifier also identifies a non-dominated set of solutions with respect to multiple objectives.

4.3.1 Proposed LP Model for Two-Phase Classifier

The set of parameters and decision variables given in Section 4.2.3 are also used in this section in addition to following decision variable.

Decision Variable

b	A real variable to capture the classification threshold/cut-off

Now, the LP model in the proposed two-phase classifier is presented. Note that in the bandwidth of boundary $[b, (b + 1)]$, b is treated as a decision variable.

The objective is to minimize the total errors:

Minimize

$$Z = \sum_{i=1}^{N} error_i \qquad (4.21)$$

subject to the following constraints, for all i: when the observation i is of category 0 (i.e.,

$$\left(\sum_{j=1}^{A} \sum_{p_j \in P_j} \left(normalize\left(a_{i,j}^{p_j}\right) \times x_{j,p_j} \right) \right.$$

$$\left. + \sum_{j'=1}^{A-1} \sum_{j''=j'+1}^{A} \sum_{q_{j'} \in Q_{j'}} \sum_{q_{j''} \in Q_{j''}} \left(normalize\left(a_{i,j'}^{q_{j'}} \times a_{i,j''}^{q_{j''}}\right) \times x'_{j',j'',q_{j'},q_{j''}} \right) + y \right)$$

should be less than or equal to b in the case of category-0 observation), we have

$$error_i \geq \left(\sum_{j=1}^{A} \sum_{p_j \in P_j} \left(normalize\left(a_{i,j}^{p_j}\right) \times x_{j,p_j} \right) \right.$$

$$\left. + \sum_{j'=1}^{A-1} \sum_{j''=j'+1}^{A} \sum_{q_{j'} \in Q_{j'}} \sum_{q_{j''} \in Q_{j''}} \left(normalize\left(a_{i,j'}^{q_{j'}} \times a_{i,j''}^{q_{j''}}\right) \times x'_{j',j'',q_{j'},q_{j''}} \right) + y \right) - b, \quad (4.22)$$

when the observation i is of category 1 (i.e.,

$$\left(\sum_{j=1}^{A} \sum_{p_j \in P_j} \left(normalize\left(a_{i,j}^{p_j}\right) \times x_{j,p_j} \right) \right.$$

$$\left. + \sum_{j'=1}^{A-1} \sum_{j''=j'+1}^{A} \sum_{q_{j'} \in Q_{j'}} \sum_{q_{j''} \in Q_{j''}} \left(normalize\left(a_{i,j'}^{q_{j'}} \times a_{i,j''}^{q_{j''}}\right) \times x'_{j',j'',q_{j'},q_{j''}} \right) + y \right)$$

should be greater than or equal to $(b + 1)$ in the case of category-1 observation), we have

$$error_i \geq (b+1) - \left(\sum_{j=1}^{A} \sum_{p_j \in P_j} \left(normalize\left(a_{i,j}^{p_j}\right) \times x_{j,p_j} \right) \right.$$

$$\left. + \sum_{j'=1}^{A-1} \sum_{j''=j'+1}^{A} \sum_{q_{j'} \in Q_{j'}} \sum_{q_{j''} \in Q_{j''}} \left(normalize\left(a_{i,j'}^{q_{j'}} \times a_{i,j''}^{q_{j''}}\right) \times x'_{j',j'',q_{j'},q_{j''}} \right) + y \right). \quad (4.23)$$

In the above constraints (4.22) and (4.23), x_{j,p_j}, $x'_{j',j'',q_{j'},q_{j''}}$, and y are unrestricted in sign, and $error_i$ and b are real variables. We have "1" in Constraint (4.23), since we have dichotomous classification. Constraints (4.22) and (4.23) capture the contribution of attributes from their higher-order polynomial degrees, and also capture the interaction effects among the attributes. The decision variable b captures the classification threshold/cut-off for the respective category. The decision variable $error_i$ captures the amount of deviation from their respective threshold for each observation (if present). Objective function (4.21) minimizes the total errors of all the observations in the training data set, so as to improve the accuracy of the prediction.

4.3.2 Algorithm for the Proposed LP-Based Two-Phase Classifier

The proposed classifier runs in two phases. In the first phase, the classifier runs the proposed LP model (LP model for the two-phase classifier) using the training data set to train the model. In the second phase, the proposed classifier uses the test data set to transform the bandwidth of boundary $[b, (b + 1)]$ into a crisp boundary $b + c$, and while determining the crisp boundary multiple objectives are considered. The proposed classifier iteratively increments the value of c from 0 to 1 by 0.05 (step size), and identifies a non-dominated set of solutions with respect to multiple objectives, while applying the trained model/expression on the test data set. Finally, for each of the solution in non-dominated set, the classifier identifies the non-dominated set of solutions with respect to objectives, Objective 1, Objective 2, and Objective 3, for the validation data set.

Phase 0:

- In this phase, the proposed classifier splits the data set into three: training data set, test data set, and validation data set.

Step 1:   Split the data set into training data set, test data set, and validation data set.

Phase 1:

- In this phase, the proposed classifier uses the training data set to train the model (i.e., to get the values of the decision variables b, x_{j,p_j}, $x'_{j',j'',q_{j'},q_{j''}}$, and y) and also identifies the bandwidth of boundary $[b, (b + 1)]$.

Step 1:   Run the proposed LP model (LP model in the proposed two-phase classifier) for the training data set to get the LP solution in terms of the value of the variable b, and also the values of the variables x_{j,p_j}, $x'_{j',j'',q_{j'},q_{j''}}$, and y, to compute the following:

$$\left(\sum_{j=1}^{A} \sum_{p_j \in P_j} \left(normalize\left(a_{i,j}^{p_j}\right) \times x_{j,p_j} \right) \right.$$

$$\left. + \sum_{j'=1}^{A-1} \sum_{j''=j'+1}^{A} \sum_{q_{j'} \in Q_{j'}} \sum_{q_{j''} \in Q_{j''}} \left(normalize\left(a_{i,j'}^{q_{j'}} \times a_{i,j''}^{q_{j''}}\right) \times x'_{j',j'',q_{j'},q_{j''}} \right) + y \right). \quad (4.24)$$

Phase 2:

- The second phase of the proposed classifier comprises two parts.

Part 1:

- In this part, the proposed classifier uses the test data set to transform the bandwidth of boundary into a crisp boundary $b + c$. In this process, the proposed classifier identifies a non-dominated set of solutions with respect to multiple objectives, and captures the corresponding set of c values.

Step 1:   Iteratively increment the value of c from 0 to 1 by 0.05 (step size), and for each value of c, do the following:

Step 1.1: For each observation in the test data set, calculate the value for the expression (4.24) and assign the value to the variable *val*.
If $val \leq b + c$
then declare the observation as category 0.
If $val \geq b + c + epsilon$
then declare the observation as category 1.

Step 1.2: Calculate the objectives, Objective 1, Objective 2, and Objective 3.

Step 1.3: Form the non-dominated set of solutions with respect to Objective 1, Objective 2, and Objective 3 and store the corresponding value of c with respect to every solution.

Part 2:

- In this part, the proposed classifier uses the validation data set to validate the model. In this process, the proposed classifier identifies a non-dominated set of solutions with respect to multiple objectives for the validation data set by evaluating the validation data set for the chosen set of c values.

Step 1:   With the values of c that are obtained (from Part 1) corresponding to the set of non-dominated solutions (see Step 1.3 of Part 1), consider the validation data set and hence obtain the set of non-dominated solutions. This set of non-dominated solutions constitutes the solutions with respect to objectives Objective 1, Objective 2, and Objective 3 for the validation data set. This set of solutions is used to benchmark/evaluate the performance of the proposed LP-based two-phase classifier.

Note:

- In the case of a single-objective optimization problem, during the second phase, the proposed classifier first identifies the best solution with respect to the objective under consideration for the test data set (see Part 1), and captures the corresponding c value; and then the proposed classifier reports the single solution with respect to the objective under consideration for the validation data set by evaluating the validation data set for the chosen c value (see Part 2).

- With the consideration of interaction among the attributes $\left(a_{i,j'}^{q_{j'}} \times a_{i,j''}^{q_{j''}}\right)$ and the contribution of attributes from their higher-order polynomial degrees $\left(a_{i,j}^{p_j}\right)$, the proposed LP-based two-phase classifier is able to capture the curvilinear/nonlinear boundary between the categories.

4.3.3 Numerical Illustration for the Constraints in Proposed LP-Based Two-Phase Classifier

In this section, we present a numerical example for proposed LP-based two-phase classifier, and for the purpose of numerical illustration, let the number of attributes be 2, and values of attributes be in the range [0, 9]. Table 4.1 represents the samples with respect to each category for the numerical illustration. We also have the power term settings according to Equations 4.15 through 4.18.

For the samples in Table 4.1, the constraints of the proposed LP model are expressed as follows:

With respect to sample 1, LP model Constraint (4.22) appears as follows:

$$
\begin{aligned}
error_i \geq (0.11 \times x_{1,1} &+ 0.01 \times x_{1,2} + 0.00 \times x_{1,3} + 0.56 \times x_{2,1} + 0.31 \times x_{2,2} \\
&+ 0.17 \times x_{2,3} + 0.06 \times x'_{1,2,1,1} + 0.03 \times x'_{1,2,1,2} + 0.01 \times x'_{1,2,2,1} \\
&+ 0.00 \times x'_{1,2,2,2} + y) - b.
\end{aligned}
\tag{4.25}
$$

With respect to sample 2, LP model Constraint (4.23) appears as follows:

$$
\begin{aligned}
error_i \geq (b+1) - (0.44 \times x_{1,1} &+ 0.20 \times x_{1,2} + 0.09 \times x_{1,3} \\
&+ 1.00 \times x_{2,1} + 1.00 \times x_{2,2} + 1.00 \times x_{2,3} + 0.44 \times x'_{1,2,1,1} + 0.44 \times x'_{1,2,1,2} \\
&+ 0.20 \times x'_{1,2,2,1} + 0.20 \times x'_{1,2,2,2} + y).
\end{aligned}
\tag{4.26}
$$

4.4 Results and Discussion

The proposed MILP model cannot be executed on large-sized data sets that commonly arise in big data analytics. The proposed MILP model can run on the training data set of size up to 100 or 200 observations, and in most cases it overfits the training data, and it fails to generalize the underlying pattern in the data set. Hence, in this section, we mainly concentrate on the performance of the proposed LP-based classifiers (LP Model 1, LP Model 2, and LP-based two-phase classifier). In Section 4.4.1, we present the comparison study of the proposed LP-based classifiers (LP Model 1, LP Model 2, and LP-based two-phase classifier) with LP classifier with fuzzy measure and the Choquet integral by Yan et al. (2006). In Section 4.4.2, we present the comparison study of the proposed LP-based classifiers (LP Model 1, LP Model 2, and LP-based two-phase classifier) with ANN. For the comparison study, we use the following settings which seem to be sufficient to produce reasonable accuracy:

$$
P_j = \{1, 2, 3, 4, 5\} \quad \forall j,
\tag{4.27}
$$

$$
Q_j = \{1\} \quad \forall j.
\tag{4.28}
$$

4.4.1 Comparison Study of the Proposed LP-Based Classifiers with the LP Classifier with Fuzzy Measure and the Choquet Integral by Yan et al. (2006)

To evaluate the performance of the proposed LP-based classifiers (LP Model 1, LP Model 2, and LP-based two-phase classifier), we use the same data set presented in the paper by Yan et al. (2006). For the purpose of numerical illustration, we take two samples from the data set, and present the constraints of the respective classifiers. This data set contains 200 observations with two attributes and the range of values with respect to each attribute is presented in Table 4.2, and the samples are presented in Table 4.3. We also have the power term settings according to Equations 4.27 and 4.28.

For the samples in Table 4.3, the constraints of the proposed LP-based classifier (LP Model 1) are expressed as follows:

With respect to sample 1, LP model Constraint (4.5) appears as follows:

$$error_i \geq \left(0.06 \times x_1'' + 0.43 \times x_2'' + y\right) - 0.5. \tag{4.29}$$

With respect to sample 2, LP model Constraint (4.6) appears as follows:

$$error_i \geq (0.5 + epsilon) - \left(0.39 \times x_1'' + 0.93 \times x_2'' + y\right). \tag{4.30}$$

For the samples in Table 4.3, the constraints of the proposed LP-based classifier (LP Model 2) are expressed as follows:

With respect to sample 1, LP model Constraint (4.11) appears as follows:

$$
\begin{aligned}
error_i \geq &(0.06 \times x_{1,1} + 0.00 \times x_{1,2} + 0.00 \times x_{1,3} + 0.00 \times x_{1,4} + 0.00 \times x_{1,5} \\
&+ 0.43 \times x_{2,1} + 0.19 \times x_{2,2} + 0.08 \times x_{2,3} + 0.04 \times x_{2,4} + 0.02 \times x_{2,5} \\
&+ 0.04 \times x_{1,2,1,1}' + y) - 0.5.
\end{aligned}
\tag{4.31}
$$

TABLE 4.2

Range of Values with Respect to Attributes in Data Set Presented in the Paper by Yan et al. (2006)

Attribute 1	Attribute 2
[0.01,0.99]	[0.01,0.99]

TABLE 4.3

Two Samples from the Data Set Presented in the Paper by Yan et al. (2006)

Sample	Attribute 1	Attribute 2	Category
1	0.07	0.43	0
2	0.39	0.92	1

With respect to sample 2, LP model Constraint (4.12) appears as follows:

$$
\begin{aligned}
error_i \geq &\, (0.5 + epsilon) - (0.39 \times x_{1,1} + 0.16 \times x_{1,2} + 0.06 \times x_{1,3} \\
&+ 0.02 \times x_{1,4} + 0.00 \times x_{1,5} + 0.93 \times x_{2,1} + 0.86 \times x_{2,2} + 0.80 \times x_{2,3} \\
&+ 0.75 \times x_{2,4} + 0.69 \times x_{2,5} + 0.45 \times x'_{1,2,1,1} + y).
\end{aligned} \tag{4.32}
$$

For the samples in Table 4.3, the constraints of the proposed LP-based two-phase classifier are expressed as follows: With respect to sample 1, LP model Constraint (4.22) appears as follows:

$$
\begin{aligned}
error_i \geq &\, (0.06 \times x_{1,1} + 0.00 \times x_{1,2} + 0.00 \times x_{1,3} + 0.00 \times x_{1,4} + 0.00 \times x_{1,5} \\
&+ 0.43 \times x_{2,1} + 0.19 \times x_{2,2} + 0.08 \times x_{2,3} + 0.04 \times x_{2,4} \\
&+ 0.02 \times x_{2,5} + 0.04 \times x'_{1,2,1,1} + y) - b.
\end{aligned} \tag{4.33}
$$

With respect to sample 2, LP model Constraint (4.23) appears as follows:

$$
\begin{aligned}
error_i \geq &\, (b + 1) - (0.39 \times x_{1,1} + 0.16 \times x_{1,2} + 0.06 \times x_{1,3} + 0.02 \times x_{1,4} \\
&+ 0.00 \times x_{1,5} + 0.93 \times x_{2,1} + 0.86 \times x_{2,2} + 0.80 \times x_{2,3} + 0.75 \times x_{2,4} \\
&+ 0.69 \times x_{2,5} + 0.45 \times x'_{1,2,1,1} + y).
\end{aligned} \tag{4.34}
$$

To calculate the objectives of the proposed LP-based classifiers (LP Model 1 and LP Model 2), we split the data into two sets, training data set (70%) and test data set (30%), and the calculated objectives with respect to test data set are listed in Table 4.4.

To calculate the objectives of the proposed LP-based two-phase classifier, we split the data into three sets, training data set (60%), test data set (10%), and validation data set (30%), and the calculated objectives of the validation data set (see Part 2 of Phase 2) are listed in Table 4.4. The results indicate that the proposed LP-based two-phase classifier is also able to give 100% accuracy for all the objectives, and it is able to predict the outcome perfectly even when the data

TABLE 4.4

Performance of the Proposed LP-Based Classifiers (LP Model 1, LP Model 2, and the LP-Based Two-Phase Classifier) for the Data Set by Yan et al. (2006)

	Objective 1 (%)	Objective 2 (%)	Objective 3 (%)
LP Model 1	48.00	100.00	1.89
LP Model 2	98.00	98.94	97.17
Two-phase classifier	100.00	100.00	100.00

Note: LP Model 1 and LP Model 2 cannot address multiobjective optimization; A single non-dominated solution of the proposed two-phase classifier (for the validation data set).

is not linearly separable. Note that Yan et al. (2006) included all the samples in the data set for training the model and the accuracy of their classifier was 100%.

4.4.2 Comparison Study of the Proposed LP-Based Classifiers for the Recommendation Data Set with Artificial Neural Networks

To evaluate the performance of the proposed LP-based classifiers, we use the data set with respect to movie recommendations (Maas et al., 2011) and note that while the data set presented by Maas et al. (2011) contains the movie reviews in text format, the last two authors of this chapter converted the text reviews into a numerical data set in their earlier work. For the purpose of numerical illustration, we take two samples from the data set, and present the constraints of the respective classifiers. This large data set contains 25,000 observations with eight attributes and the range of values with respect to each attribute is presented in Table 4.5, and the samples are presented in Table 4.6. We also have the power term settings according to Equations 4.27 and 4.28.

For the samples in Table 4.6, the constraints of the proposed LP-based classifier (LP Model 1) are expressed as follows:

With respect to sample 1, LP model Constraint (4.5) appears as follows:

$$error_i \geq (0.25 \times x_1'' + 0.09 \times x_2'' + 0.21 \times x_3'' + 0.11 \times x_4'' + 0.26 \times x_5''$$
$$+ 0.05 \times x_6'' + 0.42 \times x_7'' + 0.31 \times x_8'' + y) - 0.5. \tag{4.35}$$

With respect to sample 2, LP model Constraint (4.6) appears as follows:

$$error_i \geq (0.5 + epsilon) - (0.28 \times x_1'' + 0.22 \times x_2'' + 0.09 \times x_3'' + 0.22 \times x_4''$$
$$+ 0.14 \times x_5'' + 0.005 + 0.12 \times x_7'' + 0.06 \times x_8'' + y). \tag{4.36}$$

TABLE 4.5

Range of Values with Respect to Attributes in Recommendation Data Set

Attribute 1	Attribute 2	Attribute 3	Attribute 4	Attribute 5	Attribute 6	Attribute 7	Attribute 8
[0,36]	[0,23]	[0,47]	[0,18]	[0,35]	[0,21]	[0,26]	[0,16]

TABLE 4.6

Two Samples from the Recommendation Data Set

Sample	Attribute 1	Attribute 2	Attribute 3	Attribute 4	Attribute 5	Attribute 6	Attribute 7	Attribute 8	Category
1	9	2	10	2	9	1	11	5	0
2	10	5	4	4	5	1	3	1	1

For the samples in Table 4.6, the constraints of the proposed LP-based classifier (LP Model 2) are expressed as follows:

With respect to sample 1, LP model Constraint (4.11) appears as follows:

$$error_i \geq (0.25 \times x_{1,1} + 0.062 \times x_{1,2} + 0.02 \times x_{1,3} + 0.00 \times x_{1,4} + 0.00 \times x_{1,5}$$
$$+0.09 \times x_{2,1} + 0.01 \times x_{2,2} + 0.00 \times x_{2,3} + 0.00 \times x_{2,4} + 0.00 \times x_{2,5}$$
$$+0.21 \times x_{3,1} + 0.05 \times x_{3,2} + 0.01 \times x_{3,3} + 0.00 \times x_{3,4} + 0.00 \times x_{3,5}$$
$$+0.11 \times x_{4,1} + 0.01 \times x_{4,2} + 0.00 \times x_{4,3} + 0.00 \times x_{4,4} + 0.00 \times x_{4,5}$$
$$+0.26 \times x_{5,1} + 0.07 \times x_{5,2} + 0.02 \times x_{5,3} + 0.00 \times x_{5,4} + 0.00 \times x_{5,5}$$
$$+0.05 \times x_{6,1} + 0.00 \times x_{6,2} + 0.00 \times x_{6,3} + 0.00 \times x_{6,4} + 0.00 \times x_{6,5}$$
$$+0.42 \times x_{7,1} + 0.18 \times x_{7,2} + 0.08 \times x_{7,3} + 0.03 \times x_{7,4} + 0.01 \times x_{7,5}$$
$$+0.31 \times x_{8,1} + 0.10 \times x_{8,2} + 0.03 \times x_{8,3} + 0.01 \times x_{8,4} + 0.00 \times x_{8,5}$$
$$+0.04 \times x'_{1,2,1,1} + 0.17 \times x'_{1,3,1,1} + 0.08 \times x'_{1,4,1,1} + 0.12 \times x'_{1,5,1,1} + 0.03 \times x'_{1,6,1,1}$$
$$+0.32 \times x'_{1,7,1,1} + 0.24 \times x'_{1,8,1,1} + 0.03 \times x'_{2,3,1,1} + 0.02 \times x'_{2,4,1,1} + 0.04 \times x'_{2,5,1,1}$$
$$+0.00 \times x'_{2,6,1,1} + 0.13 \times x'_{2,7,1,1} + 0.07 \times x'_{2,8,1,1} + 0.03 \times x'_{3,4,1,1} + 0.14 \times x'_{3,5,1,1}$$
$$+0.03 \times x'_{3,6,1,1} + 0.29 \times x'_{3,7,1,1} + 0.16 \times x'_{3,8,1,1} + 0.06 \times x'_{4,5,1,1} + 0.02 \times x'_{4,6,1,1}$$
$$+0.12 \times x'_{4,7,1,1} + 0.12 \times x'_{4,8,1,1} + 0.02 \times x'_{5,6,1,1} + 0.20 \times x'_{5,7,1,1} + 0.09 \times x'_{5,8,1,1}$$
$$+0.04 \times x'_{6,7,1,1} + 0.02 \times x'_{6,8,1,1} + 0.28 \times x'_{7,8,1,1} + y) - 0.5.$$

$$(4.37)$$

With respect to sample 2, LP model Constraint (4.12) appears as follows:

$$error_i \geq (0.5 + epsilon)$$
$$-(x_{1,1} \times 0.28 + x_{1,2} \times 0.08 + x_{1,3} \times 0.02 + x_{1,4} \times 0.01 + x_{1,5} \times 0.00$$
$$+x_{2,1} \times 0.22 + x_{2,2} \times 0.05 + x_{2,3} \times 0.01 + x_{2,4} \times 0.00 + x_{2,5} \times 0.00$$
$$+x_{3,1} \times 0.09 + x_{3,2} \times 0.01 + x_{3,3} \times 0.00 + x_{3,4} \times 0.00 + x_{3,5} \times 0.00$$
$$+x_{4,1} \times 0.22 + x_{4,2} \times 0.01 + x_{4,3} \times 0.01 + x_{4,4} \times 0.00 + x_{4,5} \times 0.00$$
$$+x_{5,1} \times 0.14 + x_{5,2} \times 0.02 + x_{5,3} \times 0.00 + x_{5,4} \times 0.00 + x_{5,5} \times 0.00$$
$$+x_{6,1} \times 0.05 + x_{6,2} \times 0.00 + x_{6,3} \times 0.00 + x_{6,4} \times 0.00 + x_{6,5} \times 0.00$$
$$+x_{7,1} \times 0.12 + x_{7,2} \times 0.01 + x_{7,3} \times 0.00 + x_{7,4} \times 0.00 + x_{7,5} \times 0.00$$
$$+x_{8,1} \times 0.06 + x_{8,2} \times 0.00 + x_{8,3} \times 0.00 + x_{8,4} \times 0.00 + x_{8,5} \times 0.00$$
$$+x'_{1,2,1,1} \times 0.10 + x'_{1,3,1,1} \times 0.08 + x'_{1,4,1,1} \times 0.18 + x'_{1,5,1,1} \times 0.08 + x'_{1,6,1,1} \times 0.03$$
$$+x'_{1,7,1,1} \times 0.10 + x'_{1,8,1,1} \times 0.05 + x'_{2,3,1,1} \times 0.03 + x'_{2,4,1,1} \times 0.08 + x'_{2,5,1,1} \times 0.05$$
$$+x'_{2,6,1,1} \times 0.01 + x'_{2,7,1,1} \times 0.09 + x'_{2,8,1,1} \times 0.03 + x'_{3,4,1,1} \times 0.03 + x'_{3,5,1,1} \times 0.03$$
$$+x'_{3,6,1,1} \times 0.01 + x'_{3,7,1,1} \times 0.03 + x'_{3,8,1,1} \times 0.01 + x'_{4,5,1,1} \times 0.07 + x'_{4,6,1,1} \times 0.04$$
$$+x'_{4,7,1,1} \times 0.07 + x'_{4,8,1,1} \times 0.05 + x'_{5,6,1,1} \times 0.01 + x'_{5,7,1,1} \times 0.03 + x'_{5,8,1,1} \times 0.01$$
$$+x'_{6,7,1,1} \times 0.01 + x'_{6,8,1,1} \times 0.00 + x'_{7,8,1,1} \times 0.02 + y).$$

$$(4.38)$$

For the samples in Table 4.6, the constraints of the proposed LP-based two-phase classifier are expressed as follows:

With respect to sample 1, LP model Constraint (4.22) appears as follows:

$$
\begin{aligned}
error_i \geq (&0.25 \times x_{1,1} + 0.062 \times x_{1,2} + 0.02 \times x_{1,3} + 0.00 \times x_{1,4} + 0.00 \times x_{1,5} \\
&+0.09 \times x_{2,1} + 0.01 \times x_{2,2} + 0.00 \times x_{2,3} + 0.00 \times x_{2,4} + 0.00 \times x_{2,5} \\
&+0.21 \times x_{3,1} + 0.05 \times x_{3,2} + 0.01 \times x_{3,3} + 0.00 \times x_{3,4} + 0.00 \times x_{3,5} \\
&+0.11 \times x_{4,1} + 0.01 \times x_{4,2} + 0.00 \times x_{4,3} + 0.00 \times x_{4,4} + 0.00 \times x_{4,5} \\
&+0.26 \times x_{5,1} + 0.07 \times x_{5,2} + 0.02 \times x_{5,3} + 0.00 \times x_{5,4} + 0.00 \times x_{5,5} \\
&+0.05 \times x_{6,1} + 0.00 \times x_{6,2} + 0.00 \times x_{6,3} + 0.00 \times x_{6,4} + 0.00 \times x_{6,5} \\
&+0.42 \times x_{7,1} + 0.18 \times x_{7,2} + 0.08 \times x_{7,3} + 0.03 \times x_{7,4} + 0.01 \times x_{7,5} \\
&+0.31 \times x_{8,1} + 0.10 \times x_{8,2} + 0.03 \times x_{8,3} + 0.01 \times x_{8,4} + 0.00 \times x_{8,5} \\
&+0.04 \times x'_{1,2,1,1} + 0.17 \times x'_{1,3,1,1} + 0.08 \times x'_{1,4,1,1} + 0.12 \times x'_{1,5,1,1} + 0.03 \times x'_{1,6,1,1} \\
&+0.32 \times x'_{1,7,1,1} + 0.24 \times x'_{1,8,1,1} + 0.03 \times x'_{2,3,1,1} + 0.02 \times x'_{2,4,1,1} + 0.04 \times x'_{2,5,1,1} \\
&+0.00 \times x'_{2,6,1,1} + 0.13 \times x'_{2,7,1,1} + 0.07 \times x'_{2,8,1,1} + 0.03 \times x'_{3,4,1,1} + 0.14 \times x'_{3,5,1,1} \\
&+0.03 \times x'_{3,6,1,1} + 0.29 \times x'_{3,7,1,1} + 0.16 \times x'_{3,8,1,1} + 0.06 \times x'_{4,5,1,1} + 0.02 \times x'_{4,6,1,1} \\
&+0.12 \times x'_{4,7,1,1} + 0.12 \times x'_{4,8,1,1} + 0.02 \times x'_{5,6,1,1} + 0.20 \times x'_{5,7,1,1} + 0.09 \times x'_{5,8,1,1} \\
&+0.04 \times x'_{6,7,1,1} + 0.02 \times x'_{6,8,1,1} + 0.28 \times x'_{7,8,1,1} + y) - b.
\end{aligned}
$$

$$(4.39)$$

With respect to sample 2, LP model Constraint (4.23) appears as follows:

$$
\begin{aligned}
error_i \geq (&b+1) \\
-(&x_{1,1} \times 0.28 + x_{1,2} \times 0.08 + x_{1,3} \times 0.02 + x_{1,4} \times 0.01 + x_{1,5} \times 0.00 \\
+&x_{2,1} \times 0.22 + x_{2,2} \times 0.05 + x_{2,3} \times 0.01 + x_{2,4} \times 0.00 + x_{2,5} \times 0.00 \\
+&x_{3,1} \times 0.09 + x_{3,2} \times 0.01 + x_{3,3} \times 0.00 + x_{3,4} \times 0.00 + x_{3,5} \times 0.00 \\
+&x_{4,1} \times 0.22 + x_{4,2} \times 0.01 + x_{4,3} \times 0.01 + x_{4,4} \times 0.00 + x_{4,5} \times 0.00 \\
+&x_{5,1} \times 0.14 + x_{5,2} \times 0.02 + x_{5,3} \times 0.00 + x_{5,4} \times 0.00 + x_{5,5} \times 0.00 \\
+&x_{6,1} \times 0.05 + x_{6,2} \times 0.00 + x_{6,3} \times 0.00 + x_{6,4} \times 0.00 + x_{6,5} \times 0.00 \\
+&x_{7,1} \times 0.12 + x_{7,2} \times 0.01 + x_{7,3} \times 0.00 + x_{7,4} \times 0.00 + x_{7,5} \times 0.00 \\
+&x_{8,1} \times 0.06 + x_{8,2} \times 0.00 + x_{8,3} \times 0.00 + x_{8,4} \times 0.00 + x_{8,5} \times 0.00 \\
+&x'_{1,2,1,1} \times 0.10 + x'_{1,3,1,1} \times 0.08 + x'_{1,4,1,1} \times 0.18 + x'_{1,5,1,1} \times 0.08 + x'_{1,6,1,1} \times 0.03 \\
+&x'_{1,7,1,1} \times 0.10 + x'_{1,8,1,1} \times 0.05 + x'_{2,3,1,1} \times 0.03 + x'_{2,4,1,1} \times 0.08 + x'_{2,5,1,1} \times 0.05 \\
+&x'_{2,6,1,1} \times 0.01 + x'_{2,7,1,1} \times 0.09 + x'_{2,8,1,1} \times 0.03 + x'_{3,4,1,1} \times 0.03 + x'_{3,5,1,1} \times 0.03 \\
+&x'_{3,6,1,1} \times 0.01 + x'_{3,7,1,1} \times 0.03 + x'_{3,8,1,1} \times 0.01 + x'_{4,5,1,1} \times 0.07 + x'_{4,6,1,1} \times 0.04 \\
+&x'_{4,7,1,1} \times 0.07 + x'_{4,8,1,1} \times 0.05 + x'_{5,6,1,1} \times 0.01 + x'_{5,7,1,1} \times 0.03 + x'_{5,8,1,1} \times 0.01 \\
+&x'_{6,7,1,1} \times 0.01 + x'_{6,8,1,1} \times 0.00 + x'_{7,8,1,1} \times 0.02 + y).
\end{aligned}
$$

$$(4.40)$$

TABLE 4.7

Performance of the Proposed LP-Based Classifiers (LP Model 1 and LP Model 2) for the Recommendation Data Set

LP-Based Classifier (LP Model 1)		LP-Based Classifier (LP Model 2)	
Objective 1	49.35	Objective 1	52.22
Objective 2	50.39	Objective 2	52.11
Objective 3	48.30	Objective 3	52.32

Note: LP Model 1 and LP Model 2 cannot address multiobjective optimization.

To calculate the objectives of the proposed LP-based classifiers (LP Model 1 and LP Model 2), we split the data into two sets, training data set (70%: 17,500 observations) and test data set (30%: 7500 observations), and the calculated objectives with respect to test data set are listed in Table 4.7.

To calculate the objectives of the proposed LP-based two-phase classifier, we split the data into three sets, training data set (60%: 15,000 observations), test data set (10%: 2500 observations), and validation data set (30%: 7500 observations), and the set of non-dominated solutions for the test data set (see Part 1 of Phase 2) is listed in Table 4.8, and the same is shown in Figure 4.1,

TABLE 4.8

Performance of the Proposed LP-Based Two-Phase Classifier for the Test Data Set (2500 Observations) with Respect to Three Objectives

		Non-Dominated Set of Solutions		
Solution	c Value	Objective 1	Objective 2	Objective 3
1	0.00	72.52	93.80	50.93
2	0.05	73.84	92.77	54.63
3	0.10	74.64	91.82	57.21
4	0.15	75.76	90.95	60.35
5	0.20	77.04	89.52	64.38
6	0.25	78.44	88.64	68.09
7	0.30	78.76	86.74	70.67
8	0.35	79.32	84.83	73.73
9	0.40	79.64	83.40	75.83
10	0.45	79.96	81.41	78.49
11	0.50	80.16	79.90	80.42
12	0.55	80.16	77.12	83.24
13	0.60	80.20	75.06	85.41
14	0.65	80.32	72.99	87.75
15	0.70	80.16	71.01	89.44
16	0.75	79.52	68.39	90.81
17	0.80	78.44	64.89	92.18
18	0.85	77.40	61.87	93.15
19	0.90	76.04	58.38	93.96
20	0.95	75.08	55.76	94.68
21	1.00	74.00	52.74	95.57

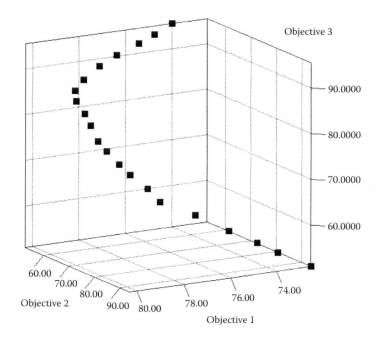

FIGURE 4.1
Non-dominated set of solutions with respect to test data set.

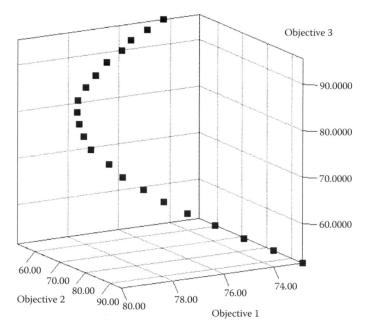

FIGURE 4.2
Non-dominated set of solutions with respect to validation data set.

and the set of non-dominated solutions for the validation data set (see Part 2 of Phase 2) is listed in Table 4.9, and the same is shown in Figure 4.2.

We can also observe from Table 4.8 that (in the case of single objective) if we choose Objective 1 as the primary objective under consideration, then the proposed classifier (in Part 1 of Phase 2) selects 0.50 as the best c value; and if we choose Objective 2 as the primary objective under consideration, then the proposed classifier (in Part 1 of Phase 2) selects 0.00 as the best c value; and if we choose Objective 3 as the primary objective under consideration, then the proposed classifier (in Part 1 of Phase 2) selects 1.00 as the best c value.

We also compare the accuracy of the proposed LP-based classifiers with the accuracy of ANN. We consider the ANN model with sigmoid function as the activation function for artificial neurons, in which the number of hidden layers is 1, and the number of input neurons is set to 8 to match with the number of attributes in the data set, and the number of output neurons is set to 2 to match with the number of categories, and the number of neurons in the hidden layer is 8. We train the ANN using training data set (70%: 17,500 observations), and we present the results of the ANN with respect to the test

TABLE 4.9

Performance of the Proposed LP-Based Two-Phase Classifier for the Validation Data Set (7500 Observations) with Respect to Three Objectives

		Non-Dominated Set of Solutions		
Solution	c Value	Objective 1	Objective 2	Objective 3
1	0.00	72.85	94.40	51.54
2	0.05	73.88	93.22	54.75
3	0.10	74.99	92.23	57.93
4	0.15	76.01	90.97	61.22
5	0.20	77.00	89.87	64.27
6	0.25	77.79	88.58	67.11
7	0.30	78.40	87.02	69.87
8	0.35	79.08	85.55	72.68
9	0.40	79.43	83.51	75.38
10	0.45	79.96	81.55	78.38
11	0.50	80.01	79.30	80.72
12	0.55	79.96	77.00	82.89
13	0.60	79.83	74.88	84.72
14	0.65	79.57	72.49	86.58
15	0.70	78.97	69.62	88.22
16	0.75	78.31	66.81	89.68
17	0.80	77.57	63.78	91.22
18	0.85	76.68	60.78	92.41
19	0.90	76.07	58.39	93.55
20	0.95	75.15	55.58	94.51
21	1.00	74.27	52.90	95.41

TABLE 4.10

Performance of the Artificial Neural Networks for the Recommendation Data Set

Solution	Objective 1	Objective 2	Objective 3
1	79.00%	81.47%	76.51%

data set (30%: 7500 observations), for the considered objectives Objective 1, Objective 2, and Objective 3 in Table 4.10. The results indicate that the proposed LP-based two-phase classifier performs well, and it is able to find a non-dominated set of solutions with respect to multiple objectives (Objective 1, Objective 2, and Objective 3). We can also observe that solution 9 (in Table 4.9) by the proposed LP-based two-phase classifier clearly dominates the solution by the ANN.

- Note that in Figures 4.1 and 4.2, Objective 1 maximizes the total accuracy with respect to both categories, Objective 2 maximizes the accuracy with respect to category 1, and Objective 3 maximizes the accuracy with respect to category 0.

4.5 Summary

This chapter proposes MILP-based classifier and LP-based classifiers for binary classification, and when we compare the accuracy of all the proposed LP-based classifiers with ANN, the results indicate that the proposed LP-based two-phase classifier is able to give better results. Consequently, the proposed LP-based two-phase classifier is able to handle data that are not inherently linearly separable, unlike the conventional MILP-based and LP-based classifiers. The salient contributions of the proposed LP-based two-phase classifier are in terms of treating the decision variables as unrestricted in sign; accounting for the contribution of attributes from their interaction effects and the contribution of attributes from their higher-order polynomial degrees; treating the classification threshold/cut-off as a decision variable; converting the bandwidth of boundary of threshold to a crisp boundary with the consideration of multiple objectives. The proposed LP-based two-phase classifier considers such multiple objectives because in the application areas such as medical diagnosis, absence of an alarm (failing to predict the category 1) is more serious than a false alarm. The proposed LP-based two-phase classifier is an efficient method in terms of the ability to solve (linear programming model) the underlying classification problem, and thus it can be effectively used in conjunction with more sophisticated and computationally demanding approaches such as random forest, support vector machines, and ANN to improve the accuracy further on the data that has high variety and high volume.

Acknowledgment

We are thankful to the reviewers and the editors for their valuable comments and suggestions to improve our chapter.

References

Freed, N. and Glover, F. 1986. Evaluating alternative linear programming models to solve the two-group discriminant problem. *Decision Sciences*, 17(2), 151–162.

Maas, A. L., Daly, R. E., Pham, P. T., Huang, D., Ng, A. Y., and Potts, C. 2011. Learning word vectors for sentiment analysis. In *Proceedings of the 49th Annual Meeting of the Association for Computational Linguistics: Human Language Technologies—Volume 1*, 142–150. Association for Computational Linguistics, Portland, Oregon.

Shi, Y., Wise, M., Luo, M., and Lin, Y. 2001. Data mining in credit card portfolio management: A multiple criteria decision making approach. In Köksalan, M. and Zionts, S. (Eds.), *Multiple Criteria Decision Making in the New Millennium*, 427–436. Springer Berlin Heidelberg, Ankara, Turkey.

Yan, N., Wang, Z., Shi, Y., and Chen, Z. 2006. Nonlinear classification by linear programming with signed fuzzy measures. In *2006 IEEE International Conference on Fuzzy Systems*, 408–413. IEEE, Vancouver, British Columbia, Canada.

5

Multicriteria Evaluation of Predictive Analytics for Electric Utility Service Management

**Raghav Goyal, Vivek Ananthakrishnan,
Sharan Srinivas, and Vittaldas V. Prabhu**

CONTENTS

5.1 Introduction

Utility companies are responsible for the infrastructure of the power delivery system and power interruptions disrupt customers as well as cause significant economic losses. In the United States, the estimated cost of power interruptions is $79 billion per year (LaCommare and Eto, 2006). The outage costs are directly proportional to the customer's dependence upon electricity during an outage. With annual electricity use in a typical U.S. home increasing 61% since 1970, it is becoming increasingly important to reduce and prevent outages (Swaminathan and Sen, 1998). Outage costs vary significantly depending upon the outage attributes such as frequency, duration, and intensity of the outage. In this chapter, an outage is considered to be a complete or total loss of service, typically resulting from a distribution-related cause or a transmission failure.

The priority of every organization is twofold: providing customers the best possible product at the lowest cost while maintaining the quality. The situation is not very different in the case of a service industry. Electric utility companies aim to optimize every form of the electric service system. Every company tries to provide the most reliable service in the form of consistent and uninterrupted service to its customers while reducing the cost of supply and maximizing profits. Companies often have to deal with problems such as downtime or outage time due to a failure in the supply system. Therefore, better recovery planning and forecasting of outages is necessary to provide reliable service to customers.

There are several reasons attributed to an electric power outage, which range from equipment failure and overloading of the line to weather-related events. Power systems are most vulnerable to storms and extreme weather events. Seasonal storms combined with wind, snow, rain, ice, etc. can cause significant outages. Data on weather-related outages have been used in the past to estimate the costs of an outage and the impact it has on consumers. According to past weather-related outage data, 90% of customer outage-minutes are owing to events which affect the local distribution systems, while the remaining 10% are from generation and transmission problems (Campbell, 2012). Electric utility companies can reduce the outages/damages resulting from severe weather conditions by enhancing the overall condition of the power delivery system and better prediction of the outage. There has been a lot of research that focuses on predicting outages as shown in Table 5.1.

However, most of the previous work does not use hourly weather forecast to predict short-term outages. Increasingly, companies are looking to tackle outages due to both local distribution systems and larger transmission systems by developing strategies to reduce or prevent outages. Short-term forecasts of an electric power outage and the cost parameters associated with the outage would help companies optimize their

TABLE 5.1

Summary of Past Research Done

Author Name	Cost Analysis of Power Outages	Predicting Power Outage for Extreme Weather Conditions	Predicting Power-Related Damages for Extreme Weather Conditions
Balijepalli et al. (2005)		✓	
Cerruti and Decker (2012)			✓
Davidson et al. (2003)		✓	
DeGaetano et al. (2008)			✓
Huang et al. (2001)			✓
J. Douglas (2000)	✓		
LaCommare and Eto (2006)	✓		
Li et al. (2010)		✓	
Liu et al. (2007)		✓	
Reed (2008)		✓	✓
Reed et al. (2010)		✓	✓
Sullivan et al. (1996)	✓		
Winkler et al. (2010)		✓	✓
Zhou et al. (2006)		✓	
Zhu et al. (2007)		✓	

manpower and resource planning. The advancement of science and technology has led to accurate short-term weather forecasts (up to 72 hours). Data from National Digital Forecast Database for 5 years of U.S. weather would be over 267 TB. The focus of this chapter is to leverage such short-term forecasts by combining it with analytics for electric utility services. These concerns have helped to stimulate research activity in the area of electric power supply system's modeling and analysis and are the motivation for this chapter.

5.2 Layout of Electric Supply Distribution Network

In a product supply chain, the product is manufactured using raw material from suppliers, then passed on to the retailers and finally to the end customer through a distribution network. A service supply chain in the form of an electric distribution system involves customer–supplier service supply relationships. Figure 5.1 shows the flow of the service, product, and information through the supply chain.

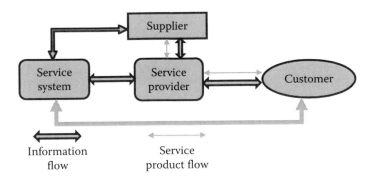

FIGURE 5.1
Service supply chain.

The electric power supply industry is an integral part of the service system industry and also important for the economy as all businesses rely heavily on electric power to operate. The Energy Information Administration predicts that there would be an increase of 29% in the electricity demand in the United States between 2012 and 2040 (Sieminski, 2014). The power generators own and operate electricity-generating facilities or power plants and sell the power produced to the utility service providers. These service providers are the key players in the network as they are responsible for providing a reliable source of electric power while ensuring uninterrupted service at an affordable cost.

The utility service provider is directly involved in the design of the service and is answerable directly to the customer. For instance, the products and services in an electric power utility service supply chain network are limited to the electric power supplied and transmission services provided. The service utility in a particular geographical area handles the transmission and distribution of electricity to the end users through a vertically integrated structure. The various decision makers in this system such as the power generators, the power suppliers, the transmitters, and the customers operate in a decentralized system. A depiction of the distribution network for electric power is shown in Figure 5.2 (Nagurney and Matsypura, 2007; Nagurney et al., 2007).

Understanding the outline of the electric power supply system is crucial for identifying the critical points in the service supply network:

- *Power generation:* The electric power generating station could be any power plant (gas, oil, nuclear, thermal, etc.) that converts fuel sources into electricity. The power generated is then stepped up to as high as 500,000 V and passed through transmission lines to the distribution lines.

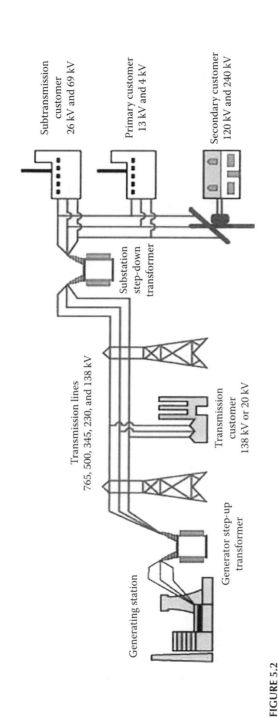

FIGURE 5.2
Basic electrical distribution system. (From U.S.–Canada Power System Outage Task Force, *Final Report on the August 14, 2003, Blackout in the United States and Canada: Causes and Recommendations,* April 2004.)

- *Transmission:* The transmission stage of the network involves transmitting electricity over long distances at very high voltages since energy transfer efficiency is high at very high voltages. Large players in the transmission market operate thousands of miles of transmission lines.

- *Distribution:* The electricity transmitted via the transmission lines is then stepped down using a step-down transformer. The distribution line networks, which are made up of the feeders and laterals, deliver electricity to commercial, industrial, and residential customers. These feeders and laterals handle current with a voltage range of 13–23 kV, while the service lines handle a voltage range of 120–480 V. The distribution costs for industrial and commercial users are less due to the high-voltage power being supplied.

Power outages could be broken down into three components: number of customers affected, total minutes of power outage, and number of interruptions. In addition, there are three customer types recognized by electric utilities—industrial, commercial, and residential. Depending on the geographic location, an electric utility company can serve a particular kind of customer or could serve all three at the same time.

Earlier, utility companies primarily focused on reducing cost. However, companies have shifted their focus to include supply assurance and risk management. One of the key metrics used to evaluate the supply assurance and risk management is system average interruption duration index, measured in units of time. The median value for North American utilities is approximately 1.50 hours (90 minutes). Typically, energy companies strive to improve this measure by directly reducing the total power outage minutes along with the number of customers affected by those outages. The predictive analytic tools used in this chapter aim to predict power outage occurrences by using the hourly weather forecasts. This information could then be used to reduce the interruption minutes by improving workforce planning of the repair crew.

As discussed in Chapter 3, analytics can be grouped into three types: descriptive analytics, predictive analytics, and prescriptive analytics. Electric utilities typically use descriptive analytics such as Pareto charts for descriptive statistics and data visualization to derive insights about the most frequent causes of power interruptions. This chapter mainly focuses on predictive analytics that uses weather forecasts to predict power interruptions. In this chapter, prescriptive analytics is limited to formulating a multiple criteria mathematical programming (MCMP) model that minimizes staffing the cost and duration of power interruption. The resulting data-driven decision process, as suggested in Chapter 3, would be as follows:

Chapter 3	Analytics for Electric Utility
1. Knowledge of relationship between KPIs and factors would then provide the decision maker with appropriate actionable items. Identify the problem or opportunity for value creation	CMI reduction
2. Identify sources of data (primary as well secondary data)	Weather forecast Outage data
3. Preprocess the data for missing and incorrect data. Prepare the data for analytics model building, if necessary transform the data	Combine the two data sources to forecast outages
4. Build the analytical models and identify the best model using model validation	MCDM problem in selecting predictive analytic technique
5. Communicate the data analysis output and decisions effectively	Ranked list
6. Implement solution/decision	Bicriteria optimization for staffing decision

5.3 Predictive Analysis

Traditionally, electric utility companies used multiple-level Pareto charts to identify the root cause of the outage and develop a strategy to minimize disruptions. However, in recent times, weather predictions have improved substantially. For example, 96–120-hour predictions of a weather-related event are nearly as accurate as the actual occurrence.

In addition to other attributes, companies could utilize hourly weather forecasts to predict power outage occurrences by the hour and use them to efficiently plan the usage of their resources.

The prediction of an outage using hourly weather forecasts can be treated as a binary classification problem (0: no outage; 1: outage). In order to predict the outage, it is necessary to have a data set that includes all the input variables (weather attributes) and the target values (either 0: no outage or 1: outage). The initial data set is split into a training set and testing set. Each instance in the training set contains one target value (either 0: no outage or 1: outage) and several input parameters (weather attributes). A machine-learning classifier uses the training data set and trains the prediction model to learn the underlying relationship between the inputs and the targets. A learned classifier uses the inputs of the testing data set to predict the outage occurrence. Finally, the classifier is evaluated by comparing the actual output and the predicted output. There are several classification techniques and algorithms that are used in the world of predictive analytics. Several prominent algorithms and their results are discussed below.

5.3.1 Decision Trees

Decision tree algorithms are one of the earliest classification algorithms that have been used in predictive modeling (Quinlan, 1986). The algorithm works on the philosophy of dividing the main problem (root node) into smaller successive decisions based on a unique node until a particular class gains a majority. This division is done until any further partitioning being done is determined to be useless. The algorithm follows a top-down, greedy search approach with the selection of the attribute that best classifies the set as the root node. The advantages of using a decision tree algorithm are (1) it can be applied to any type of data, (2) the final structure of the classifier is quite simple and can be easily interpreted by decision makers, the resulting trees can be easily used to obtain a better understanding of the phenomenon in question.

5.3.2 Logistic Regression

The logistic regression model is used to predict a dichotomous outcome using one or more independent variables (Menard, 2002). The outcome or the response variable Y is the outage occurrence (i.e., 0 or 1), while the independent variables (X) are input parameters (i.e., weather attributes). Therefore, in logistic regression, the probability of power outage and no outage are computed as shown below:

$$p(Y = 1) = \frac{e^{(\beta_0 + \beta \cdot X)}}{1 + e^{(\beta_0 + \beta \cdot X)}} = \frac{1}{1 + e^{-(\beta_0 + \beta \cdot X)}}, \tag{5.1}$$

$$p(Y = 0) = 1 - P(Y = 1). \tag{5.2}$$

5.3.3 Boosting

Boosting is one of the most recent and important developments in the domain of predictive classification techniques. It works on the principle of sequentially applying a classification algorithm to reweighted versions of the training data, and then taking a weighted majority vote of the sequence of classifiers produced as a result of this sequential process. For the two-class problem, boosting can be viewed as an approximation to additive modeling on the logistic scale using maximum Bernoulli likelihood as a criterion. This simple strategy is found to yield drastic improvements in results for many classification algorithms due to statistic principles of additive modeling and maximum likelihood estimation.

5.3.4 Random Forest

Random forest is a popular ensemble learning method for generating classification and regression models (Breiman, 2001; Ho, 1995). The method works by constructing a multitude of decision trees using the training set and outputting the mode of the class and the mean prediction of the individual trees for a classification problem and regression problem, respectively. In other words, random forests average multiple deep decision trees, trained on different parts of the same training set. The objective of reducing the variance comes at the expense of a small increase in the bias and some loss of interpretability. However, this greatly boosts the performance of the final model and it also corrects the problem of overfitting, which is a common occurrence in decision trees. The general technique of bootstrap aggregating or bagging is implemented in the training algorithm.

5.3.5 Support Vector Machines

Support vector machines (SVMs) are supervised learning methods used for classification and regression tasks that generate nonoverlapping partitions and usually employ all the attributes (Gunn, 1998; Gunn et al., 1997; Vapnik and Vapnik, 1998). The entity space is partitioned in a single pass and is based on maximum margin linear discriminants, similar to probabilistic approaches, but does not consider the dependencies among attributes. SVMs have gained popularity as they are based on the structural risk minimization (SRM) principle. This gives SVMs greater generalization ability, which is the goal in statistical learning. SVMs rely on preprocessing the data to represent patterns in a high dimension. Data from two categories can always be separated by a hyperplane when an appropriate nonlinear mapping is used. One maximizes the distance between itself and the nearest target value (optimal separating hyperplane). The basic idea behind SVM classifier is to choose the hyperplane that has the maximum margin (distance from itself to the nearest class).

5.3.6 Artificial Neural Networks

Artificial neural networks (ANNs) are used to estimate functions that can depend on a large number of inputs and are generally unknown (Boger and Guterman, 1997; Braspenning et al., 1995). ANNs assign numeric weights to the connections between the input and output variables that can be tuned based on experience. There are many different kinds of learning rules used by neural networks, with the delta rule being the most popular. The delta rule learns by updating the weights depending upon the error magnitude (i.e., the difference between the predicted output and the actual output). An initial guess and subsequent error corrections due to weight adjustments lead to a final optimal weight. ANNs provide an analytical alternative to conventional techniques which are often limited by strict assumptions of normality, linearity, variable independence, etc. ANNs can capture many kinds of relationships and thus allow the user to easily model phenomena that are not easily explainable.

5.4 Evaluating Prediction Models Using Multicriteria Decision-Making Techniques

Multicriteria decision making (MCDM) has been heavily used in supplier selection to improve the supply chain of any organization. Comprehensive literature review on different MCDM techniques used for supplier selection and evaluation has been done in the past (Ho et al., 2010). In addition to supply chain management, MDCM techniques are used in several other areas/fields such as healthcare, marketing, and financial management (Zopounidis and Doumpos, 2002). For instance, MCDM techniques are used in financial management for developing an efficient portfolio, credit-risk assessment, etc. However, none of the previous research articles use MCDM techniques to evaluate and select the best predictive model. This chapter explores how different MCDM techniques such as L_2 metric, Borda count, and rating method could be used to pick the best predictive model, given multiple selection criteria such as accuracy, specificity, and sensitivity.

5.4.1 Performance of Classification Model

The classification methods discussed in Section 5.3 are used to predict the occurrence of an interruption, given an hourly weather condition. A confusion matrix is generally used to illustrate the performance of the classification model, and is generated by comparing the predicted outcome and the actual outcome. Table 5.2 shows the layout of a confusion matrix.

The elements of the confusion matrix are as follows:

- True negatives (TN) represent the number of times the prediction model accurately predicted 0 (no interruption).
- False positives (FP) represent the number of times the prediction model predicted 1 (interruptions) when in reality there were no interruptions (represented by 0). This type of error is also called type I error.
- True positives (TP) represent the number of times the model accurately predicted 1 (interruptions).

TABLE 5.2

Confusion Matrix

		Actual Outcome	
Predicted outcome		0 (no outage)	1 (outage)
	0 (no outage)	TN	FN
	1 (outage)	FP	TP

- False negatives (FN) represent the number of times model predicted 0 (no interruptions) when in reality there were interruptions
  (represented by 1). This type of error is called type II error.

5.4.2 Criteria for Model Evaluation

The criteria considered for model evaluation are accuracy, area under the
receiver operating characteristic (ROC) curve, sensitivity, specificity, and
precision. Based on the confusion matrix, the values of the criteria for each
prediction model are determined, and are as follows:

- *Accuracy* {(TN + TP)/(TN + TP + FP + FN)}: This is a measure of
  overall accuracy of the model. This is calculated by dividing
  accurate predictions by total number of instances.
- *Area under the ROC curve (AUC)*: It is a standard measure of the
  predictive accuracy. If it is below 0.50, then the model prediction is
  worse than an unbiased flip of a coin.
- *Sensitivity* {TP/(FN + TP)}: It measures the proportion of outages that
  were correctly classified.
- *Specificity* {TN/(TN + FP)}: It measures the proportion of no outages
  that were accurately classified.
- *Precision* {(TP/(FP + TP)}: It measures the relevance of the positively
  classified (i.e., 1: outage) instances.

5.4.3 Multicriteria Ranking Techniques

The following multicriteria ranking methods are used to determine the best
classification method:

- Rating method
- Borda count
- L_2 metric method

Rating Method: In this method, the participants rate each parameter
on a scale of 1–10 (1 being least important and 10 being most important).
The weights of each criterion are then calculated by normalizing the ratings (Ravindran, 2016). The final score of each prediction model is the
weighted sum of the criteria values, and the model with the highest score is
ranked first.

Borda Count: In this method, the decision maker is asked to rank all the
criteria based on their importance. Therefore, the n criteria are ranked from
1 (most important) to n (least important). The most important criterion is
given n points, the next most important criterion gets $n–1$ points, and the

least important criterion gets 1 point. In the presence of multiple decision makers, the number of times a criterion is ranked at a particular rank on the survey is multiplied with its corresponding points to determine the overall points for each of the criteria. The weights are then computed by dividing the points of each criterion by the sum of points of all the criteria. The final score of each prediction model is the weighted sum of the criteria values, and the model with the highest score is ranked first.

L_2 *Metric Method*: It measures the distance between the vector of ideal solutions and the vector representing criteria values for each prediction model as shown below (Ravindran, 2016).

$$L_2(k) = \left[\sum_{j=1}^{n} (x_{jk} - y_j) \right]^{1/2}, \tag{5.3}$$

where
    k denotes the prediction model, j denotes the criterion, and n is the total number of criteria.

x_{jk} is the value of criterion j for prediction model k.

x_j is the ideal value of criterion j.

Therefore, each prediction model will have a score, and the model with the least L_2 score is ranked first, followed by the next smallest L_2 score, etc.

5.5 Data Description

The primary data set used in this chapter is the hourly recordings of weather conditions for all the days in 2013, at the location of the electric utility. The other data set contains time-stamped details of the complaints reported by customers in the same year. Each instance of the complaint is captured along with various other information such as equipment failure, causes of failure, parts affected, and duration of the outage. The detailed variables of these data files are described below.

5.5.1 Weather Data

The hourly weather information collected for the year 2013 is described in Table 5.3.

The variable "Event" is further divided into four binary variables, namely, fog, rain, tornado (hurricane), thunderstorm, to describe the type of event that occurred. The variable "Event" takes a value 1 if at least one of the four

TABLE 5.3

Weather Data Variables Description

Variable Name	Variable Description	Variable Type
Temperature	Temperature of the location (°F)	Numeric
Heat index	Index combining effect of temperature and humidity (°F)	Numeric
Dew point	Dew point of the location (°F)	Numeric
Humidity	Humidity of the location (%)	Numeric
Pressure	Atmospheric pressure (in)	Numeric
Visibility	Visibility of the location (miles)	Numeric
Wind speed	Speed of the wind flowing at the location (mph)	Numeric
Gust speed	Gust speed of the wind at the location (mph)	Numeric
Precipitation	Amount of precipitation at the location (in)	Numeric
Event	Special weather condition at the location	Binary

event types is 1. The final data set has over 8700 instances with 697 instances experiencing a weather-related event. A sample of the final data set is shown in Figure 5.3.

5.5.2 Power Outage Data

The power outage data set provided by an electric utility company contains data on the outages experienced by the company in the year 2013. The data set is time-stamped with each instance in this data set being referred to as a ticket (outage).

The data set contains more than 42,000 such instances with outages affecting 1,989,032 customers. The total minutes of disruption (MI) for tickets are totaled at 2,362,144 while the customer minutes interrupted (CMI) for the entire year is 48,355,553 minutes. In this chapter, binary MI, which is assigned a value of 0 when there is no outage and value of 1 when there is an outage, is selected as the response variable (output). This information has been rolled up hourly to integrate with previously mentioned weather data. The predictive models used in this chapter aim to predict the binary MI (response variable) with high accuracy. A sample of the final data table used is shown in Figure 5.4.

5.6 Experimental Results

In this section, the criteria values for each of the prediction models are obtained and the models are evaluated using multicriteria ranking techniques. The experimental results are performed by using the data set obtained from the electric utility company. The prediction models are coded

Date	Hour	Temp	Heat Index	Dew Point	Humidity	Pressure	Visibility	Wind Speed	Gust Speed	Precipitation	Event	Rain	Thunder storm	Fog	Tornado
5/2/2013	7	73.90	0.00	69.10	0.85	29.89	10.00	11.50	0.00	0.04	1	1	0	0	0
5/2/2013	8	75.55	0.00	71.35	0.87	29.91	10.00	11.55	0.00	0.00	0	0	0	0	0
5/2/2013	9	73.75	0.00	71.03	0.91	29.91	10.00	10.95	5.48	0.00	1	1	1	0	0
5/2/2013	10	75.83	0.00	71.48	0.87	29.91	10.00	9.20	0.00	0.00	1	1	0	0	0
5/2/2013	11	74.43	0.00	71.03	0.89	29.91	6.50	15.28	11.50	0.00	1	1	0	0	0
5/2/2013	12	74.43	0.00	70.13	0.92	29.91	0.85	9.80	0.00	0.26	1	1	1	0	0
5/2/2013	13	71.68	0.00	69.84	0.94	29.90	2.60	8.30	0.00	0.82	1	1	1	1	0
5/2/2013	14	71.43	0.00	69.87	0.95	29.91	5.00	3.87	0.00	0.03	1	1	1	0	0
5/2/2013	15	69.57	0.00	66.43	0.90	29.90	3.33	8.47	6.13	0.16	1	1	1	0	0

FIGURE 5.3
Sample data table for historical weather.

Temp	Heat Index	Dew Point	Humidity	Pressure	Visibility	Wind Speed	Gust Speed	Precipitation	Event	Rain	Thunderstorm	Fog	Tornado	Binary MI
73.9	0	61	0.64	30.18	10	8.1	0	0	0	0	0	0	0	0
73.2	0	63.3	0.71	30.2	8.5	8.1	0	0	1	1	0	0	0	1
70	0	66	0.87	30.21	3	6.9	0	0.02	1	1	0	0	0	1
68	0	64.9	0.9	30.18	6	4.6	0	0.05	1	1	0	0	0	0
69.1	0	66	0.9	30.15	5	5.8	0	0.02	1	1	0	0	0	0
69.27	0	65.53	0.88	30.13	4.33	5.8	0	0.06	1	1	0	0	0	1
71.1	0	66	0.84	30.12	7	0	0	0	1	1	0	0	0	0
70	0	66.9	0.9	30.11	7	0	0	0.01	1	1	0	0	0	1
71.35	0	66.1	0.84	30.11	10	3.5	0	0	0	0	0	0	0	1
69.45	0	66.1	0.89	30.12	6.5	6.35	0	0	0	1	0	0	0	0
68.9	0	66.1	0.91	30.18	10	5.8	0	0	0	0	0	0	0	1
68	0	66.13	0.94	30.14	8.67	4.23	0	0	1	1	0	0	0	1
66.9	0	64.9	0.93	30.13	10	5.8	0	0	0	0	0	0	0	0
66.9	0	66	0.97	30.13	10	4.6	0	0	0	0	0	0	0	1

FIGURE 5.4
Sample data table for prediction model.

6666666666666

66666666

6666

666666

TABLE 5.4

Criteria Values for Prediction Models

Prediction Model	Accuracy	AUC	Sensitivity	Specificity	Precision
Decision tree	0.698	0.661	0.923	0.211	0.718
Random forest	0.713	0.693	0.896	0.313	0.746
Boosting	0.710	0.720	0.906	0.288	0.734
Support vector machines	0.706	0.672	0.938	0.204	0.718
Logistic regression	0.683	0.710	0.910	0.219	0.691
Artificial neural networks	0.706	0.711	0.922	0.239	0.724

and executed in R statistical software using a computer with Intel Core i5 2.50 GHz processor with 8 GB of RAM.

The data set must be divided into training set and testing set. If there is less training data, then the parameter estimates will have higher variance. On the other hand, if there is less testing data, then the classifier performance will have higher variance. Therefore, different data splits are used for learning and validation and are as follows:

- Data used in training, 67%; data used for testing, 33%
- Data used in training, 70%; data used for testing, 30%
- Data used in training, 75%; data used for testing, 25%

The classification model is replicated 20 times with random sampling for each of the above splits, and the criteria values are recorded. It was observed that the data split did not have any impact on the performance of the prediction model because the criteria values did not have any substantial deviation. Therefore, the average of the results is taken and compiled as shown in Table 5.4.

In order to ensure easy comparison and avoid any bias in estimating the overall score for each prediction model, the criteria values are scaled using the ideal value method, where the criteria values are divided by their ideal (best) value. For example, the ideal (best) value for accuracy is 0.713. The accuracy value for each of the prediction models is divided by the ideal value for accuracy. Therefore, the accuracy of decision tree will be scaled to 0.979 (i.e., 0.698/0.713). Note that the scaled criteria values will always be ≤1, and the best value of each criterion is 1. Table 5.5 presents the scaled criteria values for all the prediction models.

It can be observed from Table 5.5 that none of the prediction models perform the best with respect to all the criteria. For instance, random forest is the best prediction model with respect to accuracy, specificity, and precision. However, boosting and SVMs perform the best with respect to AUC and sensitivity, respectively. Therefore, the rating, Borda count, and L_2 metric method are used to determine the best model. The criteria weights for rating method and Borda count method are shown in Table 5.6.

TABLE 5.5

Summary of Scaled Results

Prediction Model	Accuracy	AUC	Sensitivity	Specificity	Precision
Decision tree	0.979	0.918	0.984	0.676	0.963
Random forest	1.000	0.962	0.955	1.000	1.000
Boosting	0.997	1.000	0.966	0.921	0.983
Support vector machines	0.990	0.934	1.000	0.652	0.963
Logistic regression	0.958	0.986	0.970	0.699	0.927
Artificial neural networks	0.991	0.988	0.983	0.764	0.971

TABLE 5.6

Criteria Weights Using Rating and Borda Count Method

Criteria	Weights Using Rating Method	Weights Using Borda Count Method
Accuracy	0.17	0.35
AUC	0.22	0.32
Sensitivity	0.22	0.07
Specificity	0.15	0.09
Precision	0.24	0.17

The overall score and the corresponding rank for each prediction model obtained using the three MCDM ranking methods are presented in Table 5.7.

Note that the final scores of each prediction model obtained using rating method and Borda count method are calculated by multiplying the scaled weights of the criterion with the corresponding criterion value.

Based on the analysis of data presented in Table 5.7, it is evident that random forest, boosting, and ANNs are the best methods as they are consistently

TABLE 5.7

Results from L_2 Metric

	MCDM Ranking Methods					
	Rating Method		Borda Count		L_2 Metric	
Prediction Model	Final Score	Rank	Final Score	Rank	Final Score	Rank
Decision tree	0.9174	6	0.9298	6	0.540	5
Random forest	0.9817	1	0.9847	2	0.8240	1
Boosting	0.9761	2	0.9866	1	0.7650	2
Support vector machines	0.9227	4	0.9378	5	0.537	6
Logistic regression	0.9205	5	0.9392	4	0.560	4
Artificial neural networks	0.9497	3	0.9657	3	0.637	3

ranked in the top three. Since random forest is ranked first under the rating method and L_2 metric method, it is regarded as the best classifier to predict outages for the given data set.

5.7 Smart Staffing

The machine-learning classifiers can predict future interruptions and their intensity using predictors such as weather forecasts and history of maintenance. The advancements in science and technology have led to accurate short-term weather forecasts (up to 72 hours) and, therefore, enable the development of good prediction models. These predictions can be used to plan the workforce (crew) for power restoration. Low staffing levels decrease the staffing cost. However, it leads to high repair time resulting in higher restoration time. On the other hand, higher staffing levels substantially decrease the restoration times at the expense of increased staffing cost. Therefore, the staffing problem involves conflicting criteria, and hence an MCMP model is used to formulate the staffing problem.

Set and indices

$w \in W$	Set of all worker types
$f \in F$	Set of all failure types

Parameters

R_{wf}	Total man-hours required by worker of type w to repair failure type f
A_w	Total workers of type w available for repairs
Q_{wf}	1 if failure type f requires worker of type w; 0 otherwise
M	Large positive number

Decision variables

x_{wf}	Number of workers of type w hired to repair failure type f

---

5.7.1 Objective Functions

5.7.1.1 Objective 1: Minimize Staffing Costs

A stable repair crew with essential skills is necessary to restore power in case of outages. However, the increase in the number of workers results in an increase in cost for the electric utility company. Therefore, the first objective seeks to minimize the total workers hired as shown below:

$$\text{Minimize } z_1 = \sum_{f \in F} \sum_{w \in W} x_{wf}. \tag{5.4}$$

5.7.1.2 Objective 2: Minimize Power Restoration Time

The time taken to restore power depends on the intensity of failure and the number of workers assigned to restore power. Assigning few workers for a high-intensity failure would increase the restoration time. Therefore, the second objective seeks to minimize the power restoration times as shown in

$$\text{Minimize } z_2 = \sum_{f \in F} \sum_{w \in W} \left(\frac{R_{wf}}{x_{wf}} \right). \tag{5.5}$$

5.7.2 Model Constraints

5.7.2.1 Restriction on Workforce Capacity

The total number of workers of a specific type is finite (A_w). Therefore, constraint (5.6) ensures that the total number of workers hired is always within the finite capacity restriction for all the worker types:

$$\sum_f x_{wf} \leq A_w \quad \forall w \in W. \tag{5.6}$$

5.7.2.2 Worker Type Requirement

Each failure type requires a crew with different worker types. For instance, a particular failure type may require two different worker types such as lineman and foreman. It is essential to predict the type of failure and have the crew prepared in advance. Constraints (5.7) and (5.8) ensure that the worker type requirement is satisfied. If there is a requirement for a particular worker type (i.e., $Q_{wf} = 1$), then constraint (5.7) ensures that at least one worker of that particular type is assigned to the failure, while constraint (5.8) becomes redundant. However, if a failure does not require a particular worker type (i.e., $Q_{wf} = 0$), the constraints (5.7) and (5.8) ensure that no workers of that type are assigned to the failure:

$$x_{wf} \geq Q_{wf} \quad \forall w \in W, f \in F, \tag{5.7}$$

$$x_{wf} \leq MQ_{wf} \quad \forall w \in W, f \in F. \tag{5.8}$$

5.7.2.3 Nonnegativity Restriction

Constraint (5.9) ensures that the number of workers hired is always a positive integer:

$$x_{wf} \geq 0, integer \quad \forall w \in W, f \in F. \tag{5.9}$$

The model can be solved using solution techniques such as goal programming and ε-constraint method to obtain a set of efficient solutions that provide a trade-off between staffing costs and restoration time. The set of efficient solutions can then be presented to a decision maker to obtain the best compromise solution that meets the need of the electric utility company.

5.8 Conclusions

In the United States, power outages cause billions of dollars in losses. This chapter aims at predicting the power outage occurrences accurately for an electric utility company that serves over 9 million people in the United States. Several machine-learning classifiers are used to predict the outages using the hourly weather forecasts. The machine-learning classifiers are later ranked based on different metrics, namely, accuracy, AUC, sensitivity, specificity, and precision using MCDM techniques.

Based on the analysis using multicriteria ranking methods, it was evident that random forest was the best method to predict the power outage occurrences as it was ranked first by two out of the three MCDM ranking techniques. In addition, an MCMP model was presented to determine the appropriate staffing levels using the outputs of the prediction models. In future work, we plan to develop prediction models to measure the intensity of the outage, and solve the MCMP model to obtain the set of efficient solutions that presents the trade-off between staffing costs and restoration time.

References

Balijepalli, N., Venkata, S. S., Richter Jr, C. W., Christie, R. D., and Longo, V. J. 2005. Distribution system reliability assessment due to lightning storms. *IEEE Transactions on Power Delivery*, 20(3), 2153–2159.

Boger, Z. and Guterman, H. 1997. Knowledge extraction from artificial neural network models (Vol. 4, pp. 3030–3035). *IEEE International Conference on Systems, Man, and Cybernetics. Computational Cybernetics and Simulation*, IEEE, Orlando, FL.

Braspenning, P. J., Thuijsman, F., and Weijters, A. J. M. M. 1995. *Artificial Neural Networks: An Introduction to ANN Theory and Practice* (Vol. 931). Springer Science & Business Media, Berlin, Germany.

Breiman, L. 2001. Random forests. *Machine Learning*, 45(1), 5–32.

Campbell, R. J. 2012. Weather-related power outages and electric system resiliency. Congressional Research Service, Library of Congress, Washington, DC.

Cerruti, B. J. and Decker, S. G. 2012. A statistical forecast model of weather related damage to a major electric utility. *Appl Meteor Climatol*, 51(2), 191–204.

Clemmensen, J. 1993. Estimating the cost of power quality. *IEEE Spectr*, 30(6), 40–41.

Davidson, R. A., Liu, H., Sarpong, I. K., Sparks, P., and Rosowsky, D. V. 2003. Electric power distribution system performance in Carolina hurricanes. *Natural Hazards Review*, 4(1), 36–45.

DeGaetano, A. T., Belcher, B. N., and Spier, P. L. 2008. Short-term ice accretion forecasts for electric utilities using the weather research and forecasting model and a modified precipitation-type algorithm. *Weather and Forecasting*, 23(5), 838–853.

Douglas, J. 2000. *Research Analyst to EPRI*. Personal Communication to J. Eto. Lawrence Berkeley National Laboratory, Berkeley, California.

Gunn, S. R. 1998. *Support Vector Machines for Classification and Regression*. ISIS Technical Report, 14.

Gunn, S. R., Brown, M., and Bossley, K. M. 1997. Network performance assessment for neurofuzzy data modelling. In Liu, X., Cohen, P., Berthold, M. R. (Eds.), *Advances in Intelligent Data Analysis Reasoning about Data* (pp. 313–323). Springer, London, UK.

Ho, T. K. 1995. Random decision forests (Vol. 1, pp. 278–282). *Proceedings of the Third International Conference on Document Analysis and Recognition*, IEEE, Montreal, Canada.

Ho, W., Xu, X., and Dey, P. K. 2010. Multi-criteria decision making approaches for supplier evaluation and selection: A literature review. *European Journal of Operational Research*, 202(1), 16–24.

Huang, Z., Rosowsky, D., and Sparks, P. 2001. Hurricane simulation techniques for the evaluation of wind-speeds and expected insurance losses. *Journal of Wind Engineering and Industrial Aerodynamics*, 89(7), 605–617.

LaCommare, K. H. and Eto, J. H. 2006. Cost of power interruptions to electricity consumers in the United States (US). *Energy*, 31(12), 1845–1855.

Li, H., Treinish, L. A., and Hosking, J. R. 2010. A statistical model for risk management of electric outage forecasts. *IBM Journal of Research and Development*, 54(3), 8–1.

Liu, H., Davidson, R. A., and Apanasovich, T. V. 2007. Statistical forecasting of electric power restoration times in hurricanes and ice storms. *Power Systems, IEEE Transactions on Power Systems*, 22(4), 2270–2279.

McNulty, S. and Solutions, E. 2001. *The Cost of Power Disturbances through Industrial and Digital Economy Companies*. EPRI's Consortium for Electric Infrastructure for a Digital Society (CEIDS), USA, Tech. Rep, 108829.

Menard, S. 2002. *Applied Logistic Regression Analysis* (Vol. 106). Sage, Thousand Oaks, California.

Nagurney, A., Liu, Z., Cojocaru, M.-G., and Daniele, P. 2007. Dynamic electric power supply chains and transportation networks: An evolutionary variational inequality formulation. *Transportation Research Part E: Logistics and Transportation Review*, 43(5), 624–646.

Nagurney, A. and Matsypura, D. 2007. A supply chain network perspective for electric power generation, supply, transmission, and consumption. In Kontoghiorghes, E. and Gatu, C. (Eds.), *Optimisation, Econometric and Financial Analysis* (pp. 3–27). Springer, Germany.

Quinlan, J. R. 1986. Induction of decision trees. *Machine Learning*, 1(1), 81–106.

Ravindran, A. R. 2016. Managing supply chains: An introduction. *Multiple Criteria Decision Making in Supply Chain Management*, 1, 1–14.

Reed, D. A. 2008. Electric utility distribution analysis for extreme winds. *Journal of Wind Engineering and Industrial Aerodynamics*, 96(1), 123–140.

Reed, D. A., Powell, M. D., and Westerman, J. M. 2010. Energy supply system performance for Hurricane Katrina. *Journal of Energy Engineering*, 136(4), 95–102.

Sieminski, A. 2014. *International Energy Outlook*. Energy Information Administration (EIA), https://www.eia.gov/pressroom/presentations/sieminski_11182014.pdf.

Sullivan, M. J., Suddeth, B. N., Vardell, T., and Vojdani, A. 1996. Interruption costs, customer satisfaction and expectations for service reliability. *IEEE Transactions on Power Systems*, 11(2), 989–995.

Swaminathan, S. and Sen, R. K. 1998. *Review of Power Quality Applications of Energy Storage Systems*. Sandia National Laboratories, Livermore, California.

Vapnik, V. N. and Vapnik, V. 1998. *Statistical Learning Theory* (Vol. 1). Wiley, New York.

Winkler, J., Duenas-Osorio, L., Stein, R., and Subramanian, D. 2010. Performance assessment of topologically diverse power systems subjected to hurricane events. *Reliability Engineering & System Safety*, 95(4), 323–336.

Zhou, Y., Pahwa, A., and Yang, S.-S. 2006. Modeling weather-related failures of overhead distribution lines. *IEEE Transactions on Power Systems*, 21(4), 1683–1690.

Zhu, D., Cheng, D., Broadwater, R. P., and Scirbona, C. 2007. Storm modeling for prediction of power distribution system outages. *Electric Power Systems Research*, 77(8), 973–979.

Zopounidis, C. and Doumpos, M. 2002. Multicriteria classification and sorting methods: A literature review. *European Journal of Operational Research*, 138(2), 229–246.

6

Multiobjective Forecasting: Time Series Models Using a Deterministic Pseudo-Evolutionary Algorithm

Nagulapally Venkat Ramarao, P. Y. Yeshwanth Babu,
Sankaralingam Ganesh, and Chandrasekharan Rajendran

CONTENTS

6.1 Introduction

Autoregressive integrated moving average (ARIMA) method is a generalization of autoregressive moving average (ARMA) models developed by Box and Jenkins (1970). ARIMA method provides a parsimonious description of the stationary data in terms of two polynomials, one for the autoregression and the other for the moving average. In case of nonstationary data, an initial differencing step (corresponding to the integrated part of the model) is applied to reduce nonstationarity. Nonseasonal ARIMA models are denoted by ARIMA (p, d, q) where parameters p, d, and q are nonnegative integers, p is the order of the autoregressive part, d is the degree of differencing, and q is the order of the moving average part of ARIMA model. Seasonal ARIMA models are denoted as ARIMA (p, d, q) $(P, D, Q)_m$, where m refers to the number of periods in each season, and P, D, Q refer to the autoregressive, differencing, and the moving average terms for the seasonal part of the ARIMA model.

In big data analytics, we encounter the forecasting problem that deals with large data sets. In most business scenarios, the older data might be less useful in building the forecast models and more weightage has to be given to the immediate past; this is where ARIMA, seasonal ARIMA, and hybrid ARIMA models come into play. Using batch-processing technologies such as Apache's Hadoop to handle the volume component of big data and using stream-processing technologies such as Apache's Spark to handle the velocity component of big data, we could convert the unstructured data into structured data. Once the data is structured, a forecasting model can be used to arrive at forecasts, either with a single objective or with respect to multiple objectives.

The usual industry requirement for time series forecasting has objectives such as reducing the average error in predictions for a time period, as well as restricting the maximum error that can pop up in any given period. A typical example would be inventory management where the objective of the time series model would be that the average inventory of products in any quarter should be minimum and at any point of time, the inventory should not go beyond a threshold based on the capacity of the warehouse. At times, such objectives are conflicting since a model, which has the least mean absolute percentage error (MAPE), does not necessarily give the least maximum absolute percentage error (MaxAPE; true in most of cases). This chapter addresses the multiple criteria decision analysis involved in the time series forecasting, by using the ARIMA model. The algorithm detailed in Section 6.2 is scalable to any number of time series, and it can cater to present industrial requirement of forecasting the sales at stock-keeping unit level (which is very large in fast moving consumer goods [FMCG] or any other consumer goods sector).

6.1.1 Seasonality

Seasonal components consist of effects that are reasonably stable with respect to timing, direction, and magnitude. Seasonality in a time series can be identified by regularity in crests and troughs, which have consistent direction and magnitude, relative to the trend. Commonly employed approaches to modeling seasonal patterns include the Holt–Winters exponential smoothing model and ARIMA model.

6.1.2 Stationarity and Nonstationarity

A stationary time series is the one whose statistical properties such as mean, variance, and autocorrelation remain constant over time. In other words, the joint probability distribution of the series does not change when shifted in time. If a time series does not hold the above property, then it is said to be nonstationary. Seasonality is a characteristic of a time series in which the data experiences regular and predictable changes recurring at certain intervals. The data can be checked for stationarity by plotting the time series and by Dickey–Fuller test. Sometimes, differencing of the data is needed to arrive at a stationary series.

We could arrive at the values for the p, d, q and P, D, Q parameters via the autocorrelation factor (ACF) and the Partial ACF (PACF) of the time series. The ACF explains the correlation between values of the series at different points in time. Given measurements Y_1, Y_2, Y_3, ..., Y_N at time $X_1, X_2, X_3, ..., X_N$, respectively, the lag k autocorrelation function is defined as

$$r_k = \frac{\sum_{i=1}^{N-k}(Y_i - \bar{Y})(Y_{i+k} - \bar{Y})}{\sum_{i=1}^{N}(Y_i - \bar{Y})^2}. \tag{6.1}$$

The PACF gives the partial correlation of a time series with its own lagged values, controlling for the values of the time series at all shorter lags.

6.1.3 Overview of ARMA Models

The autoregressive or AR model can be written in the form:

$$y_t = c + \alpha_1 y_{t-1} + \cdots + \alpha_p y_{t-p} + e_t; \text{ or } y_t = c + \sum_{i=1}^{p}\alpha_i y_{t-i} + e_t, \tag{6.2}$$

where the terms in α are autocorrelation coefficients with respect to lags 1, 2, ..., p, and e_t is a residual error term. Note that this error term specifically relates to the current time period t.

The moving average or MA part of the model can be written as follows:

$$y_t = c + \beta_0 e_t + \beta_1 e_{t-1} + \cdots + \beta_q e_{t-q}; \text{ or } y_t = c + \sum_{i=0}^{q} \beta_i e_{t-i}, \tag{6.3}$$

where the β_i terms are the weights applied to prior error values in the time series, and it is usual to define $\beta_0 = 1$ without loss of generality.

ARMA(p, q) model can be expressed as below:

$$y_t = \alpha_0 + \alpha_1 y_{t-1} + \cdots + \alpha_p y_{t-p} - \beta_0 e_t - \beta_1 e_{t-1} - \cdots - \beta_q e_{t-q}. \tag{6.4}$$

Seasonal ARIMA (that is, SARIMA(p, d, q)(P, D, Q)$_m$) model can be expressed as below, where p, d, q, P, D, and Q are nonnegative integers and m is periodicity:

$$\varphi(B)\Phi(B^m)(1-B)^d(1-B^m)^D(Z_t - \mu) = \theta(B)\Theta(B^m)a_t, \tag{6.5}$$

where

$\varphi(B) = 1 - \varphi_1 B - \varphi_2 B^2 - \cdots - \varphi_p B^p$, φ_1, φ_2, ..., φ_p are coefficients;

$\Phi(B^m) = 1 - \Phi_1 B^m - \Phi_2 B^{2*m} - \cdots - \Phi_p B^{P*m}$, Φ_1, Φ_2, ..., Φ_P are coefficients;

$\theta(B) = 1 + \theta_1 B + \theta_2 B^2 + \cdots + \theta_q B^q$, θ_1, θ_2, ..., θ_q are coefficients; and

$\Theta(B^m) = 1 + \Theta_1 B^m + \Theta_2 B^{2*m} + \cdots + \Theta_Q B^{Q*m}$, Θ_1, Θ_2, ..., Θ_Q are coefficients.

As shown above, the four functions are polynomials in B of degrees p, q, P, and Q, where B is the backward shift operator, that is, $By_t = y_{t-1}$, $B^2 y_t = y_{t-2}$, $B^3 y_t = y_{t-3}$, etc., d is the order of regular differences, D is the order of seasonal differences, and Z_t denotes the observed value of time series data. For detailed explanation of these equations, refer to Hyndman and Athanasopoulos (2013).

6.1.4 Brief Literature Review on ARIMA Models

Zou and Yang (2004) proposed an algorithm to assign weights and combine ARIMA models for a better performance of prediction. Cools et al. (2009) used ARIMAX and SARIMAX models to forecast the daily traffic counts. Lee and Hamzah (2010) developed an ARIMAX model to forecast monthly sales of Muslim boys' clothes in Indonesia. This model combines ARIMA model and calendar variation effect during Eid holidays using linear regression, and had better forecast results than decomposition method, SARIMA, and Artificial Neural Network (ANN). The regression–SARIMA modeling framework captures important drivers of electricity demand. Nie et al. (2012) introduced a hybrid model which integrates ARIMA with support vector machines (SVM) to forecast short-term load forecasting for energy management. Chikobvu and Sigauke (2012) predicted the daily peak electricity demand in South Africa using SARIMA and regression–SARIMA. The performance of the developed models was evaluated by comparing them with Winter's triple exponential smoothing model. Empirical results from the study showed that the SARIMA model produced more accurate short-term forecasts.

In this chapter, we attempt to improve on the basic ARIMA model with the consideration of multiple objectives such as the minimization of MAPE and MaxAPE. We develop offspring time series from the best parent ARIMA models by considering the fitness values of parent ARIMA models. These fitness values are computed using the appropriate primary objective function, namely, MAPE/MaxAPE; and considering the parent ARIMA models, coupled with their relative fitness values, we deterministically generate the offspring time series.

6.2 Proposed Multiobjective Deterministic Pseudo-Evolutionary Algorithm

In this chapter, we have considered sales data of two retail segments. The real-life sales data of a beverage company and the real-life sales data of an online fashion store are being used here. Since the data has granularity (Section 6.3) at day level, we observe seasonality for every 7 days. The details of the data sets are not furnished here for the purpose of confidentiality, and for making the chapter concise we present the application of the algorithm considering one data set only.

We fit 288 parent SARIMA models to the time series with possible combinations of the parameters $p \in \{0, 1, 2, 3, 4, 5\}$, $d \in \{0, 1\}$, $q \in \{0, 1, 2, 3, 4, 5\}$, $(P,D,Q) \in \{(0,0,0),(0,1,1),(1,1,0),(1,1,1)\}$, and seasonality m as 7. Once these models are obtained, we then proceed to calculate MAPE and MaxAPE for each of these parent SARIMA models and arrange them in nondecreasing order of the respective objective function. Fitness Values are calculated for each time series generated from the parent SARIMA model, as explained in step 3 in phases 1 and 2 of the step-by-step procedure of the proposed multiobjective deterministic pseudo-evolutionary algorithm (MDPEA). Once the relative fitness values are obtained for each set of parent time series, we obtain deterministically the offspring time series corresponding to the consideration of these parent time series. Subsequently, we compare offspring and parent time series to get the netfront in the training period, followed by their use in the generation of time series in the test period.

6.2.1 Step-by-Step Procedure of the Proposed MDPEA

MAPE and MaxAPE are chosen as the criteria or objectives for the following reason. The MaxAPE provides the scenario of the worst-case model error and the MAPE provides the scenario of the average model error. Also, MAPE and MaxAPE are better metrics to judge the forecast model's performance in real life, compared to other theoretical likelihood metrics such as Akaike information criteria (AIC) or Bayesian information criteria (BIC).

The algorithm has four phases. In the first phase, we fit SARIMA models to the time series, and find the top parent models with MAPE as the primary objective. The parent time series from the respective SARIMA models are assigned relative fitness values, according to their corresponding MAPE values. Thereafter we generate offspring time series by using respective parent time series.

In the second phase, we choose the top parent SARIMA models with MaxAPE as the primary objective and generate offspring time series; here, the relative fitness values are assigned to parent time series according to the MaxAPE of the corresponding SARIMA models, and thereafter we generate offspring time series by using respective parent time series.

In the third phase, we find the nondominated netfront with respect to MAPE and MaxAPE by considering parent time series and offspring time series, with respect to the training period.

In the fourth phase, we find the nondominated netfront with respect to MAPE and MaxAPE by considering those parent SARIMA models and offspring SARIMA time series (that have entered the netfront corresponding to the training period) in the test period. Note that while generating these offspring time series in the test period, we make use of the corresponding and respective parent-series models (include their fitness values) that have been obtained in the training period. Let I' be the size of training set and I'' be that of test set.

6.2.1.1 Phase 1: MAPE Being the Primary Objective

Step 1: Set

$$p \in \{0,1,2,3,4,5\},$$
$$q \in \{0,1,2,3,4,5\},$$
$$d \in \{0,1\},$$
$$(P,D,Q) \in \{(0,0,0), (0,1,1), (1,1,0), (1,1,1)\}, \text{ and}$$
$$m = 7.$$

Step 2: Run the SARIMA model for every combination of $(p, d, q_r)(P, D, Q)_m$, and choose the best N parent models with respect to MAPE and arrange them in the nondecreasing order of their respective MAPE.

Denote the following:

- Actual data point of the time series as y_i, where $i = 1, 2, 3, \ldots, I'$, corresponding to the training set.
- Predicted value of nth parent model, with respect to data point i (i.e., time period i) corresponding to training set, as $\widehat{y1}_{n,i}^{p,I'}$ for $n = 1, 2, 3, \ldots, N$, and $i = 1, 2, 3, \ldots, I'$.

- The MAPE with respect to nth parent time series considering the training set is given as follows:

$$E1_n^{p,l'} = \left(\sum_{i=1}^{l'} \frac{\left| \widehat{y1}_{n,i}^{p,l'} - y_i \right|}{y_i} \times 100 \right) \times \frac{1}{l'}. \tag{6.6}$$

- The MaxAPE with respect to nth parent time series (chosen with the primary objective of minimizing the MAPE) considering the training set is given as follows:

$$\xi1_n^{p,l'} = \max_i \left(\frac{\left| \widehat{y1}_{n,i}^{p,l'} - y_i \right|}{y_i} \times 100 \right). \tag{6.7}$$

Step 3: Do the following to generate offspring time series with the primary objective of minimizing the MAPE, by considering $(n' - 1)$ parent time series at a time to generate a corresponding offspring time series, and by considering the relative fitness of these chosen parents:

Step 3.1: Set $n' = 3$.

Step 3.2: Calculate the fitness of nth parent model (i.e., parent time series), where $1 \leq n \leq n' - 1$, as follows, given n':

$$f1_{n,n'} = E1_{n'}^{p,l'} - E1_n^{p,l'}. \tag{6.8}$$

Step 3.3: Calculate the relative fitness of nth parent time series, where $1 \leq n \leq n' - 1$:

$$f1'_{n,n'} = \frac{f1_{n,n'}}{\sum_{n''=1}^{n'-1} f1_{n'',n'}}. \tag{6.9}$$

Step 3.4: Generate the predicted values with respect to offspring, that is, the offspring time series n'' with respect to training set, where $n'' = n' - 2$:

$$\widehat{y1}_{n'',i}^{o,l'} = \sum_{n=1}^{n'-1} \left(\widehat{y1}_{n,i}^{p,l'} \times f1'_{n,n'} \right), \quad \text{for } i = 1, 2, 3, \ldots, l'. \tag{6.10}$$

Step 3.5: Set $n' = n' + 1$ and repeat step 3.2 through 3.5 up to $n' = N$.

Step 4: Calculate MAPE and MaxAPE values for n''th offspring time series with respect to the training set:

$$E1_{n''}^{o,I'} = \left(\sum_{i=1}^{I'} \frac{\left| \widehat{y1}_{n'',i}^{o,I'} - y_i \right|}{y_i} \times 100 \right) \times \frac{1}{I'}, \tag{6.11}$$

$$\xi 1_{n''}^{o,I'} = \max_i \left(\frac{\left| \widehat{y1}_{n'',i}^{o,I'} - y_i \right|}{y_i} \times 100 \right) \quad \text{for } n'' = 1,2,3,\ldots,(N-2). \tag{6.12}$$

6.2.1.2 Phase 2: MaxAPE Being the Primary Objective

Step 1: Set

$$p \in \{0,1,2,3,4,5\},$$
$$q \in \{0,1,2,3,4,5\},$$
$$d \in \{0,1\},$$
$$(P,D,Q) \in \{(0,0,0), (0,1,1), (1,1,0), (1,1,1)\}, \text{ and}$$
$$m = 7.$$

Step 2: Run SARIMA models for every combination of $(p, d, q)(P, D, Q)_m$ and choose the best N parent models with respect to MaxAPE and arrange them in nondecreasing order of their respective MaxAPE.

Denote the following:

- Actual data point of time series as y_i, where $i = 1, 2, 3, \ldots, I'$ corresponds to the training set.
- Predicted value of nth parent model with respect to data point i corresponding to training set as $\widehat{y2}_{n,i}^{p,I'}$ for $n = 1, 2, 3, \ldots, N$, and $i = 1, 2, 3, \ldots, I'$ (note: these parents are obtained with the consideration of MaxAPE).
- The MAPE with respect to nth parent time series considering the training set (chosen with the primary objective of minimizing MaxAPE) is given as follows:

$$E2_n^{p,I'} = \left(\sum_{i=1}^{I'} \frac{\left| \widehat{y2}_{n,i}^{p,I'} - y_i \right|}{y_i} \times 100 \right) \times \frac{1}{I'}. \qquad (6.13)$$

- The MaxAPE with respect to nth parent time series considering the training period is:

$$\xi 2_n^{p,I'} = \max_i \left(\frac{\left| \widehat{y2}_{n,i}^{p,I'} - y_i \right|}{y_i} \times 100 \right). \qquad (6.14)$$

Step 3: Do the following to generate offspring time series with the primary objective of minimizing the MaxAPE, by considering $(n' - 1)$ parent time series:

Step 3.1: Set $n' = 3$.

Step 3.2: Calculate the fitness of nth parent time series, $1 \le n \le n' - 1$, as follows, given n':

$$f2_{n,n'} = \xi 2_{n'}^{p,I'} - \xi 2_n^{p,I'}. \qquad (6.15)$$

Step 3.3: Calculate the relative fitness of nth parent model, $1 \le n \le n' - 1$, as follows, given n':

$$f2'_{n,n'} = \frac{f2_{n,n'}}{\sum_{n''=1}^{n'-1} f2_{n'',n'}}. \qquad (6.16)$$

Step 3.4: Generate the predicted values with respect to offspring, that is, the offspring time series n'' with respect to training set, where $n'' = n' - 2$:

$$\widehat{y2}_{n'',i}^{o,I'} = \sum_{n=1}^{n'-1} \left(\widehat{y2}_{n,i}^{p,I'} \times f2'_{n,n'} \right), \quad \text{for } i = 1,2,3,\ldots,I'. \qquad (6.17)$$

Step 3.5: Set $n' = n' + 1$ and repeat step 3.2 through 3.5 up to $n' = N$.

Step 4: Calculate MAPE and MaxAPE values for n''th offspring time series with respect to training set:

$$E2_{n''}^{o,I'} = \left(\sum_{i=1}^{I'} \frac{\left| \widehat{y2}_{n'',i}^{\,o,I'} - y_i \right|}{y_i} \times 100 \right) \times \frac{1}{I'}, \tag{6.18}$$

$$\xi 2_{n''}^{o,I'} = \max_i \left(\frac{\left| \widehat{y2}_{n'',i}^{\,o,I'} - y_i \right|}{y_i} \times 100 \right) \quad \text{for } n'' = 1,2,3,\ldots,(N-2). \tag{6.19}$$

6.2.1.3 Phase 3: Generating the Combined Netfront with Respect to the Training Period

Dominated model: A model is referred to as dominated if both MAPE and MaxAPE are bettered by or equal to at least one other model in the lot (consisting of the parents and offspring).

Nondominated model: A model is referred to as nondominated if its MAPE or MaxAPE is better than any other model in the lot. Eliminating all the dominated models in the lot will result in the nondominated netfront.

We compare the performance of parent and offspring time series with respect to both MAPE and MaxAPE in the training period, and identify the nondominated sets of parent and offspring time series. We proceed as follows considering this nondominated netfront:

- Let the performance of top N parent SARIMA time series with respect to MAPE be denoted by $\left(E1_n^{p,I'}, \xi 1_n^{p,I'} \right)$, $n = 1, 2, 3, \ldots, N$; From these time series, the nondominated parents, which enter the netfront corresponding to training period, are denoted by the set $\Psi 1^P$.

- Let the performance of top parent SARIMA time series with respect to MaxAPE denoted by $\left(E2_n^{p,I'}, \xi 2_n^{p,I'} \right)$, $n = 1, 2, 3, \ldots, N$. From these time series, the nondominated parents, which enter the netfront corresponding to training period, are denoted by the set $\Psi 2^P$.

- Let the performance of offspring time series derived in phase 1 be denoted by $\left(E1_{n''}^{o,I'}, \xi 1_{n''}^{o,I'} \right)$, $n'' = 1, 2, 3, \ldots, (N-2)$; From these offspring, the nondominated offspring, which enter the netfront corresponding to training period, are denoted by the set $\Psi 1^o$. Note that these offspring are generated with the consideration of the corresponding $f1_{n,n'}'$ associated with the respective parents.

- Let the performance of offspring time series derived in phase 2 be denoted by $\left(E2_{n''}^{o,I'}, \xi 2_{n''}^{o,I'} \right)$, $n'' = 1, 2, 3, \ldots, (N-2)$. From these offspring, the nondominated offspring, which enter the netfront corresponding to training period, are denoted by $\Psi 2^o$. Note that these offspring are generated with the consideration of the corresponding $f2'_{n,n'}$ associated with the respective parents.

- Note that when we carry over offspring in the sets $\Psi 1^o$ and $\Psi 2^o$, we also carry over the respective parent's parameters (p, d, q) $(P, D, Q)_m$ and their relative fitness values that have led to the generation of the offspring, along with the offspring's n'' and n'. This information is used, while evaluating the offspring in the test period.

6.2.1.4 Phase 4: Generating the Combined Netfront with Respect to the Test Data Set (of Size *l″* Time Periods) from the Models Which Form the Netfront Corresponding to the Training Period

- The performance of the parent time series models in the test data set (with MAPE as primary objective), which have earlier entered the netfront corresponding to the training data set, is denoted by $\left(E1_n^{p,l''}, \xi 1_n^{p,l''}\right)$, $n \in \Psi 1^p$. Note that using the parent time series models and their corresponding (p, d, q) $(P, D, Q)_m$ values obtained from the training data set, we now generate the parent time series (forecasts) for the test data set, and their performance is denoted as above. The generated parent time series are checked for possible entry into the nondominated netfront corresponding to the test period. The nondominated parents which enter the netfront corresponding to test period are denoted by their respective performance $\left(E1_n^{p,l''}, \xi 1_n^{p,l''}\right)$, $n \in \Omega 1^p$.

- The performance of the parent time series models in the test data set (with MaxAPE as primary objective), which have earlier entered the netfront corresponding to the training data set, is denoted by $\left(E2_n^{p,l''}, \xi 2_n^{p,l''}\right)$, $n \in \Psi 2^p$. Note that using the parent time series models and their corresponding (p, d, q) $(P, D, Q)_m$ values obtained from the training data set, we generate the parent time series (forecasts) for the test data set. The generated parent time series are checked for possible entry into the nondominated netfront corresponding to the test period. The nondominated parents which enter the netfront corresponding to test period are denoted by $\left(E2_n^{p,l''}, \xi 2_n^{p,l''}\right)$, $n \in \Omega 2^p$.

The following steps give the procedure to obtain the offspring time series with respect to the test data set.

Step 1: MAPE as the primary objective:

Considering the offspring time series in $\Psi 1^o$ one by one, and using the corresponding set of parent time series models and their fitness values that have led to the generation of this offspring time series, we do the following.

Step 1.1: Generate the forecasted time series for the test data set by considering the parent time series model in the set of parent time series models that have led to the generation of this offspring time series; denote such a forecasted time series given by every such parent model as $\widehat{y1}_{n,i}^{p,I''}$ with respect to the time period i in the test data set.

Step 1.2: Using the above and the parents' corresponding fitness values (that have been obtained in the training period) with respect to this offspring in $\Psi1^o$ and this offspring's corresponding n'' and n', we obtain the corresponding forecasted offspring time series for the test data set as follows:

$$\widehat{y1}_{n'',i}^{o,I''} = \sum_{n=1}^{n'-1}\left(\widehat{y1}_{n,i}^{p,I''} \times f1'_{n,n'}\right), \quad \text{for } i = 1,2,3,\ldots,I''. \tag{6.20}$$

Note that $f1'_{n,n'}$ is inherited from the training data set (Equation 6.9).

Step 1.3: Calculate:

$$E1_{n''}^{o,I''} = \left(\sum_{i=1}^{I''}\frac{\left|\widehat{y1}_{n'',i}^{o,I''} - y_i\right|}{y_i}\times 100\right)\times\frac{1}{I''}; \tag{6.21}$$

$$\xi1_{n''}^{o,I''} = \max_i\left(\frac{\left|\widehat{y1}_{n'',i}^{o,I''} - y_i\right|}{y_i}\times 100\right). \tag{6.22}$$

By repeating the above for every offspring in $\Psi1^o$, we get the corresponding set of offspring time series with respect to the test period. This set of offspring time series with respect to the test period is checked for possible entry into the set of nondominated solutions with respect to the test period. Let the set of nondominated offspring time series in the test period be denoted by $\Omega1^o$. Let the performance of the nondominated offspring which enter the netfront corresponding to the test period be denoted by $\left(E1_{n''}^{o,I''}, \xi1_{n''}^{o,I''}\right)$, $n'' \in \Omega1^o$.

Step 2: MaxAPE as the primary objective:

Considering the offspring time series in $\Psi2^o$ one by one, and using the corresponding set of parent time series models and their fitness values that have led to the generation of this offspring time series, we do the following.

Step 2.1: Generate the forecasted time series for the test data set by considering every parent time series model in the set of parent time series models that have led to the generation of this offspring time series, denote such a forecasted time series given by every such parent model as $\widehat{y2}_{n,i}^{p,I''}$ with respect to the time period i in the test data set.

Step 2.2: Using the above and the parents' fitness values (that have been obtained in the training period) with respect to this offspring from $\Psi2^o$ and this offspring's corresponding n'' and n', we obtain the corresponding forecasted offspring time series for the test data set as follows:

$$\widehat{y2}_{n'',i}^{o,I''} = \sum_{n=1}^{n'-1}\left(\widehat{y2}_{n,i}^{p,I''} \times f2'_{n,n'}\right), \quad \text{for } i = 1,2,3,\dots,I''. \tag{6.23}$$

Note that $f2'_{n,n'}$ is inherited from the training data set (Equation 6.16).

Step 2.3: Calculate:

$$E2_{n''}^{o,I''} = \left(\sum_{i=1}^{I''}\frac{\left|\widehat{y2}_{n'',i}^{o,I''} - y_i\right|}{y_i} \times 100\right) \times \frac{1}{I''}; \tag{6.24}$$

$$\xi2_{n''}^{o,I''} = \max_i\left(\frac{\left|\widehat{y2}_{n'',i}^{o,I''} - y_i\right|}{y_i} \times 100\right). \tag{6.25}$$

By repeating the above for every offspring in $\Psi2^o$, we get the corresponding set of offspring time series with respect to the test period. This set of offspring time series with respect to the test period is checked for possible entry into the set of nondominated solutions with respect to the test period. Let the set of nondominated offspring time series in the test period be denoted by $\Omega2^o$. Let the performance of the nondominated offspring which enter the netfront corresponding to the test period be denoted by $\left(E2_{n''}^{o,I''}, \xi2_{n''}^{o,I''}\right)$, $n'' \in \Omega2^o$.

6.2.1.5 Phase 5: Stop

The algorithm is terminated and the time series denoted by $\Omega1^p$, $\Omega2^p$, $\Omega1^o$, and $\Omega2^o$ form the nondominated time series with respect to the test period.

6.3 Computational Evaluation of the Proposed MDPEA

We consider one data set (Figure 6.1) from a retailer segment. The details are masked for confidentiality. In a generic sense, the time series can be considered as sales of consumer products, which are durable and conveniently available in market, and have a weekly seasonality since the data is at daily granularity.

6.3.1 Data Set

Figure 6.1 details the real-life sales data of a beverage company.

6.3.2 Multiobjective Netfront for a Retail Segment Sales Data (90:10 with Respect to the Split of Training Data Set: Test Data Set)

Multiple data set SARIMA models are fitted to the sales time series and arranged in the increasing order of MAPE and MaxAPE separately. These two sets are considered as the parent models to derive the offspring models. Table 6.1 shows the parent and offspring nomenclature, and their respective description. Note that we report 20 top parent time series with respect to MAPE and 20 top parents time series with respect to MaxAPE, because the next 10 parent time series with respect to MAPE and the next 10 parent time series with respect to MaxAPE have not entered the nondominated front in the training data set. However, we consider the top 32 parent time series with respect to MAPE and top 32 parent time series with respect to MaxAPE to generate 30 offspring time series with respect to MAPE and 30 offspring time series with respect to MaxAPE, respectively.

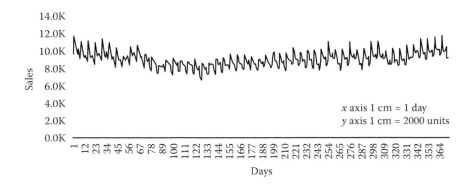

FIGURE 6.1
Real-life sales data of a retail segment; time period: 1 year at a day-level granularity.

TABLE 6.1

Labels for Parent and Offspring Time Series

Label	Top 20 Parents with MAPE as Primary Objective	Label	Top 20 Parents with MaxAPE as Primary Objective
P-1	ARIMA(5, 0, 3)(1, 1, 1)	P-21	ARIMA(0, 0, 0)(0, 1, 1)
P-2	ARIMA(4, 1, 5)(0, 1, 1)	P-22	ARIMA(0, 0, 0)(1, 1, 1)
P-3	ARIMA(5, 1, 4)(0, 1, 1)	P-23	ARIMA(0, 0, 4)(0, 1, 1)
P-4	ARIMA(5, 0, 2)(0, 1, 1)	P-24	ARIMA(0, 0, 4)(1, 1, 1)
P-5	ARIMA(5, 0, 5)(1, 1, 1)	P-25	ARIMA(1, 0, 4)(0, 1, 1)
P-6	ARIMA(5, 1, 5)(1, 1, 1)	P-26	ARIMA(2, 0, 3)(1, 1, 1)
P-7	ARIMA(4, 0, 4)(1, 1, 1)	P-27	ARIMA(2, 0, 4)(0, 1, 1)
P-8	ARIMA(5, 0, 2)(1, 1, 1)	P-28	ARIMA(2, 0, 4)(1, 1, 1)
P-9	ARIMA(4, 0, 5)(0, 1, 1)	P-29	ARIMA(2, 0, 5)(1, 1, 1)
P-10	ARIMA(4, 0, 3)(0, 1, 1)	P-30	ARIMA(3, 0, 4)(0, 1, 1)
P-11	ARIMA(5, 0, 5)(0, 1, 1)	P-31	ARIMA(4, 0, 2)(0, 1, 1)
P-12	ARIMA(5, 1, 3)(0, 1, 1)	P-32	ARIMA(4, 0, 2)(1, 1, 1)
P-13	ARIMA(5, 1, 5)(0, 1, 1)	P-33	ARIMA(4, 0, 3)(1, 1, 1)
P-14	ARIMA(5, 1, 4)(1, 1, 1)	P-34	ARIMA(4, 1, 5)(1, 1, 0)
P-15	ARIMA(4, 0, 4)(0, 1, 1)	P-35	ARIMA(4, 1, 5)(1, 1, 1)
P-16	ARIMA(5, 1, 3)(1, 1, 1)	P-36	ARIMA(5, 0, 4)(0, 1, 1)
P-17	ARIMA(3, 1, 4)(1, 1, 1)	P-37	ARIMA(5, 0, 4)(1, 1, 0)
P-18	ARIMA(5, 1, 1)(1, 1, 1)	P-38	ARIMA(5, 0, 4)(1, 1, 1)
P-19	ARIMA(5, 1, 2)(1, 1, 1)	P-39	ARIMA(5, 0, 5)(1, 1, 0)
P-20	ARIMA(4, 1, 3)(1, 1, 1)	P-40	ARIMA(5, 0, 5)(1, 1, 1)

Label	Parent description (MAPE as primary objective)
P-1 to P-20	Top 20 parent SARIMA time series with respect to MAPE as primary objective

Label	Parent description (MaxAPE as primary objective)
P-21 to P-40	Top 20 parent SARIMA models with respect to MaxAPE as primary objective

Label	Offspring time series description (MAPE as primary objective)
O-41	Offspring obtained from top 3 parents with MAPE as primary objective
O-42	Offspring obtained from top 4 parents with MAPE as primary objective
...	...
O-70	Offspring obtained from top 32 parents with MAPE as primary objective

Label	Offspring time series description (MaxAPE as primary objective)
O-71	Offspring obtained from top 3 parents with MaxAPE as primary objective
O-72	Offspring obtained from top 4 parents with MaxAPE as primary objective
...	...
O-100	Offspring obtained from top 32 parents with MaxAPE as primary objective

Scenario 1:

In this scenario, we see the formation of nondominated solutions when we consider only 20 offspring, with respect to MAPE and MaxAPE. We compare O-41 to O-60 and O-71 to O-90 offspring time series with P-1 to P-40 parent time series considering the training period and we create a netfront of nondominated time series (Figure 6.2). The parent and offspring time series, which enter this netfront, are evaluated in the test period and the final netfront is created as shown in netfront test period chart (Figure 6.3). We refrain from labeling all the data points due to congestion of data points.

As we can see from Figures 6.2 and 6.3, both parent and offspring time series have entered the netfront of nondominated solutions in the training set; however, of those time series that have entered the netfront in the training set, only one offspring time series has entered the netfront of nondominated

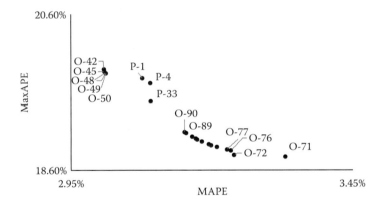

FIGURE 6.2
Netfront corresponding to the training period with consideration of (O-41 to O-60) and (O-71 to O-90) offspring time series, (P-1 to P-20) and (P-21 to P-40) parent time series.

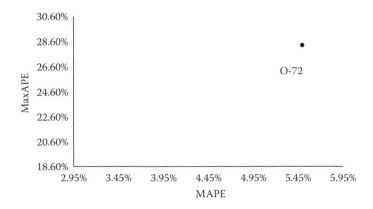

FIGURE 6.3
Netfront corresponding to the test period.

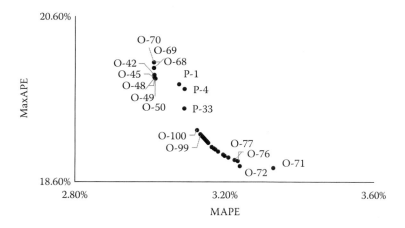

FIGURE 6.4
Netfront corresponding to the training period with consideration of (O-41 to O-70) and (O-71 to O-100) offspring time series, and (P-1 to P-20) and (P-21 to P-40) parent time series.

solutions in the test set, when dealing with multiple objectives, that is, MAPE and MaxAPE.

Scenario 2:
In this scenario, we see the formation of nondominated solutions when we consider a set of 30 offspring, with respect to MAPE and MaxAPE. We compare O-41 to O-70 and O-71 to O-100 offspring time series with P-1 to P-40 parent time series considering the training period and create a netfront of nondominated time series as shown in netfront training period chart (Figure 6.4). The parent and offspring time series which enter this netfront are evaluated in the test period and the final netfront is created as shown in netfront test period chart (Figure 6.5).

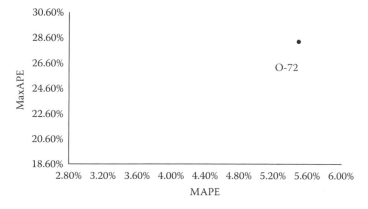

FIGURE 6.5
Netfront corresponding to test period for offspring.

As we can see from Figures 6.4 and 6.5, both parent and offspring time series have entered the netfront of nondominated solutions in the training set; however, of those time series that have entered the netfront in the training set, only one offspring time series has entered to the netfront of nondominated solutions in the test set, when dealing with multiple objectives, that is, MAPE and MaxAPE.

6.4 Summary and Conclusions

The proposed MDPEA offers the development of offspring time series from the parent time series, deterministically by considering the fitness values of parent SARIMA models. These fitness values are computed using the appropriate primary objective function, namely, MAPE/MaxAPE. Considering the parent SARIMA models deterministically, we generate these offspring time series by considering the respective parents whose fitness values are separately calculated with respect to the MAPE and MaxAPE.

The salient contributions of the proposed MDPEA are as follows:

1. *Accuracy*: It is evident from the computational experiments that the MDPEA is able to produce the offspring time series that perform better than the parent time series with respect to the multiple objectives of minimizing the MAPE and MaxAPE.
2. *Speed*: MDPEA gets the best of both worlds, that is, it exploits the simplicity of a deterministic model and uses the inheritance feature from the evolutionary algorithms.
3. *Compatibility*: The offspring forecasts from this algorithm can be used as inputs to ANN to improve the performance.

Acknowledgments

The authors are grateful to the editors and reviewers for their suggestions and comments to improve the earlier version of the chapter.

References

Box, G.E.P. and Jenkins, G.M. 1970. *Time Series Analysis Forecasting and Control.* Holden-Day, San Francisco, CA.

Chikobvu, D. and Sigauke, C. 2012. Regression-SARIMA modelling of daily peak electricity demand in South Africa. *Journal of Energy in Southern Africa* 23(3): 23–30.

Cools, M., Moons, E., and Wets, G. 2009. Investigating variability in daily traffic counts using ARIMA X and SARIMA X models: Assessing impact of holidays on two divergent site locations. In *Transportation Research Board (TRB) 88th Annual Meeting Compendium of Papers DVD*, Washington, DC, January 11–15, 2009.

Hyndman, R.J. and Athanasopoulos, G. 2013. Forecasting principles and practice. Available via: www.OTexts.com/fpp, accessed December 26, 2015.

Lee, M. and Hamzah, N. 2010. Calendar variation model based on ARIMAX for forecasting sales data with Ramadhan effect. In *Proceedings of the Regional Conference on Statistical Sciences*, Kuala Lumpur, Malaysia, June 2010, pp. 349–361.

Nie, H., Liu, G., Liu, X., and Wang, Y. (2012). Hybrid of ARIMA and SVMs for short-term load forecasting. *Energy Procedia* 16: 1455–1460.

Zou, H. and Yang, Y. 2004. Combining time series models for forecasting. *International Journal of Forecasting* 20(1): 69–84.

7

A Class of Models for Microgrid Optimization

Shaya Sheikh, Mohammad Komaki, Camelia Al-Najjar,
Abdulaziz Altowijri, and Behnam Malakooti

CONTENTS

7.1 Introduction

Microgrid is defined as a set of local energy generators, energy transmitters, energy storages, and users that can work either independent (islanded mode) from the main grid or in connection with the main grid. In practice, main grids could function as a backup system for microgrids. Microgrid concept was originally presented as a solution for creating a sustainable, green, and efficient energy model; see Zamora and Srivastava (2010).

Renewable energy resources have had increasing and accelerating penetration rate in microgrids since the turn of the century. However, the intermittent nature of renewables such as solar and wind has always been a challenge for their integration in microgrids. For example, electricity generated from solar panels is impacted by factors such as weather conditions and time of day. Fortunately, energy storages can decrease this impact by adding more flexibility to the system and by alleviating the imbalance between energy supply and demand. There has been promising progress toward building and deploying energy storages in microgrid scale and with reasonable costs.

Microgrids are becoming more and more intelligent and connected to each other. Generators, storages, and users are equipped with the sensors that can measure and monitor parameters of the system at every spot and in every second. Entities are also connected to each other through internet or intranet where they can communicate to each other. A controller (such as heating, ventilation, and air conditioning [HVAC] in buildings) can then receive these data and adjust the parameters of the system (e.g., energy usage) in real time. These set of sensors generate huge amount of data that can grow exponentially within a short span of time. Energy big data and renewable resources amplify the urgency for designing more efficient and flexible microgrids.

Energy big data contains variety of data sets, including but not limited to weather data, energy market data, geographical data (e.g., global positioning system [GPS] information), and field measurement data (e.g., device status, electricity consumption, storage level, etc.). These sets of data are blended together and then classified in order to generate useful information and insights about the whole system. For example, GPS data help to visualize locational marginal price in a geographic context. Common sources of big data in microgrids are smart appliances and metering points throughout the grid. Metering points can generate new set of data as frequent as every 5–15 minutes. These points generate large volume of data which makes communication, complexity, and data storage an inefficient and costly process; see Diamantoulakis et al. (2015). Data is also being generated from generators, transformers, and local distributors. In addition, information from grid monitoring and maintenance are generated on a regular basis. The volume of collected data from a grid can easily reach to terabytes over a year. As a result, data-mining and machine-learning tools are required to refine the data, discover meaningful patterns in data, and generate useful information from it. Information and insights are then applied to analyze energy price fluctuations in real time, to analyze energy generators status, and to plan and monitor energy system's status.

According to Zhou et al. (2016), seven steps for managing big energy data are: data collection, data cleaning, data integration, data mining, visualization, intelligent decision making, and smart energy management. Given the large volume of generated data and its dynamic nature, the data processing would be a daunting and challenging task.

Data-mining methods such as regression, clustering, neural network, and support vector machines are important tools for extracting useful and relevant information from a large pool of data. Regression models, time series models, and state–space models are among the most popular short-term forecasting methods; see Kyriakides and Polycarpou (2007). Also, forecasting can predict the demand accurately and optimization tool can achieve the optimal energy generation and distribution. Implementing these tools together can result in effective dynamic pricing, planning, and operations policies in energy systems.

Dynamic nature of data in microgrids highlights the importance of short-term, very short-term, and ultrashort-term forecasting methods; see Diamantoulakis et al. (2015). Very short-term and ultrashort-term forecasting are mainly used for immediate forecasting. However, there are fewer methods (such as artificial neural network) for these two categories. All aforementioned data-mining tools are mainly used for averaging out the measures in energy systems and are not useful for individual meters. Therefore, customized methods such as empirical mode decomposition and extreme learning with kernel are suggested for analyzing data from single meters; see Yoo et al. (2011).

In this chapter, we concentrate on the modeling of microgrids in their simplest form. We assume that data and energy are allowed to flow between users, generators, and storages at any time. Therefore, all entities of a microgrid communicate to each other and a controller is constantly collecting and processing data from all entities in the system. One of the presented models, called multiperiod energy model, utilizes the temporal data such as energy prices as the input and generates the best energy generation schedule among renewable generators, and best distribution among storages, and consumers.

We present a number of operational and design energy models for solving microgrids problems. The first model, called energy operation model, is used as a base to represent a microgrid. The second model, energy design optimization, offers the best design for a microgrid with specified demand range. The extension of first and second model that considers multiperiods is also presented. Finally, the last model combines two important criteria of cost and environmental impact in energy operation model. For further information about presented models, see Malakooti (1986, 2014), Malakooti et al. (2013), and Ravindran and Warsing (2012). A multicriteria energy operation model is also explained in Sheikh et al. (2014).

Factories, residential buildings, office buildings, and hybrid cars are some examples of microgrids. A basic microgrid is composed of energy generators, energy users, and energy storages; see Carmona and Ludkovski (2010). These entities are linked together in order to supply, store, and consume energy; see Figure 7.1. A system of interconnected microgrids is presented in Figure 7.2.

There are always different alternatives for generating or procurement of energy, but there are trade-offs among these alternatives. The multicriteria microgrid approach distinguishes the most preferred different sources with their associated generation and transportation costs. For further information about multicriteria energy system models, see Sheikh et al. (Sheikh and Malakooti, 2011, 2012; Sheikh, 2013; Sheikh et al., 2014, 2015), Ren et al. (2010), Liu et al. (2010), Fazlollahi et al. (2012), Fadaee and Radzi (2012),

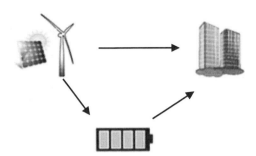

FIGURE 7.1
A simple microgrid.

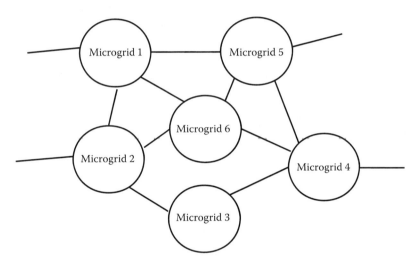

FIGURE 7.2
System of microgrids.

Alarcon-Rodriguez et al. (2010), Shabanpour and Seifi (2015), and Sharafi and ELMekkawy (2014).

7.2 Energy Models for Microgrids

Microgrid consists of three connected entities: users, energy storages, and generators. A user such as a factory machinery receives energy from either a generator or a storage. Storage, such as a battery, is a physical entity which stores energy for later use. If the rate of energy production is higher than the rate of energy consumption, the excess energy can be stored in the storage for later use or be sold to the main grid. A generator such as a wind turbine generates energy. Demand of each user needs to be satisfied such that it reduces energy costs and stores some energy for future use. Major costs consist of generating and transportation (transmission and distribution) costs.

Users, generators, and storages are coordinating to minimize the total cost and to satisfy the needed demands of the users. The energy stored by the storages is paid for by the users of the system for future use. Stored energy at the end of the period is left for the next period's use. The amount in the storage is set depending on the prediction of the next period's energy prices. If the prediction shows that generator's prices will be higher, then more energy is stored in storages for future use and vice versa. Δs_p is defined as the predicted change in the future price of energy. The storage can also act as a hub in which all input energy from the generator is transported to the user without storing energy in the storage.

Supply and demand behavior of storages: Depending on the price of the energy in the future, it may be economical to store or release energy from storages; see Escribano et al. (2011), Marckhoff and Wimschulte (2009). If the future price of energy will be higher, then more energy should be stored in the storage now. On the other hand, if the future price of energy will be lower, then the storage should sell as much energy as it can now, and the energy level of the storage should be set to the minimum; see Conejo et al. (2005).

7.2.1 Energy Operations Model

A linear programming optimization approach for the operation of microgrids is presented in this section. This model finds the optimal amount of energy generated by each energy generator, stored in each storage for future use, and transported on each link. Decision variables and costs for formulating the energy operations model are presented in Table 7.1. Net flow is defined as the difference between all incoming energy flows and all outgoing energy flows:

Net flow = All outgoing energy flows – All incoming energy flows

The sign of net flow for each entity signifies the nature of that entity. Specifically,

Net flow > 0	Generator
Net flow < 0	User
Net flow = 0	Storage

Nomenclature: By using the notation provided in Table 7.1, the microgrids model for a given period is formulated as follows. For further explanation

TABLE 7.1

Notations and Symbols Used for Microgrids

Parameter	Index
K_i: Maximum capacity of gen. i	i: Origin entity
D: Set of energy demand for users	j: Destination entity
g_i: Cost per unit of energy for gen. i	
$q_{p, max}$: Maximum capacity of storage p	**Decision Variable**
$q_{p, min}$: Minimum capacity of storage p	x_{ij}: Amount of energy from
Δs_p: Projected gain in dollars of energy price for next period	entity i to entity j
c_{ij}: Energy transportation cost from entity i to entity j	q_p: Energy level of storage p

and applications of Model 1, see Malakooti et al. (2013), Malakooti (2014), Sheikh and Malakooti (2011, 2012), Sheikh (2013), and Sheikh et al. (2014, 2015).

Model 1: Energy operation optimization

$$\text{Minimize } f_1 = \sum_{\text{for all i}} \sum_{\text{for all j}} c_{ij} x_{ij} \tag{7.1}$$

$$\sum_{\text{for all i}} g_i \sum_{\text{for all j}} x_{ij} \tag{7.2}$$

$$G \sum_{\text{for all p}} \Delta s_p q_p. \tag{7.3}$$

Subject to:

$$\sum_{\text{for all i}} x_{ij} \geq D_j \quad \text{for all j demand constraint for users} \tag{7.4}$$

$$\sum_{\text{for all j}} x_{ij} \leq k_i \quad \text{for all i capacity constraint for generators} \tag{7.5}$$

$$q_{p,0} + \sum_{\text{for all j}} x_{ij} - \sum_{\text{for all i}} x_{ij} = q_p \quad \text{for all p (energy level of storages)} \tag{7.6}$$

$$q_p \leq q_{p,max} \quad \text{for all (maximum allowable level for storages)} \tag{7.7}$$

$$q_p \geq q_{p,min} \quad \text{for all (minimum required level for storages)} \tag{7.8}$$

$$x_{ij} \geq 0, q_p \geq 0 \quad \text{for all i, j, and p.} \tag{7.9}$$

Equation 7.1 calculates transportation cost from generators to users, from generators to storages, and from storages to users. Cost of purchasing energy from generators to users and storages are shown with Equation 7.2. G in Equation 7.3 is a given large positive number (e.g., set $G = 100{,}000$). This equation finds total monetary gain of the system by storing energy at current time. The objective function minimizes the total cost of generating, transferring, and storing energy. The marginal gain or loss for energy storages is defined such that if $\Delta s_p > 0$, then the energy price of the next period will be higher. Therefore, store (or buy) energy for future use (i.e., q_p is maximized) and if $\Delta s_p < 0$, then the energy price of the next period will be lower. Therefore, sell the existing energy in the storage (i.e., q_p is minimized).

The demand constraints (7.4) ensure that the total amount of energy being transferred to the users from the generators and the storages is greater than or equal to the energy demand by the users. The production constraints (7.5) show that the amount of energy being transferred from each generator to the users and the storages is not more than the maximum capacity of the given generator. Constraints (7.6) show the amount of energy that is present in the storages at the end of the period. This amount is equal to the amount of energy that is initially in the storage plus the amount that is transferred from the generators to the storage minus the amount that is transferred from the storage to users. Constraints (7.7) show that the amount of energy in storage p at the end of the period should be less than or equal to the maximum allowable capacity of the storage. Constraints (7.8) ensure that the amount of energy in storage p at the end of the period is more than or equal to the required minimum level of the storage for emergency purposes. Based on the prediction of energy prices, either more energy is stored in the storage or it is sold to result in less energy in the storage. The nonnegativity constraints (7.9) ensure that all amounts of energy generated, transferred, or stored will be nonnegative values. See microgrid operations example.

7.2.1.1 Microgrid Operations Optimization Example

There exist three users, three generators, and two storages shown in Figure 7.3. The demands for users 1, 2, and 3 and 4 are 400, 450, and 350 units of energy, respectively. The generating energy cost per unit of energy for generators 1, 2, and 3 are \$44.5, \$44, and \$43.5, respectively. Maximum capacities of generators 1, 2, and 3 are 1000, 950, 1200 units of energy, respectively. The initial energy levels in storages 1 and 2 are 150 and 150 and the maximum capacity of storages 1 and 2 are 250 and 240 units of energy, respectively. The minimum capacity of storages 1 and 2 are 50 and 20 units of energy, respectively. The data for this example is presented in Table 7.2. Suppose that $\Delta s_1 = +1$ and

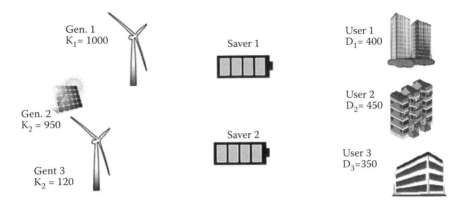

FIGURE 7.3
Illustration for systems operations example with $\Delta s_1 = +1$ and $\Delta s_2 = +2$.

TABLE 7.2

Data for Microgrid Operations

c_{ij}		User j		
		1	2	3
Gen. i	1	6	5	8
	2	7	4	5
	3	5	4	6

c_{ij}		User j		
		1	2	3
Storage i	1	1	2	1
	2	2	0.5	1.5
Demand D_j		400	450	350

c_{ij}		Storage j	
		1	2
Gen.i	1	4.5	5
	2	6	4
	3	5	6

Generator	g_i ($)	K_i
1	44.5	1000
2	44	950
3	43.5	1200

Storage	$q_{p,0}$	$q_{p,min}$	$q_{p,max}$
1	150	50	250
2	150	20	240

$\Delta s_2 = +2$. The next period's energy price is predicted to be higher for storages 1 and 2 (i.e., buy energy or $\Delta s_p > 0$ for all storages).

Formulation for the microgrids operation is shown below. This formulation is used to find the optimal solution that minimizes the total cost for the energy operation example.

Optimal distribution of energy for above microgrid is as follows. Lingo 13 was used to find the optimal solution for this problem.

Energy Trans. x_{ij}	User j 1	2	3	Tot.
Gen. i 1	0	0	0	0
2	0	0	350	350
3	400	450	0	850
Tot.	400	450	350	1200

Energy Trans. y_{ip}	Storage p 1	2	Tot.
Gen. i 1	0	0	0
2	0	90	90
3	100	0	100
Tot.	100	90	190

Energy Trans. z_{pj}	User j 1	2	3	Tot.
Storage 1	0	0	0	0
P 2	0	0	0	0
Tot.	0	0	0	0

According to the solution, generator 2 should generate 350 and 90 units of energy for user 3 and storage 2, respectively, and generator 3 should generate 400, 450, and 100 units of energy for users 1, 2, and storage 1, respectively. In total, generator 2 generates $(350 + 90) = 440$ units of energy and generator 3 generates $(400 + 450 + 100) = 950$ units. At the end of the period, the energy levels of storages 1 and 2 are $q_1 = 150 + 100 = 250$ and $q_2 = 150 + 90 = 240$, respectively. The solution is presented in Figure 7.4. The objective function value, f_1, and total benefit of this example is $66,365.

Energy Trans. x_{ij}	User j 1	2	3	Tot.
Gen. i 1	0	0	0	0
2	0	0	120	120
3	400	450	0	850
Tot.	400	450	120	970

Energy Trans. y_{ip}	Storage p 1	2	Tot.
Gen. i 1	0	0	0
2	0	0	0
3	0	0	0
Tot.	0	0	0

Energy Trans. z_{pj}	User j 1	2	3	Tot.
Storage 1	0	0	100	100
P 2	0	0	130	130
Tot.	0	0	230	230

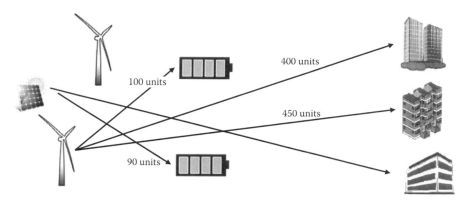

100 units
90 units
400 units
450 units

FIGURE 7.4
Solution for systems operations example.

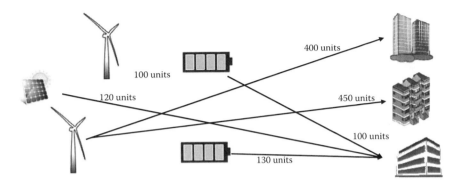

FIGURE 7.5
Solution for systems operations example with $\Delta s_1 = -1$ and $\Delta s_2 = -2$.

If the future prices were predicted to be lower than the current prices, $\Delta s_1 = -1$ and $\Delta s_2 = -2$, the solution as presented in Figure 7.5 would be: $q_1 = 50$, $q_2 = 20$, and $f_1 = \$47{,}040$.

7.2.2 Microgrids Design Optimization

In Section 2.1, an optimization model for planning and operations of existing microgrids was discussed. This section presents the design of microgrids. As an example, in a community, there are choices to buy different energy generators and storages. In this model, both operational and design costs are taken into consideration. The formulation of this model is based on energy operation model discussed earlier.

Generator- and storage-purchasing costs need to be considered in the objective function (7.1 through 7.3). The formulations for purchasing generators or storage are similar. Expression (7.10) considers the total purchasing cost for generators which needs to be added to the objective function. u_i is a binary variable and c_i is the cost of purchasing generator i.

$$+ \sum_{i=1}^{I} c_i u_i. \tag{7.10}$$

The constraints of microgrids design are the same as energy operation model except that binary variables for each generator are added, where 1 means the item is purchased, and 0 means it is not purchased. Generator constraints (7.5) need to be replaced with constraints (7.11).

$$\sum_{\text{for all } j} x_{ij} \le K_i u_i \quad \text{for } i = 1, \ldots, I \text{ (generator investment constraints)}, \tag{7.11}$$

where

$$u_i : \begin{cases} 1 & \text{If generator i is chosen to be purchased} \\ 0 & \text{otherwise} \end{cases}.$$

For further explanation regarding Models 1 and 2, see Malakooti et al. (2013, 2014).

Model 2: Microgrid design optimization

Minimize objective function (7.1 through 7.3) + Expression (7.10)

Constraints (7.4), (7.6) through (7.9) and (7.11)

7.2.2.1 Example for Microgrids Design

The microgrid is illustrated in Figure 7.6. For simplicity, the arrows between entities are not shown in the figure. Suppose that there is no generator and it is possible to buy one or more of three possible generators. The users and storages are existing entities as given in energy operation example. The input parameters are as follows (Table 7.3).

The optimal design and operation plan for this example are found by CPLEX 13. The solution is shown in Table 7.4.

In the final solution, $u_1 = 0$, $u_2 = 1$, $u_3 = 1$, and $u_4 = 1$ which means that the best decision is to purchase generators 2, 3, and 4. The total cost is \$249,192.5. The solution is shown in Figure 7.7.

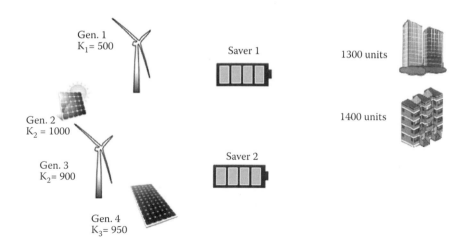

FIGURE 7.6
Energy systems design.

TABLE 7.3

Information for Microgrids Design

c_{ij}		User j	
		1	2
Gen. i	1	4	4
	2	5	3
	3	5	4
	4	4	6

c_{pj}		User j	
		1	2
Storage p	1	3.5	3
	2	4	3.5
Demand D_j		1300	1400

c_{ip}, Trans. Cost from i to p		Storage p	
		1	2
Gen. i	1	1	2
	2	2	0.75
	3	2	0.8
	4	1	1

Gen	Purchase Cost in $1000	Gen. Costs, g_i ($)	Gen. Cap., K_i
1	25	60.5	500
2	25.5	60.75	1000
3	25.3	60.25	900
4	24.9	60.5	950

Storage	Init. Enr $q_{p,0}$	Min. Enr $q_{p,min}$	Max. Enr $q_{p,max}$
1	50	50	300
2	50	50	350

TABLE 7.4

Solution for Microgrids Design

Energy Trans. x_{ij}		User j		
		1	2	Tot
Gen. i	1	0	0	0
	2	0	1000	1000
	3	0	400	400
	4	950	0	950
	Tot	950	1400	2350

Energy Trans. y_{ip}		Storage p		
		1	2	Tot.
Gen. i	1	0	0	0
	2	0	0	0
	3	0	350	350
	4	0	0	0
	Tot.	0	350	350

Energy Trans. z_{pj}		User j		
		1	2	Tot.
Storage p	1	0	0	0
	2	350	0	350
	Tot.	350	0	350

7.2.3 Multiperiod Energy Model

Energy price in many microgrids are set as a variable rate where this rate is a function of supply and demand. As a result, the market value of energy changes during a given day. Variable energy price is addressed by energy operations optimization for multiple periods. Suppose that there are T periods, where $t = 1, 2, \ldots, T$. For multiperiod models, the formulation is similar to Model 1. However, index t is added to all decision variables. Also, Equation 7.6 changes to:

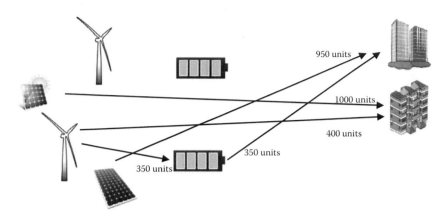

FIGURE 7.7
Solution for energy design example.

$$q_{p,t-1} + \sum_{\text{for all } j} x_{ijt} - \sum_{\text{for all } i} x_{ijt} = q_{p,t} \quad \text{for all } p \text{ (energy level of storages),} \quad (7.12)$$

where $q_{p,t-1}$ and $q_{p,t}$ represent energy level in storage p at the end of period t – 1 (or start of period t) and the end of period t, respectively. The total gain for future savings is not used in this formulation since future prices of given periods are known.

We introduce two approaches for solving multiperiod energy model: (1) Period-by-period approach which is used when energy price and demands per period are only known at the beginning of that period. In this case, the energy model is solved at the beginning of each period; (2) Aggregate multiperiod approach which is used when energy price and demands of a given number of future periods are known at the beginning of the first period. If the data for all upcoming periods are known at the beginning, multiperiod energy operation can be solved in one step. In this case, energy stored in storages at the end of period t – 1 will be considered as the initial energy level for period t.

7.2.3.1 Multiperiod Energy Planning Optimization Example

Consider energy operations example with a day and night period. Suppose that the information for both periods is known at the beginning of the first period. Different costs of buying and selling energy are considered. The demands and cost information for day period are shown in Table 7.5. The initial energy levels in storages 1 and 2 are 100 and 150, respectively.

Data for day period is provided in Table 7.6.

The formulation is similar to energy operation model except that Constraints (7.13) show input/output of the different periods and their connection together. The solution would be as shown in Table 7.7.

TABLE 7.5

Information for Night-Period Energy Operation

Period	Demand User 1	Demand User 2	Demand User 3	Predicted Change in Energy Price for Storage 1, $\Delta s_{1,t}$	Predicted Change in Energy Price for Storage 2, $\Delta s_{2,t}$	Generator 1 Costs $(g_{1,t})$	Generator 2 Costs $(g_{2,t})$	Generator 3 Costs $(g_{3,t})$
1, Night	200	250	100	4	3	44.5	44	43.5

TABLE 7.6

Information for Day-Period Energy Operation

Period	Demand User 1	Demand User 2	Demand User 3	Predicted Change in Energy Price for Storage 1, $\Delta s_{1,t}$	Predicted Change in Energy Price for Storage 2, $\Delta s_{2,t}$	Generator 1 Costs $(g_{1,t})$	Generator 2 Costs $(g_{2,t})$	Generator 3 Costs $(g_{3,t})$
2, Day	500	600	300	−4	−3	44.5	44	43.5

Energy levels for storages in period 1 are $q_{1,2,1} = 60$, $q_{2,2,1} = 150$. In period 2, energy level for storage 2 will remain unchanged. However, storage 2 decreases its energy to 24. These energy levels are in accordance with the expected behavior from storages.

TABLE 7.7

Solution for Multiperiod Energy Problem

Energy Trans. x_{ij1}	User j 1	2	3	Tot.
Gen. i 1	0	0	0	0
2	0	0	60	60
3	200	250	0	450
Tot.	200	250	60	510

Energy Trans. y_{ip1}	Storage p 1	2	Tot.
Gen. i 1	0	0	0
2	0	0	0
3	0	0	0
Tot.	0	0	0

Energy Trans. z_{pj1}	User j 1	2	3	Tot.
Storage p 1	0	0	40	40
2	0	0	0	0
Tot.	0	0	40	40

Energy Trans. x_{ij2}	User j 1	2	3	Tot.
Gen. i 1	0	0	0	0
2	0	0	174	174
3	500	600	0	1100
Tot.	500	600	174	1274

Energy Trans. y_{ip2}	Storage p 1	2	Tot.
Gen. i 1	0	0	0
2	0	0	0
3	0	0	0
Tot.	0	0	0

Energy Trans. z_{pj2}	User j 1	2	3	Tot.
Storage p 1	0	0	0	0
2	0	0	126	126
Tot.	0	0	126	126

7.2.4 Bicriteria Optimization of Microgrids

Energy cost is considered an important criterion in microgrids. Another important criterion is the environmental impact of the microgrid. Different energy generators have different environmental impacts. This section introduces bicriteria microgrids operations optimization with two criteria functions: energy cost and environmental impact. The environmental impact factor per unit of energy for each energy generator is presented by m_i for $i = 1, \ldots , I$. The bicriteria model is as follows:

Model 3: Bicriteria of energy operations

$$\text{Minimize total cost } f_1 = \text{Use (7.1 through 7.3)}$$

$$\text{Minimize total environmental impact } \quad f_2 = \sum_{\text{for all } i} m_i \sum_{\text{for all } j} x_{ij}. \quad (7.13)$$

Subject to Constraints (7.4) through (7.9)

$\sum_{\text{for all } j} x_{ij}$ in (7.13) is the generated energy by generator i. This amount is multiplied by environmental factor per unit of energy (m_i) to find the total environmental impact of generator i. The constraints of Model 3 are the same as the constraints of energy operation model. To find the best alternative for a given system, use the following Multi-Criteria Decision Making (MCDM) process. In the first step, an upper and lower bound is found for each criteria using the following:

	Solution	
Problem I: Minimize f_1 subject to Constraints of energy operation model	$f_{1,min}$	$f_{2,max}$
Problem II: Minimize f_2 subject to Constraints of energy operation model	$f_{1,max}$	$f_{2,min}$

Finally, an evenly distributed set of values for f_2 is selected and used as a constraint in the following bicriteria operations model.

Minimize	$f_1 =$ Equations 7.1 through 7.3
Subject to:	$f_2 =$ Equation 7.13
	$f_2 \le$ the selected f_2 value and Constraints of Model 1

Varying f_2 over its range will result in many alternatives with a wide range of solutions. The decision maker will be able to select the best alternative by specifying a value for f_2.

7.2.4.1 Bicriteria Microgrid Operations Example

Consider the following microgrid where the first generator is operated by wind energy and the second generator is operated by coal with environmental factors of $m_1 = 0.1$ and $m_2 = 0.9$, respectively. A set of alternatives are

TABLE 7.8

Information for Bicriteria Microgrid Example

c_{ij}, Trans. Cost from i to j		User j			
		1	2	3	4
Gen. i	1	6	5	8	4
	2	7	4	5	7

c_{pj}, Trans. Cost from p to j		User j			
		1	2	3	4
Storage p	1	1	2	1	2
	2	2	0.5	1.5	1
Demand D_j		600	800	400	300

c_{ip}, Trans. Cost from i to p		Storage p	
		1	2
Gen.	1	4.5	5
i	2	6	4

Gen.		Gen. Costs, g_i ($)	Gen. Cap. K_i	Env. Imp. m_i
	1	40	1900	0.1
	2	34	1800	0.9

Storage	Init. Enr $q_{p,0}$	Min. Enr $q_{p,min}$	Max. Enr $q_{p,max}$
1	50	50	300
2	50	50	350

presented for this example. Note that environmental impacts of storages are not considered in this example (Table 7.8).

First, the range of the criteria must be found. Then, the criteria must be calculated when minimizing only f_1, the total cost. This model is as follows:

$$\text{Minimize } f_1 \quad \text{Subject to : Constraints of Model 3}$$

The solution to this example is shown as Problem I in Table 7.9.

Then, the criteria must be calculated when minimizing only f_2, the environmental impact. This problem is as follows:

$$\text{Minimize } f_2 \quad \text{Subject to : Constraints of Model 3}$$

The solution to this problem is shown as Problem II in Table 7.9.

Minimize	f_1
Subject to:	$f_2 \leq 1651.6$
	Constraints of Model 1

In the next step, an evenly distributed range of values for f_2 is found. The following five evenly distributed values of f_2 are used: 384.4, 637.5, 1021.5, 1405.5, and 1651.6.

TABLE 7.9

Optimal Solution Considering Single Objectives

	Solution	
Problem I: Minimize f_1 subject to Constraints of Energy Operation Model	83,886	1651.6
Problem II: Minimize f_2 subject to Constraints of Energy Operation Model	94,806	384.4

TABLE 7.10

Solution for Bicriteria Microgrid Example

Enr. Trans. x_{ij}	User j					Enr. Trans. y_{ip}	Storage p			Enr Trans. z_{pj}	User j				
	1	2	3	4	Tot.		1	2	Tot.		1	2	3	4	Tot.
Gen. i 1	0	0	0	300	300	Gen. i 1	16	0	16	Storage p 1	26	0	0	0	26
2	0	800	400	0	1200	2	0	600	600	2	574	0	0	0	574
Tot	0	800	400	300	1500	Tot.	16	600	616	Tot	600	0	0	0	600

Storage Energy Level, q_p	1	2
	50	50

For example, for $f_2 = 1651.6$, the following model is solved to find the optimal solution for bicriteria energy operations example.

The solution to this bicriteria microgrids example is shown in Table 7.10. The total cost (f_1) and total environmental impact (f_2) for this solution are $f_1 = \$83,886$, and $f_2 = 1651.6$, respectively. This solution shows that generator 1 generates a total of 316 units of energy while generator 2 generates 1800 units of energy.

7.3 Conclusion

Integrating renewable energy resources with microgrids requires developing flexible and efficient energy systems. In this chapter, energy operation optimization problem was used as base for developing other applicable models such as multiple period microgrid operation model and bicriteria operation model. Linear structure of these models enables us to solve medium- to large-scale problems in a matter of few seconds to few minutes. All presented models can be applied as a base for more complex microgrid operation and design problems where parameter values are subject to change every few minutes. Bicriteria energy model can also be expanded to multicriteria model by incorporating additional objectives such as reliability of energy system or thermal comfort of energy users; see Sheikh (2013); Sheikh et al. (2014).

References

Alarcon-Rodriguez, A., Ault, G., and Galloway, S. 2010. Multi-objective planning of distributed energy resources: A review of the state-of-the-art. *Renewable and Sustainable Energy Reviews*, 14(5), 1353–1366.

Carmona, R. and Ludkovski, M. 2010. Valuation of energy storage: An optimal switching approach. *Quantitative Finance*, 10(4), 359–374.

Conejo, A. J., Fernandez-Gonzalez, J. J., and Alguacil, N. 2005. Energy procurement for large consumers in electricity markets. *IEE Proceedings-Generation, Transmission and Distribution*, 152(3), 357–364.

Diamantoulakis, P. D., Kapinas, V. M., and Karagiannidis, G. K. 2015. Big data analytics for dynamic energy management in smart grids. *Big Data Research*, 2(3), 94–101.

Escribano, A., Peña, J. I., and Villaplana, P. 2011. Modelling electricity prices: International evidence. *Oxford Bulletin of Economics and Statistics*, 73(5), 622–650.

Fadaee, M. and Radzi, M. A. M. 2012. Multi-objective optimization of a stand-alone hybrid renewable energy system by using evolutionary algorithms: A review. *Renewable and Sustainable Energy Reviews*, 16(5), 3364–3369.

Fazlollahi, S., Mandel, P., Becker, G., and Maréchal, F. 2012. Methods for multi-objective investment and operating optimization of complex energy systems. *Energy*, 45(1), 12–22.

Kyriakides, E. and Polycarpou, M. (eds.) 2007. Short term electric load forecasting: A tutorial. In *Trends in Neural Computation* (pp. 391–418). Springer, Berlin.

Liu, P., Pistikopoulos, E. N., and Li, Z. 2010. A multi-objective optimization approach to polygeneration energy systems design. *AIChE Journal*, 56(5), 1218–1234.

Malakooti, B. 1986. Implementation of MCDM for the glass industry energy system. *IIE Transactions*, 18(4), 374–379.

Malakooti, B. 2014. *Production and Operations Systems with Multi-Objectives*. John Wiley.

Malakooti, B., Sheikh, S., Al-Najjar, C., and Kim, H. 2013. Multi-objective energy aware multiprocessor scheduling using bat intelligence. *Journal of Intelligent Manufacturing*, 24, 805–819.

Marckhoff, J. and Wimschulte, J. 2009. Locational price spreads and the pricing of contracts for difference: Evidence from the Nordic market. *Energy Economics*, 31(2), 257–268.

Ravindran, A. R. and D. P. Warsing, Jr. 2012. *Supply Chain Engineering: Models and Applications*. Boca Raton, FL: CRC Press.

Ren, H., Zhou, W., Nakagami, K. I., Gao, W., and Wu, Q. 2010. Multi-objective optimization for the operation of distributed energy systems considering economic and environmental aspects. *Applied Energy*, 87(12), 3642–3651.

Shabanpour-Haghighi, A. and Seifi, A. R. 2015. Multi-objective operation management of a multi-carrier energy system. *Energy*, 88, 430–442.

Sharafi, M. and ELMekkawy, T. Y. 2014. Multi-objective optimal design of hybrid renewable energy systems using PSO-simulation based approach. *Renewable Energy*, 68, 67–79.

Sheikh, S. 2013. Optimization and Risk Scenario Analysis of Energy Planning Problem. PhD dissertation. Advisor: Professor Behnam Malakooti, Case Western Reserve University.

Sheikh, S., Komaki, M., and Malakooti, B. 2014. Multi-objective energy operation problem using Z utility theory. *International Journal of Advanced Manufacturing Technology*, 74(9), 1303–1321.

Sheikh, S., Komaki, M., and Malakooti, B. 2015. Integrated risk and multi-objective optimization of energy systems. *Computers & Industrial Engineering*, 90, 1–11.

Sheikh, S. and Malakooti, B. 2011. Integrated energy systems with multi-objective. *IEEE Energy Tech Conference*, Cleveland, OH, May 25–26, 2011, pp. 1–5.

Sheikh, S. and Malakooti, B. 2012. Design, operation, and efficiency of energy systems. *IEEE Energy Tech Conference*, Cleveland, OH.

Yoo, P. D., Ng, J. W., and Zomaya, A. Y. 2011. An energy-efficient kernel framework for large-scale data modeling and classification. In *IEEE International Symposium on Parallel and Distributed Processing Workshops and PhD Forum (IPDPSW)*, pp. 404–408.

Zamora, R. and Srivastava, A. K. 2010. Controls for microgrids with storage: Review, challenges, and research needs. *Renewable and Sustainable Energy Reviews*, 14(7), 2009–2018.

Zhou, K., Fu, Chao, and Yang, Shanlin. 2016. Big data driven smart energy management: From big data to big insights. *Renewable and Sustainable Energy Reviews*, 56, 215–225.

8

A Data-Driven Approach for Multiobjective Loan Portfolio Optimization Using Machine-Learning Algorithms and Mathematical Programming

Sharan Srinivas and Suchithra Rajendran

CONTENTS

8.1 Introduction

In the United States, one in three Americans experiences financial problems and may not be able to completely fulfill his/her needs using his/
her savings or income (Soergel, 2015). Financial institutions, such as banks,
lend money to qualified borrowers to help them meet their credit needs and
the borrowers are expected to repay the loan in installments over a certain
time period. Irrespective of the size of the bank, loans contribute a major
portion of the bank's total asset. For large commercial banks, such as Wells
Fargo, JP Morgan, Citibank, and Bank of America, loans as a percentage of
asset size is nearly 45%. For medium-sized banks, such as Capital One and
PNC, loans contribute about half of their total asset (Perez, 2015). Therefore,
banks use a major portion of deposited funds to issue different types of
loans (e.g., credit card loans, mortgage loans, auto loans) and earn a profit
by collecting interest on the loaned amount. The profit earned from a loan
is the difference between the total amount of money collected from the borrower throughout the loan repayment period and the amount lent to the
borrower.

8.1.1 Impact of Banks on the Society

Banks and societies are mutual constitutions. Banks generate value by pro-
moting the well-being and fulfillment of the global population. In other
words, banks generate value by supporting the global population in financing, savings, electronic payment systems, and asset management. For example, customers can safely deposit their savings in banks and avoid the major
risks involved in protecting them. Therefore, the society's well-being ensures
the well-being of the banks and hence, banks must work in the best interest
of the community.

8.1.2 Risks Associated with Lending

As discussed earlier, revenue generated from loans is one of the largest assets
for any bank. However, there is also a huge risk involved because a percentage of the total approved loans may result in bad loans (debts that are not
recovered in time). Bad loans are one of the most common causes for bank
revenue losses and sometimes bankruptcy. Many national and international

banks suffered huge losses due to late payment or loan default resulting in financial instability. For instance, the net loss for Bank of India in the second quarter of 2015 due to bad loans was about $170 million (Tripathy, 2015). In late 2000s, banks issued a large sum of money toward home and personal loans without considering the risk of bad loans. Due to this, banks could not derive profits from all the approved loans leading to a financial crisis and a significant decrease in the loan approval rate. The decrease in the loan approval rate led to a fall in house prices and as a result, the borrowers had to sell more assets to repay their loans. This was one of the major causes of the recession in 2009 and could have been avoided if the banks had carefully chosen their borrowers.

8.1.3 Loan Approval Process

The sequence of events involved in approving a loan starting from the submission of loan applications by the potential borrowers to the final outcome is referred to as the *loan approval process*. Figure 8.1 illustrates the steps involved in the loan approval process as described by Power (2002). The loan application process begins when the applicant initiates a loan request by first submitting the preapproval documents (e.g., proof of employment, driver's license) requested by the bank. These documents are then verified by the lender in the preapproval process. If the documents are valid, then the loan preapproval statement is sent to the applicant and additional supporting documents (e.g., number of dependents, number of existing loans) are requested to make a final decision. The lender's team of underwriters processes these documents and inputs the borrower's attributes into the loan approval model (a decision support system for loan approvals). Some of the criteria that are given as inputs to the loan approval model are income, open line of credits and age, and these criteria may vary from one bank to another. Based on the outcome of the decision support system and the strategic goals of the company (e.g., limit on the number of loans approved), the underwriter decides whether to issue a loan to the applicant and he/she is notified of the bank's decision.

8.1.4 Problem Statement and Objective

Even though the largest asset of banks is the revenue generated from loans, it is necessary to consider the huge risk of approving bad loans to avoid bankruptcy. Therefore, it is evident that a robust loan approval system is essential for efficient operation of the banks. Banks have certain strategic goals (e.g., policies, targets) for their loan portfolio. In addition, the loan portfolio must be efficient to achieve the optimal risk–return tradeoff (i.e., a portfolio that offers the highest expected return for a specified risk level or lowest risk level for a specified level of expected return). Hence, the objectives of this chapter are to:

approvals may vary depending on the type of loan. For instance, factors such as loan amount, total income, number of dependents, and real estate securities are major factors that impact the approval of auto loans (Hill, 2014), and factors such as history of late payments, credit scores, payment for existing debt, and loan tenure impact the approval of mortgage loans (Choi, 2011).

8.2.1 Predictive Analytics Approach for Loan Approvals

Zurada and Barker (2011) suggested that the accuracy of predicting the risk type of the borrower (good or bad borrower) plays a significant role in the bank's financial stability. Due to this, machine-learning techniques, such as logistic regression and discriminant analysis, are studied (Bell et al., 1990; Khandani et al., 2010; Neelankavil, 2015). Ince and Aktan (2009) compared different techniques such as discriminant analysis, logistic regression, neural networks, and classification and regression trees (CART) for credit scoring in banking. Historical data from a financial institution were used to evaluate the techniques, and the authors observed that CART and neural networks outperformed other techniques. The authors also considered the probability of making an error of granting loan for a bad customer and the probability of not granting loan for a good customer. A detailed review of the methods used to classify the "good" and "bad" risk classes in the literature is given by Hand and Henley (1997). The methods discussed in their paper classified the applicants based on their repayment behavior.

Banks are faced with the challenge of using good classification models to make sound decisions and gain competitive advantage. Several methods used for classifying good and bad risks are given below:

- Discriminant analysis (Durand, 1941; Meyers and Forgy, 1963; Lane, 1972; Apilado et al., 1974)
- Regression (Orgler, 1970; Fitzpatrick, 1976; Wiginton, 1980; Srinivasan and Kim, 1987; Leonard, 1993; Henley, 1994; Lucas, 2004)
- Mathematical programming methods (Hand, 1981; Showers and Chakrin, 1981; Kolesar and Showers, 1985)
- Artificial neural network (ANN; Davis et al., 1992; Ripley, 1994; Rosenberg and Gleit, 1994)
- Time-varying models (Bierman and Hausman, 1970; Dirickx and Wakeman, 1976; Srinivasan and Kim, 1987)
- Random forest (RF; Brown and Mues, 2012; Wang et al., 2012; Kruppa et al., 2013)
- Stacking (Wang and Ma, 2012; Koutanaei et al., 2015)

8.2.2 Motivation

In reality, financial institutions offer different types of loans (such as mortgage loans, commercial loans) to serve the varying customer needs. The return varies depending on the type of loan and the risk type of the borrower. However, most of the existing researches primarily focus on developing a recommender system for loan approvals and do not take into account the different types of loans. Moreover, the strategic goals of the financial institution (e.g., restriction on the total approvals for a particular type of loan) are seldom taken into consideration. Also, most of the previous researches consider only single loan applications and ignore the possibility of joint loan applications (application with coborrower). If the applicant has a bad credit score, then the probability of loan acceptance can be increased by having a cosigner with a good credit score because most banks use the better of the two credit scores to determine the eligibility for loan approval (Somers and Hollis, 1996). Therefore, the present work aims to address these gaps in the literature. The proposed two-stage decision support system involves the integration of data analytics with multiple criteria decision making (MCDM). In the first stage, we evaluate different machine-learning algorithms to classify the risk type of the applicants. In the second stage, we propose an integer programming model with multiple objectives to construct the loan portfolio. The MCMP model uses the risk type of the borrower as one of the inputs and constructs the portfolio by considering the different types of loans, strategic goals of the banks, and both single and joint applications.

8.3 Methodology

The proposed methodology is illustrated using Figure 8.2.

In the first stage, the borrowers are categorized as high risk or low risk by different machine-learning classifiers using factors such as borrower's age and late payment history. The best machine-learning classifier is selected based on the output performance measures discussed in Section 8.3.3. It is important for banks to differentiate between high- and low-risk borrowers because the interest rate for all borrowers depends on their risk type. High-risk borrowers are often charged more interest rate to compensate for the high probability of loan defaulting (Diette, 2000). In the second stage, a diversified loan portfolio is developed with the objective of achieving higher interest rate returns and lower risk using an MCMP model. Since the risk type of the potential borrowers impacts the return and the risk of the portfolio, the solutions obtained in the first stage using machine-learning techniques are used as an input to the second stage.

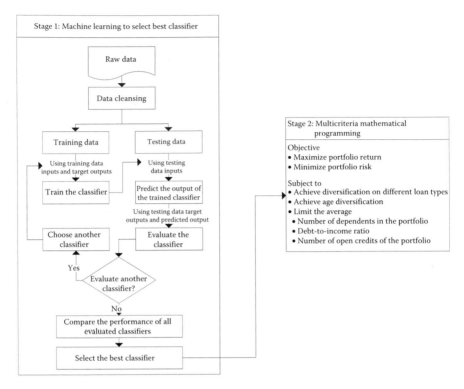

FIGURE 8.2
Overview of research methodology.

8.3.1 Predicting the Risk Associated with Each Applicant

In this section, the tasks involved in determining the risk associated with applicants are discussed.

8.3.1.1 Data Description

The data used for determining the risk type of an applicant (high risk or low risk) were downloaded from the Kaggle competition site (Kaggle, 2011). A detailed description of the downloaded data is presented in Table 8.1.

The variable "Risk" is the output variable and all the other variables are treated as input variables or features or predictors. Therefore, the data has ten predictors and one response variable. The input variables are not highly correlated, and hence all the variables can be used to predict the response variable. The downloaded raw data contains 150,000 samples. As in any real-world data, these data also suffer from user input errors and missing values. The user input errors are values that appear to be meaningless for a variable under consideration. For example, the age of some of the borrowers was entered as 0 and the monthly income of some of the borrowers was

TABLE 8.1

Description of Data

Variable Name	Description	Type
Risk	Risk associated with the borrower	Binary (high risk or low risk)
Age	Age of the borrower (in years)	Integer
Debt ratio	Ratio of monthly debt payments to monthly gross income	Continuous between 0 and 1
LOC	Number of open loans and lines of credit	Integer
Income	Monthly income of the borrower	Continuous
MREL	Number of mortgage and real estate loans	Integer
Dependents	Number of dependents of the borrower	Integer
Utilization	Ratio of total balance on lines of credit to the total credit limits	Continuous between 0 and 1
30 days	Number of times the borrower has been 30–59 days past the due date in the last 2 years	Integer
60 days	Number of times the borrower has been 60–89 days past the due date in the last 2 years	Integer
90 days	Number of times the borrower has been equal to or more than 90 days past the due date	Integer

entered as $1. In addition, there were 29,731 instances in which the monthly income was missing, and 3924 instances in which the number of dependents was missing. The presence of erroneous or missing data can significantly impact the results of the analyses, and therefore, it is important to cleanse the data before using them as inputs to the machine-learning algorithm. Section 8.3.1.2 describes the various data-cleansing techniques used to handle the user input errors and missing values.

8.3.1.2 Data Cleansing

The process of detecting any inconsistencies in the data and replacing them with suitable values and making them usable is called *data cleansing or scrubbing*. In order to handle the user input errors and missing values in the data, several alternatives are considered, and are as follows:

Substitute with a Unique Value: The data entry error and missing data are coded with a value that never occurs in the data set. For example, all the missing values and the user input errors can be substituted with –1.

Substitute with Median: The data entry error and missing data are substituted with the median value of that feature.

Discard Variable and Substitute with a Unique Value: The features with too many missing values are removed and the data entry error and missing values for the remaining features are then coded as –1.

Discard Variable and Substitute with Median: The features with too many missing values are removed and the data entry error and missing data for the remaining features are then substituted with the median value of that feature.

Discard Incomplete Rows: The rows containing error or missing values are removed from the data set.

8.3.2 Training and Testing Using Machine-Learning Algorithms

Several classification techniques are evaluated using the data obtained from Kaggle (2011). The data is split into training data and testing data. Each algorithm or classifier uses the training data to learn the underlying relationship between the input features and the response variable, and the testing data to evaluate the strength of the trained classifier. Therefore, the classifier automatically learns to classify the risk type of the potential borrowers using the features in the training phase, and the trained classifier uses the features of the testing data to predict the output (i.e., risk type of the potential borrower).

Five different machine-learning techniques are used and the best method for classifying the risk type is identified. The five methods considered are: (1) logistic regression, (2) RFs, (3) neural networks, (4) gradient boosting, and (5) stacking. A supervised learning procedure is used in the five machine-learning algorithms, and hence the training of the methods involves the use of known inputs and outputs.

8.3.2.1 Logistic Regression

The probability that the borrower will be of high risk or low risk is computed using the input features and is shown in Equations 8.1 and 8.2. b_0 is a constant and b_is are the regression coefficients of the input parameters. The training data set is used to estimate the regression constant and regression coefficients that best fit the observed data.

$$P(\text{Risk Type} = \text{High}) = \frac{1}{1 + e^{-(b_0 + b_1 * \text{Age} + b_2 * \text{Debt Ratio} + b_3 * \text{LOC} + b_4 * \text{Income} + \cdots + b_{10} * 90 \text{ Day})}} \quad (8.1)$$

$$P(\text{Risk Type} = \text{Low}) = 1 - P(\text{Risk Type} = \text{High}). \quad (8.2)$$

A schematic representation of the relationship between the input features and the response variable with a logistic function is shown in Figure 8.3.

8.3.2.2 Artificial Neural Network

ANN is a machine-learning algorithm that is inspired by the biological neural network. As shown in Figure 8.4, ANN includes three different types

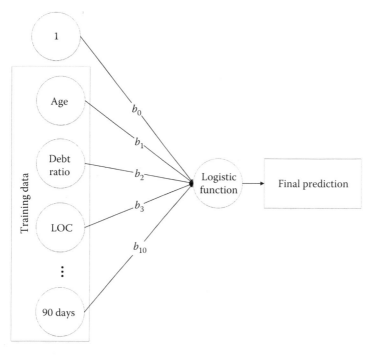

FIGURE 8.3
Representation of the logistic regression algorithm.

of layers, namely, an input layer, hidden layer, and output layer and each layer has a certain number of nodes. Each node (i) in a given layer (l) is connected to each node (j) in the next layer ($l + 1$) by a connection weight (w_{ij}). In order to train the classifier, the features are given as inputs to the input layer. Each input value is multiplied by a weight at the input layer, and the weighted input is relayed to each node in the hidden layer. Each node in the hidden layer will combine the weighted inputs that it receives, use it with the activation function (e.g., sigmoid activation), and relay the value to the nodes in the output layer. The output layer then determines the network output (risk type) by performing a weighted sum of the outputs of the hidden layer. The process of using the training inputs, hidden layers, and activation function to compute the response variable (risk type) is called *feed-forwarding* or *forward pass*. Initially, the training process begins with random weights. At the end of each feedforward step, the predicted output is compared with the actual output. If the predicted risk type is same as the actual risk type, then the neural network's weights are reinforced. On the other hand, if the predicted risk type is incorrect, then the neural network's weights are adjusted based on a feedback. This process is called *backward pass*. The forward pass and the backward pass are repeated for different

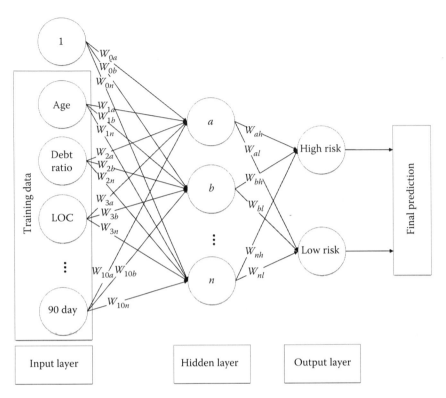

FIGURE 8.4
Representation of a three-layered neural network algorithm.

training samples until the ANN classifier is fully trained. ANN can be useful to uncover complex relationship between the features and the output. However, it requires more parameters to be estimated, and therefore, may require more time for training.

8.3.2.3 Random Forests

RF is an ensemble of decision trees proposed by Breiman (2001). Decision trees use a tree-like structure to obtain the output class and have nodes at each level of the tree. Each node splits into two nodes in the next level and the data set is divided among the two nodes based on a test (e.g., is the feature "age" greater than 50?). This process is repeated until the output class (risk type) is reached. At each level, it is necessary to select the feature that is most useful for classifying the response variable, and information gain is a metric to measure the usefulness of a feature at that level. Information gain is the expected reduction in entropy due to sorting on a given node. Entropy is the measure of impurity and a higher entropy indicates more information content. If p_i indicates the probability of class i, then Equations 8.3 and 8.4 give the entropy and information gain, respectively.

$$\text{Entropy} = \sum_i - p_i \log_2 p_i \tag{8.3}$$

$$\text{Information gain} = \text{Entropy (parent)} - \text{Weighted average}$$
$$[\text{Entropy (children)}]. \tag{8.4}$$

Each decision tree provides an output class and the final output class is the plurality voting of all the decision trees. Figure 8.5 is a schematic representation of the RF algorithm.

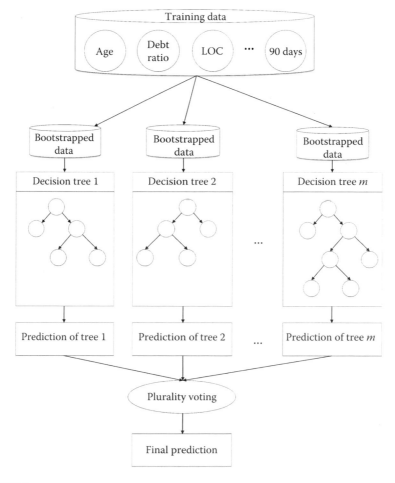

FIGURE 8.5
Representation of the random forests algorithm.

8.3.2.4 *Stochastic Gradient-Boosted Decision Trees*

Stochastic gradient boosting is an ensemble method introduced by Friedman (2002). It iteratively trains shallow decision trees with the objective of minimizing a loss function (e.g., negative log-likelihood) and sequentially learns from the errors of the previous trees. Therefore, the trees are trained one at a time and cannot be trained in parallel. A schematic representation of stochastic gradient-boosted decision trees (SGBDT) algorithm is shown in Figure 8.6.

The samples that are wrongly classified by a decision tree are upweighted, and the samples that are correctly classified are downweighted. This procedure is continued iteratively for all the decision trees leading to a higher weight for observations with correctly classified outputs and a lower weight for observations with wrongly classified outputs. It is to be noted that

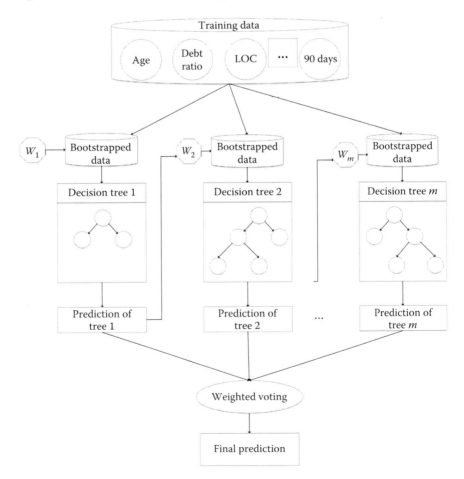

FIGURE 8.6
Representation of the SGBDT algorithm.

the SGBDT randomly samples a subset of the training data to train each decision tree. The final output would be a weighted voting of the decision trees.

8.3.2.5 Stacking

Stacking is the idea of combining the predictions of multiple classifiers and it involves two phases. In the first phase, the features are independently trained using different classifiers (e.g., logistic regression, RFs). The classifiers used to train the features are called *base-level classifiers*. In the second phase, the predicted outputs obtained from the base-level classifiers are used as inputs to the second-phase classifier (e.g., ANN, RFs) called *meta-level classifier*. In other words, the second phase combines the individual predictions of the base-level classifiers. The final class is the output obtained from the meta-level classifier. A schematic representation of the stacking algorithm is shown in Figure 8.7.

8.3.3 Evaluating a Learning Algorithm

Given the set of input values, the trained classifier provides the probability value for the risk type of borrowers. A threshold is then required to convert

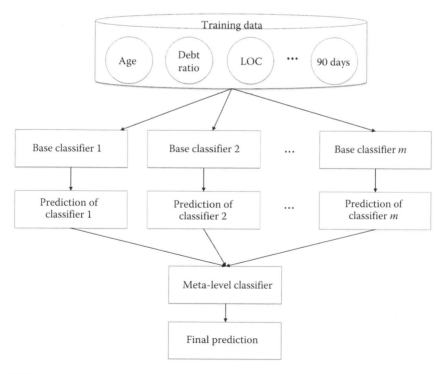

FIGURE 8.7
Representation of stacking.

the probability value to either high risk or low risk. For instance, a threshold of 0.50 indicates that any value below 0.50 is low risk and above 0.50 is high risk. A classifier may have any threshold value (e.g., 0.70) depending on the data. Therefore, the area under the curve (AUC) value for receiver operating characteristics (ROC) is used to obtain the best threshold value for a given classification problem. The AUC value quantifies the overall ability of the classifier to discriminate between the risk types of the applicant. A random classifier has an AUC value of 0.50 and a perfectly accurate classifier has an AUC value of 1.0. The ROC curve plots the true-positive rate (TPR) versus the false-positive rate (FPR) for various threshold settings as shown in Figure 8.8. Therefore, each point on the ROC curve represents the TPR/FPR value corresponding to a particular threshold. The dotted line has an AUC value of 0.50 and is the performance of a random classifier. A perfect classifier would yield a point in the upper left corner or the coordinate (0,1) of the ROC curve. Hence, using the ROC curve, a decision maker can choose a threshold value that is closest to the (0,1) coordinate. In other words, the threshold should be chosen in such a way that the TPR is high and the FPR is low. Once the threshold value is obtained, the probability values below the threshold are categorized as low-risk borrowers and values above the threshold are categorized as high-risk borrowers.

The accuracy of the classifier is determined by using the actual output and the predicted output by constructing a confusion matrix. A model has high accuracy if the actual outputs are same as the predicted outputs for many instances. As shown in Figure 8.9, the confusion matrix has four categories, namely, true positive, true negative, false positive, and false negative.

If a classifier predicts the risk type of a potential borrower as high risk and if the actual risk type of the potential borrower is low risk, then it is a *false positive* or *Type I error*. Type I error may result in loss of customers to the financial institution because the borrower may go to a competitor who

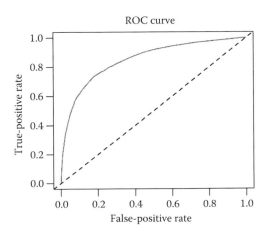

FIGURE 8.8
A sample diagram of the ROC curve.

		Actual output	
		High risk	Low risk
Predicted output	High risk	True positive	False positive (type I error)
	Low risk	False negative (type II error)	True negative

FIGURE 8.9
Elements of a confusion matrix.

offers a lower interest rate. If a classifier predicts the risk type of a potential borrower as low risk and if the actual risk type of the borrower is high risk, then it is a *false negative* or *Type II error*. Type II error may result in loss of revenue for the bank since the borrower has a higher chance to default on his loan. It is important to avoid both the Type I and Type II errors because the loan portfolio model seeks to reduce the risk and increase the return. If a classifier predicts the risk type of a potential borrower as high risk and if the actual borrower is high risk, then it is a *true positive*. If a classifier predicts the risk type of a potential borrower as low risk and if the actual borrower is low risk, then it is a *true negative*. True positive and true negative both improve the accuracy of the classifier. Therefore, three metrics, TPR (sensitivity), true negative rate (specificity) and accuracy, are derived from the confusion matrix and are used in this present work to evaluate a classifier, and are given below:

$$\text{TPR (Sensitivity)} = \frac{\sum \text{True Positive}}{\sum \text{True Positive} + \sum \text{False Negative}} \tag{8.5}$$

$$\text{True Negative Rate (Specificity)} = \frac{\sum \text{True Negative}}{\sum \text{True Negative} + \sum \text{False Positive}} \tag{8.6}$$

$$\text{Accuracy} = \frac{\sum \text{True Positive} + \sum \text{True Negative}}{\sum \text{True Positive} + \sum \text{False Negative} + \sum \text{False Positive} + \sum \text{True Negative}}. \tag{8.7}$$

8.3.4 MCMP Model

In this section, the proposed multiobjective integer programming model for constructing the loan portfolio is discussed. The notations used in the model are given below:

Sets and Indices

$c \in C$	Set of all age categories
$t \in T$	Set of all loan types
$a \in A$	Set of all loan applications
$a \in N(t)$	Set of all loan applications of loan type t
$b \in B(a)$	Set of all applicants for application a $B(a) \in \{1,2\}$
	Note: $B(a) = 1$ if single loan application and 2 if joint loan application

Input Parameters Associated with Loan Applicant

$E_{a,b,c}$	1 if age of bth borrower of application a falls within age category c, 0 otherwise
$D_{a,b}$	Total number of dependents for bth borrower of application a
$DR_{a,b}$	Debt ratio for bth borrower of application a
$R_{a,b}$	Risk associated with bth borrower of application a (output of Stage 1)
$L_{a,b}$	Number of late payments made by bth borrower of application a
$O_{a,b}$	Number of open credits for bth borrower of application a

Input Parameters Associated with Financial Institution

T_c^E	Lower bound on the percentage of total borrowers for age category c
T^D	Upper bound on number of dependents
T^{DR}	Upper bound on debt ratio
T^L	Upper bound on late payment
T^O	Upper bound on open credit
LP_t	Lower bound on percentage of loan issued to loan type t
UP_t	Upper bound on percentage of loan issued to loan type t
I_a	Interest rate of loan issued to application a

Decision Variables

Δ_a	1 if application a is approved; 0 otherwise
$\delta_{a,b}$	1 if b is the primary borrower for application a; 0 otherwise

8.3.4.1 Model Objectives

Objective 1: Maximize the total return due to loan interest

Bank's stability is essential for a bank to function effectively. The increase in revenue results in the improved capital base of banks, thereby increasing the stability. The main source of revenue for banks is from interest generated from loans. Therefore, it is necessary to maximize the revenue generated from loans and hence first objective considered in the loan portfolio optimization model is to maximize the return generated from the loan as given below:

$$\text{Max } Z_1 = \sum_{a \in A}(I_a \times \Delta_a). \tag{8.8}$$

Objective 2: Minimize the portfolio's risk

If many borrowers fail to meet the financial obligations mentioned in the loan agreement, then the asset of the bank becomes less than its liability and may even reach a state of insolvency. Therefore, the bank must evaluate the risk of the applicant defaulting on a loan and hence, objective 2 is to minimize the total risk incurred by the bank as given below:

$$\text{Min } Z_2 = \sum_{a \in A} \sum_{b \in B(a)} (R_{a,b} \times \delta_{a,b}). \tag{8.9}$$

8.3.4.2 Model Constraints

8.3.4.2.1 Limit the Average Debt-to-Income Ratio Associated with the Borrowers

Debt-to-income ratio is a measure of the proportion of gross income being paid toward debt. A high debt-to-income ratio indicates that the borrower either has a high debt or does not have sufficient income to repay the debt. Evidence from the literature suggests that a borrower with a high debt-to-income ratio is more likely to have trouble in repaying the loan (Lanza, 2014). Hence, financial institutions consider debt-to-income ratio as an important factor when deciding to lend money or approve credit. Therefore, constraint (8.10) ensures that, on an average, the debt ratio of the borrowers in the portfolio must be within the bank-specified debt ratio value (T^{DR}).

$$\frac{\sum_{a \in A} \sum_{b \in B(a)} (DR_{a,b} \times \delta_{a,b})}{\sum_{a \in A} \sum_{b \in B(a)} \delta_{a,b}} \leq T^{DR}. \tag{8.10}$$

8.3.4.2.2 Diversify the Types of Loan

Borrowers have a variety of purposes for requesting loans, such as purchasing a new house or funding for a new business (Curtis, 2007). Bank loans can be categorized into low interest generating loan types, such as commercial loans or commercial real estate loans, and high interest generating loan types, such as consumer loans and credit card loans. A diversified portfolio helps to maintain a steady revenue when the types of loans are not highly correlated. Therefore, it is necessary to consider different types of loans to achieve loan diversification ensuring customer satisfaction. Constraints (8.11) and (8.12) achieve the loan diversity across the different types of loans.

$$\sum_{a \in N(t)} \Delta_a \geq LP_t \times (|A|) \quad \forall t \in T \tag{8.11}$$

$$\sum_{a \in N(t)} \Delta_a \leq UP_t \times (|A|) \quad \forall t \in T. \tag{8.12}$$

8.3.4.2.3 Limit the Average Late Payment Frequency of the Borrowers

Avoiding a payment, missing a payment, or late payment may result in the reduction of the net revenue generated by the bank. In addition to the lost revenue, there are other costs, such as the cost involved in contacting the borrower to reclaim the loan through debt collection agency (Sullivan, 2014). Therefore, it is necessary to choose borrowers who are less likely to make late payments, and financial institutions use the borrower's late payment history to determine the likelihood of making late payments in the future. Constraint (8.13) ensures that, on an average, the late payments of the borrowers in the portfolio must be within the bank-specified late payment value (T^L).

$$\frac{\sum_{a \in A} \sum_{b \in B(a)} (L_{a,b} \times \delta_{a,b})}{\sum_{a \in A} \sum_{b \in B(a)} \delta_{a,b}} \leq T^L. \tag{8.13}$$

8.3.4.2.4 Limit the Average Number of Dependents Associated with the Borrowers

Studies have proven that the number of dependents is a significant factor to be considered when approving loans (Riungu, 2014). Borrower with many dependents may experience increased financial burden and, therefore, increases his/her likelihood of not paying the loan on time. Constraint (8.14) ensures that, on an average, the number of dependents of the borrowers in the portfolio must be within the bank-specified number of dependents value (T^D).

$$\frac{\sum_{a \in A} \sum_{b \in B(a)} (D_{a,b} \times \delta_{a,b})}{\sum_{a \in A} \sum_{b \in B(a)} \delta_{a,b}} \leq T^D. \tag{8.14}$$

8.3.4.2.5 Limit the Average Open Credits Associated with the Borrowers

Opening a new line of credit indicates that the borrower is unable to repay his/her debts and too many open credits may be riskier for borrowers who do not have an established credit history. Hence, financial institutions do not prefer borrowers with many open credits (Dykstra and Wade, 1997). Constraint (8.15) ensures that, on an average, the number of open credits of the borrowers in the portfolio must be within the bank-specified open credits value (T^O).

$$\frac{\sum_{a \in A} \sum_{b \in B(a)} (O_{a,b} \times \delta_{a,b})}{\sum_{a \in A} \sum_{b \in B(a)} \delta_{a,b}} \leq T^O. \tag{8.15}$$

8.3.4.2.6 Achieve Age Diversification among the Borrowers

According to Chapman (1940), bad loan experience for a certain age category is significantly different from the other age categories. Also, older borrowers are more likely to make wise decisions and do not run into bankruptcy (Carr, 2013). Therefore, in the optimal loan portfolio, it is necessary to consider at least a certain percentage of borrowers to fall within that bank-specified age category to ensure risk reduction and loan diversification. Therefore, constraint (8.16) is introduced to have at least a certain percentage of the loan application $\left(T_c^E\right)$ to be accepted with borrowers in that age category.

$$\frac{\sum_{a \in A} \sum_{b \in B(a)} (E_{a,b,c} \times \delta_{a,b})}{\sum_{a \in A} \sum_{b \in B(a)} \delta_{a,b}} \geq T_c^E \quad \forall c \in C. \tag{8.16}$$

8.3.4.2.7 Hard Constraints

Constraint (8.17) ensures that in a joint loan application, only one of the two borrowers can be a primary borrower, and Constraint (8.18) forces Δ_a to be 1 if application a is approved. Constraint (8.19) forces binary restrictions on the model.

$$\sum_{b \in B(a)} \delta_{a,b} \leq 1 \quad \forall a \in A \tag{8.17}$$

$$\sum_{b \in B(a)} \delta_{a,b} = \Delta_a \quad \forall a \in A \tag{8.18}$$

$$\delta_{a,b} \in \{0,1\} \quad \forall a \in A, b \in B(a). \tag{8.19}$$

8.3.4.3 Solution Approach: Multiobjective Optimization Using ε-Constraint Method

In a multicriteria problem, all objectives cannot be optimized simultaneously. Therefore, the main focus is to find the set of efficient solutions. A solution is said to be efficient (Pareto-optimal or nondominated) if it cannot be improved any further without sacrificing the performance of at least one of the other objective function value (Masud and Ravindran, 2008). The MCMP model described in Sections 8.3.4.1 and 8.3.4.2 can be solved using the ε-constraint method.

The multicriteria problem under consideration has two objectives:

- Objective 1: maximize return (Max $f_1(x)$)
- Objective 2: minimize risk (Min $f_2(x)$)

In the ε-constraint method, objective 1 is maximized (Equation 8.20) and objective 2 is given as a constraint (Equation 8.21) to the model along with the other regular constraints discussed in Section 8.3.4.2 (Equation 8.22).

$$\text{Max } f_1(x) \tag{8.20}$$

subject to

$$f_2(x) \leq \varepsilon \tag{8.21}$$

$$x \in S, \text{where } S \text{ is the feasible region.} \tag{8.22}$$

The RHS of the constrained objective function (ε) in Equation 8.21 is varied to obtain the efficient frontier.

First, the model is solved as a single objective problem considering only maximizing return and ignoring the risk constraint. The resulting total risk and return are the upper bounds on the value of risk and return, respectively. Next, the model is again solved as a single objective problem considering only minimizing risk and ignoring the return. The resulting total risk and return are the lower bounds on the value of risk and return, respectively. Finally, the return is maximized as the risk is varied (the value of ε is varied) from its lower bound to its upper bound. The model using ε-constraint method is given below:

Objective

$$\text{Max } Z_1 = \sum_{a \in A} \sum_{b \in B(a)} (I_a \times \delta_{a,b}).$$

Subject to:
Constraints (8.10) through (8.19), and

$$\sum_{a \in A} \sum_{b \in B(a)} (R_{a,b} \times \delta_{a,b}) \leq \varepsilon.$$

8.3.4.4 Demand Uncertainty in Loan Application Requests

The total number of loan application requests is unknown for any given time period. Therefore, to study the effect of the variability of the total number of loan applications on the portfolio's return and risk, n different scenarios are considered. For each scenario, the total number of loan applications is sampled from a known distribution and therefore known *a priori*, and then the mathematical formulation is solved by varying the ε values. Finally, the mean and the standard deviation of the output parameters across all scenarios for each ε value are obtained and analyzed.

8.4 Experimental Results

In this section, the input parameters are defined and the best classifier for the given input parameters is selected. Model accuracy, specificity, sensitivity, and training time defined in Section 8.3.3 are the performance measures used to identify the best classifier. The output of the best trained classifier is given as inputs to the MCMP model.

8.4.1 Input Parameters of Machine-Learning Algorithms

As discussed in Section 8.3.1.1, the data set used for classifying the risk type of the potential borrowers is downloaded from Kaggle (Kaggle, 2011), and it has historical data for 150,000 borrowers. The historical data have both the feature values (inputs) and the corresponding risk type (outputs). In addition to the historical data with both inputs and outputs, Kaggle also provided data for 100,000 borrowers that have only the feature values (inputs), and in this work, they are assumed to represent the population of the potential borrowers. The best classifier is used to predict the risk type of these potential borrowers. However, the loan portfolio cannot be generated only based on the risk type of potential borrowers. It is essential to consider various strategic plans of banks, such as the diversification of the loan types, achieving higher returns, maintaining low risks, and clearly these goals are conflicting in nature. Therefore, the predicted risk type of the potential borrower is used in the MCMP model to develop a loan portfolio that aims to achieve the bank-specified goals. The summary of the data parameters for machine-learning algorithms is given in Table 8.2.

TABLE 8.2

Summary of Data Parameters for Machine-Learning Algorithms

Data	Value
Training data	100,000
Testing data	50,000
Population of potential borrowers	100,000
Number of features	10
Number of outputs	1
Number of classifiers evaluated	5
Number of base-level classifiers used in stacking	2
Base-level classifiers in stacking	RF and SGBDT
Meta-level classifier in stacking	ANN
Types of loans considered	5
Percentage of individual loan applications	80%
Percentage of joint loan applications	20%

8.4.2 Analysis

The five classifiers are trained and tested using the caret package in R and the MCMP model was coded in Visual C++ and executed in IBM CPLEX®12.4.0.0 optimizer. The experiments were conducted on a computer with 8 GB RAM, Intel i5 2.50 GHz processor running Windows 10.

Section 8.3.1.2 discusses five different alternatives for data cleansing: substitute with −1 (A1), substitute with median (A2), discard variable and substitute with unique value (A3), discard variable and substitute with median value (A4), and discard incomplete rows (A5). Each alternative is then used to train and test each of the five classifiers. Finally, the AUC values for each alternative for the five classifiers are obtained and are shown in Table 8.3. The best approach for data cleansing is the alternative with the highest AUC value.

Based on the values in Table 8.3, A1 has the highest AUC value for all the classifiers except neural network. Neural network has the highest AUC value for the alternative A3. However, the difference between the AUC value for A3 and A1 obtained using the neural network is very small. Therefore, A1 is chosen as the best alternative for data cleansing and error and missing values are substituted with −1. Two-thirds of the cleansed data is used for training and the remaining is used for testing. The training data was resampled using fourfold cross validation technique in which the training data is internally split to train/test runs to determine the optimal parameters of the classifier. The AUC value is then obtained for each classifier for the testing set. It is important to note that the entries in the training set and testing set are randomly sampled and the AUC value changes depending on the entries in the training and testing set. Therefore, to estimate the accuracy value of each classifier, the procedure of estimating the AUC values is repeated 50 times. Figure 8.10 shows the average AUC value over 50 replications along with its standard deviation for the five classifiers.

It is observed that the standard deviation of the AUC values of the classifiers is very low and therefore, all the classifiers are robust in their classification. The average AUC value for logistic regression is very low compared with the other classifiers. Hence, logistic regression is no longer considered

TABLE 8.3

AUC Values for Different Alternatives

	Logistic Regression	Neural Network	Random Forest	SGBDT	Stacking
A1	0.69522	0.85519	0.84529	0.86210	0.86351
A2	0.69319	0.84344	0.84526	0.86148	0.86192
A3	0.68583	0.85856	0.84084	0.86082	0.86126
A4	0.69343	0.84344	0.84527	0.86148	0.86188
A5	0.68190	0.83376	0.83534	0.85182	0.85332

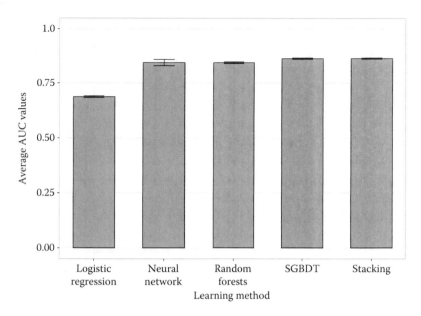

FIGURE 8.10
Average AUC values for each learning method.

in choosing the best classifier. In order to choose the best classifier among the other four, the class probabilities should be converted to class labels by using a specified threshold value. The ROC curve for the classifiers is shown in Figure 8.11 and can be used to select an appropriate threshold value. The threshold value impacts the sensitivity, specificity, and accuracy, and hence, different threshold values are evaluated to obtain a trade-off between sensitivity, specificity, and accuracy.

Using the ROC curve and by experimenting different threshold values, a threshold of 0.82 is chosen for RF and a threshold of 0.80 is chosen for SGBDT and Neural Network to obtain the respective class labels. Since stacking uses the neural network as its meta-level classifier, its threshold is also set to 0.80. The class label is low risk if the class probability is less than the threshold and is high risk if it is greater than the threshold.

The sensitivity, specificity, and accuracy values for each classifier are obtained using the confusion matrix and the time required to train each classifier is recorded as shown in Table 8.4. It is to be noted that the classifiers must be trained periodically to learn any new patterns emerged during environmental changes. An ideal method will have a sensitivity, specificity, and accuracy values close to 1 and will require low-training time.

All the methods perform well in classifying the risk type and have accuracy values close to each other. It can be observed that stacking yields better results in terms of sensitivity, specificity, and accuracy when compared with neural network, RFs, and SGBDT. However, the total training

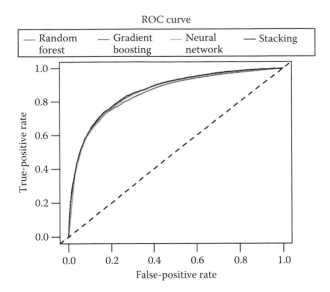

FIGURE 8.11
ROC curve for comparing the learning algorithms.

TABLE 8.4

Performance of Classification Algorithms

Algorithm	Sensitivity	Specificity	Accuracy	Training Time (in sec)
Neural network	0.7154	0.8308	0.8200	172
Random Forests	0.7212	0.8313	0.8200	229
SGBDT	0.7431	0.8231	0.8278	130
Stacking	0.7433	0.8314	0.8296	648

time (training time of base-level classifier + training time of meta-level classifier) is very high, compared with the other three classifiers. SGBDT has better sensitivity, accuracy, and training time values, and has slightly worse specificity value when compared with neural network and RFs. Therefore, for this particular data set, SGBDT is preferable over neural network and RFs. Further, sensitivity, specificity, and accuracy values of SGBDT are close to those of the stacking method, and its training time is almost five times less than that of the stacking method. Hence, SGBDT is chosen as the best classifier to classify the risk type of the potential borrowers.

8.4.3 Input Parameters of MCMP Model

Table 8.5 gives the summary of input parameters used in MCMP model. Five different types of loans are considered: mortgage loans, credit

TABLE 8.5

Summary of Data Parameters for Mathematical Programming Model

Data	Value
Bound on percentage of time the borrowers must be above 50 years of age	25%
Bound on number of dependents	2
Bound on debt ratio	0.4
Bound on late payments	5
Bound on open credits	10
Percentage of loan issued to loan type 1	29%
Percentage of loan issued to loan type 2	16%
Percentage of loan issued to loan type 3	13%
Percentage of loan issued to loan type 4	12%
Percentage of loan issued to loan type 5	30%
Interest rate of loan issued to loan type 1	5%
Interest rate of loan issued to loan type 2	3%
Interest rate of loan issued to loan type 3	2.5%
Interest rate of loan issued to loan type 4	2.5%
Interest rate of loan issued to loan type 5	5%

card loans, consumer loans, commercial real estate loans, and commercial loans. The interest rate and percentage of these five different types of loans considered are obtained from the literature (Dilworth, 2015; Issa, 2015).

8.4.4 MCMP Model Analysis

Figure 8.12 illustrates risk–return curve for different scenarios. The bounds for the risk and return vary for each scenario depending on the total number of loan applications. The risk objective is treated as ε-constraint and the value of ε is varied from the lower bound to the upper bound with a step size of 100. Across all scenarios, the return increases steeply when the underwriter is willing to accept up to 100 high-risk borrowers (i.e., ε = 100) and the slope of the line gradually decreases because the maximum revenue generating high-risk borrowers are already granted loan when ε = 100. Using the efficient frontier, the underwriter can estimate the maximum return generated for each scenario for a bank-specified risk value.

8.4.4.1 Scenario Analysis

Figures 8.13a through e illustrate the impact of portfolio risk on the percentage of different types of loans. From the error bars, it can be observed that the percentage of each type of loan issued varies as the portfolio risk increases;

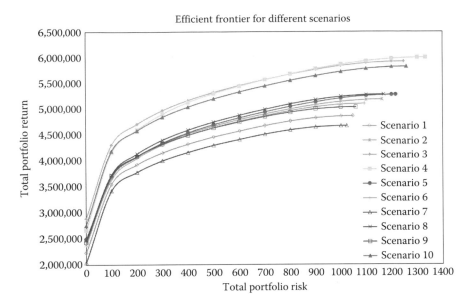

FIGURE 8.12
Efficient frontiers for different scenarios.

however, the variation is not statistically significant. Hence, the portfolio risk does not significantly impact the percentage of different types of loans. Also, within scenarios, the variation in the percentage of a particular type of loan issued is not significant. Therefore, uncertainties in the total number of applications received in a planning horizon do not impact the loan diversification with respect to the type of loans.

Figure 8.14a shows the impact of the portfolio risk on the loan approval rate. It can be observed that the loan approval rate increases as the portfolio risk increases. This is so because, if the financial institution is willing to tolerate more risk, then more high-risk borrowers are accepted since the objective of the model is to maximize the return. This results in an increase in the loan approval rate. Due to the increase in the loan approval rate, the portfolio's return also increases as risk increases as shown in Figure 8.14b.

Also, based on the various risk settings tested, the underwriter can decide on the portfolio return and loan approval rate. For example, consider a setting in which the bank is willing to accept up to 500 high-risk borrowers. From Figure 8.14a and b, it can be concluded that the return generated is on average around 4.75 million and the average loan approval rate is about 70%. Hence, for this multiobjective problem, the model serves as a decision support tool for determining the portfolio return and loan approval rate for a bank-specified risk value.

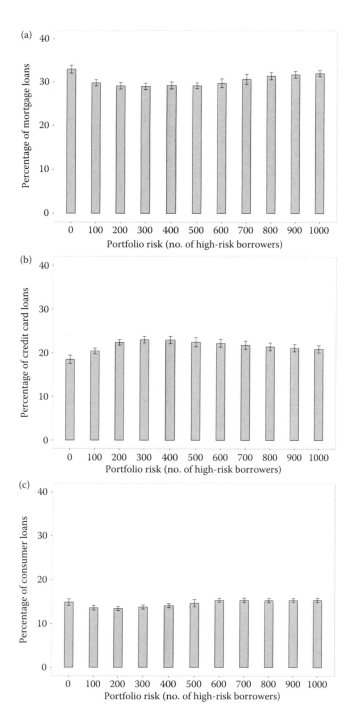

FIGURE 8.13
Impact of portfolio risk on percentage of different types of loans. *(Continued)*

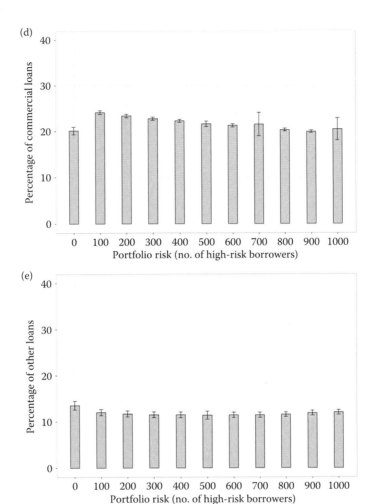

FIGURE 8.13 (*Continued*)
Impact of portfolio risk on percentage of different types of loans.

8.5 Conclusions

Financial institutions, such as banks, cater to the needs of various customers by offering a variety of loans, and the revenue generated by loans is one of the largest assets for any bank. However, there is also a large risk associated with the approval of bad loans and may sometimes even lead to bankruptcy. Therefore, in this work, a decision support recommender system is developed for loan approvals taking into account the different types of loans.

The developed recommender system is a two-stage system. In the first stage, the best machine-learning classifier is selected among the five classifiers

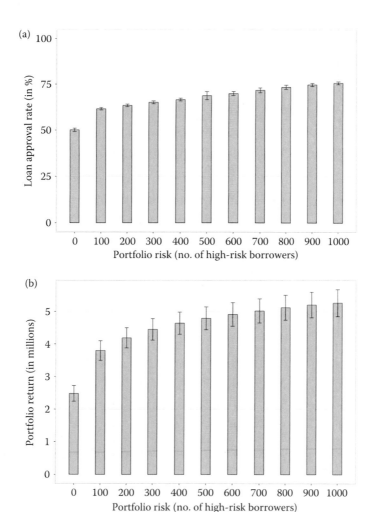

FIGURE 8.14
Impact of the portfolio risk on the loan approval rate and portfolio return.

considered to classify the potential borrowers as high risk or low risk. Several factors of the applicant, such as age, late payment history, and number of open credits, are given as inputs to the classifiers. The five different classifiers used are logistic regression, RFs, neural networks, gradient boosting, and stacking. The sensitivity, specificity, and accuracy values for each classifier are obtained using the confusion matrix and the time required to train each classifier is recorded. The method which has a sensitivity, specificity, and accuracy values close to 1 and least training time is selected as the best classifier. The output of the best classifier in the first stage along with other attributes of the applicants is given as input to the second stage of the model. In the second stage, a diversified loan portfolio is developed considering the types of loans, return, and risk associated with the borrower. The strategic goals and constraints of the bank are also given as input to the second stage.

The data of about 150,000 samples with ten predictors and one response variable were downloaded from the Kaggle competition site (Kaggle, 2011). SGBDT performs the best for the given data set. The findings indicate that the portfolio risk does not significantly impact the percentage of different types of loans as well as the uncertainties in the total number of applications received in a planning horizon. The loan approval rate and return increase as the portfolio risk increases since the financial institution is willing to tolerate more high-risk borrowers resulting in an increased loan approval rate and portfolio return. Finally, the efficient frontier is developed with the objective of maximizing the return and minimizing the risk, using which the underwriter can estimate the maximum return possible generated for each scenario for a bank-specified risk value.

References

Altman, E. I. 1980. Commercial bank lending: Process, credit scoring, and costs of errors in lending. *Journal of Financial and Quantitative Analysis*, Vol. 15(4), pp. 813–832.

Apilado, V. P., Warner, D. C., and Dauten, J. J. 1974. Evaluative techniques in consumer finance—Experimental results and policy implications for financial institutions. *Journal of Financial & Quantitative Analysis*, Vol. 9(2), pp. 275–283.

Bell, T. B., Ribar, G. S., and Verchio, J. 1990. Neural nets versus logistic regression: A comparison of each model's ability to predict commercial bank failures. In *Proceedings of the 1990 Deloitte and Touche/University of Kansas Symposium of Auditing Problems*, Lawrence, KS, pp. 29–58.

Bierman, Jr, H. and Hausman, W. H. 1970. The credit granting decision. *Management Science*, Vol. 16(8), pp. B-519–B532.

Breiman, L. 2001. Random forests. *Machine Learning*, Vol. 45(1), pp. 5–32.

Brown, I. and Mues, C. 2012. An experimental comparison of classification algorithms for imbalanced credit scoring data sets. *Expert Systems with Applications*, Vol. 39(3), pp. 3446–3453.

Burke, A. E. and Hanley, A. 2003. How do banks pick safer ventures? A theory relating the importance of risk aversion and collateral to interest margins and credit rationing. *The Journal of Entrepreneurial Finance*, Vol. 8(2), pp. 13–24.

Carr, D. 2013. Why older minds make better decisions. http://www.forbes.com/sites/nextavenue/2013/04/29/why-older-minds-make-better-decisions/, accessed on December 6, 2015.

Chapman, J. M. 1940. Factors affecting credit risk in personal lending. In *Commercial Banks and Consumer Installment Credit*. New York: National Bureau of Economic Research, pp. 109–139.

Choi, C. 2011. Mortgage approved: 5 factors that lenders consider on home loan applications in tighter financial market. http://www.postandcourier.com/article/20110703/PC05/307039941/, accessed on December 2, 2015.

Clemente, Jr, D. A. 1980. Prediction of agricultural loan repayment performance. *Philippine Review of Economics*, Vol. 17(1 & 2), pp.31–60.

Curtis, G. 2007. Different needs, different loans | Investopedia. http://www.investopedia.com/articles/pf/07/loan_types.asp, accessed on October 10, 2015.

Davis, R. H., Edelman, D. B., and Gammerman, A. J. 1992. Machine-learning algorithms for credit-card applications. *IMA Journal of Management Mathematics*, Vol. 4(1), pp. 43–51.

Diette, M. D. 2000. How do lenders set interest rates on loans? https://www.minneapolisfed.org/publications/community-dividend/how-do-lenders-set-interest-rates-on-loans, accessed on December 6, 2015.

Dilworth, K. 2015. Credit card interest rates slide to 14.89%. http://www.creditcards.com/credit-card-news/interest-rate-report-10715-down-2121.php, accessed on August 6, 2016.

Dirickx, Y. M. and Wakeman, L. 1976. An extension of the Bierman-Hausman model for credit granting. *Management Science*, Vol. 22(11), pp. 1229–1237.

Durand, D. 1941. Risk elements in consumer installment financing. In: *NBER Books*. National Bureau of Economic Research, Inc, Cambridge, MA, number dura41-2, June.

Dykstra, D. R. and Wade, P. M. 1997. U.S. Patent No. 5,611,052. Washington, DC: U.S. Patent and Trademark Office.

Fama, E. F. and Miller, M. H. 1972. *The Theory of Finance*. New York: Holt Rinehart & Winston.

Fitzpatrick, D. 1976. An analysis of bank credit card profit. *Journal of Bank Research*, Vol. 7, pp. 199–205.

Friedman, J. H. 2002. Stochastic gradient boosting. *Computational Statistics and Data Analysis*, Vol. 38(4), pp. 367–378.

Gerlach, S. and Peng, W. 2005. Bank lending and property prices in Hong Kong. *Journal of Banking & Finance*, Vol. 29(2), pp. 461–481.

Greenbaum, S. I. and Thakor, A. V. 2007. *Contemporary Financial Intermediation*. Boston: Academic Press, an imprint of Elsevier.

Hand, D. J. 1981. *Discrimination and Classification*. Wiley Series in Probability and Mathematical Statistics, Chichester: Wiley.

Hand, D. J. and Henley, W. E. 1997. Statistical classification methods in consumer credit scoring: A review. *Journal of the Royal Statistical Society. Series A (Statistics in Society)*, Vol. 160(3), pp. 523–541.

Henley, W. E. 1994. Statistical aspects of credit scoring. Doctoral dissertation. Open University. http://ethos.bl.uk/OrderDetails.do?uin=uk.bl.ethos.241084, accessed on August 6, 2016.

Hill, T. 2014. How lenders decide your auto loan rate. http://www.dallasnews.com/business/personal-finance/headlines/20140815-how-lenders-decide-your-auto-loan-rate.ece, accessed on December 15, 2015.

Ince, H. and Aktan, B. 2009. A comparison of data mining techniques for credit scoring in banking: A managerial perspective. *Journal of Business Economics and Management*, Vol. (3), pp. 233–240.

Issa, E. 2015. American household credit card debt statistics: 2015—NerdWallet. https://www.nerdwallet.com/blog/credit-card-data/average-credit-card-debt-household/, accessed on December 15, 2015.

Kaggle. 2011. Data—Give me some credit. https://www.kaggle.com/c/GiveMe SomeCredit/data, accessed on September 15, 2015.

Khandani, A. E., Kim, A. J., and Lo, A. W. 2010. Consumer credit-risk models via machine-learning algorithms. *Journal of Banking & Finance*, Vol. 34(11), pp. 2767–2787.

Kolesar, P. and Showers, J. L. 1985. A robust credit screening model using categorical data. *Management Science*, Vol. 31(2), pp. 123–133.

Koutanaei, F. N., Sajedi, H., and Khanbabaei, M. 2015. A hybrid data mining model of feature selection algorithms and ensemble learning classifiers for credit scoring. *Journal of Retailing and Consumer Services*, Vol. 27, pp. 11–23.

Kruppa, J., Schwarz, A., Arminger, G., and Ziegler, A. 2013. Consumer credit risk: Individual probability estimates using machine learning. *Expert Systems with Applications*, Vol. 40(13), pp. 5125–5131.

Lane, S. 1972. Submarginal credit risk classification. *Journal of Financial and Quantitative Analysis*, Vol. 7(1), pp. 1379–1385.

Lanza, A. 2014. Afford a mortgage with student loan debt. http://www.usnews.com/ education/blogs/student-loan-ranger/2014/10/29/afford-a-mortgage-with-student-loan-debt, accessed on December 15, 2015.

Leonard, K. J. 1993. Empirical Bayes analysis of the commercial loan evaluation process. *Statistics & Probability Letters*, Vol. 18(4), pp. 289–296.

Lovati, J. M. 1975. The changing competition between commercial banks and thrift institutions for deposits. Review July, pp. 2–8.

Lucas, A. 2004. Updating scorecards: Removing the mystique. In: *Readings in Credit Scoring: Foundations, Developments, and Aims*. Thomas, L. C., Edelman, D. B., and Crook, J. N. (eds.). New York: Oxford University Press, pp. 93–109.

Masud, A. and Ravindran, A. 2008. Multiple criteria decision making, Chapter 5. In *Operations Research and Management Science Handbook*, ed. A. Ravindran. Boca Raton, FL: CRC Press Taylor and Francis Group, pp. 1–35.

Meyers, J. H. and Forgy, E. W. 1963. The development of a numerical credit evaluation system. *Journal of the American Statistical Association*, Vol. 58(303), pp. 799–806.

Neelankavil, J. P. 2015. International Business Research. New York: Routledge.

Orgler, Y. E. 1970. A credit scoring model for commercial loans. *Journal of Money, Credit and Banking*, Vol. 2(4), pp. 435–445.

Perez, S. 2015. Welcome to market realist. http://marketrealist.com/2015/03/loan-assets-gaining-importance-banking-sector/, accessed on December 15, 2015.

Power, D. J. 2002. *Decision Support Systems: Concepts and Resources for Managers*. Westport, CT: Greenwood Publishing Group.

Ripley, B. D. 1994. Neural networks and related methods for classification. *Journal of the Royal Statistical Society. Series B (Methodological)*, Vol. 56(3), pp. 409–456.

Riungu, M. K. 2014. Factors influencing loan repayment in micro-finance institutions: A case of South Imenti District. *Doctoral dissertation*.

Rosenberg, E. and Gleit, A. 1994. Quantitative methods in credit management: A survey. *Operations Research*, Vol. 42(4), pp. 589–613.

Saunders, D., Xiouros, C., and Zenios, S. 2007. Credit risk optimization using factor models. *Annals of Operations Research*, Vol. 152(1), pp. 49–77.

Sharpe, W. F. 1970. Portfolio Theory and Capital Markets. McGraw-Hill, New York.

Showers, J. L. and Chakrin, L. M. 1981. Reducing uncollectible revenue from residential telephone customers. *Interfaces*, Vol. 11(6), pp. 21–34.

Sirignano, J., Tsoukalas, G., and Giesecke, K. 2015. Large-scale loan portfolio selection. Available at SSRN: http://dx.doi.org/10.2139/ssrn.2641301

Soergel, A. 2015. 1 in 3 Americans near financial disaster. http://www.usnews.com/news/blogs/data-mine/2015/02/23/study-suggests-1-in-3-americans-flirting-with-financial-disaster/, accessed on December 15, 2015.

Somers, P. and Hollis, J. M. 1996. Student loan discharge through bankruptcy. *American Bankruptcy Institute Law Review*, Vol. 4, pp. 457.

Srinivasan, V. and Kim, Y. H. 1987. Credit granting: A comparative analysis of classification procedures. *Journal of Finance*, Vol. 42(3), pp. 665–681.

Sullivan, B. 2014. 4 student loan debt collection tricks. http://www.cbsnews.com/media/4-student-loan-debt-collection-tricks/, accessed on December 15, 2015.

Taylor, W. F. 1980. Meeting the equal credit opportunity act's specificity requirement: Judgmental and statistical scoring systems. *Buffalo Law Review*, Vol. 29, pp. 73.

Tripathy, D. 2015. Bank of India sinks to Q2 loss as bad debts jump. http://in.reuters.com/article/bank-of-india-q2-results-idINKCN0SY1A020151109, accessed on December 15, 2015.

Van Leuvensteijn, M., Bikker, J. A., Van Rixtel, A. A., and Kok Sorensen, C. 2007. A new approach to measuring competition in the loan markets of the euro area. Available at SSRN: http://ssrn.com/abstract=991604

Wang, G. and Ma, J. 2012. A hybrid ensemble approach for enterprise credit risk assessment based on Support Vector Machine. *Expert Systems with Applications*, Vol. 39(5), pp. 5325–5331.

Wang, G., Ma, J., Huang, L., and Xu, K. 2012. Two credit scoring models based on dual strategy ensemble trees. *Knowledge-Based Systems*, Vol. 26, pp. 61–68.

Wiginton, J. C. 1980. A note on the comparison of logit and discriminant models of consumer credit behavior. *Journal of Financial and Quantitative Analysis*, Vol. 15(3), pp. 757–770.

Zurada, J. and Barker, R. M. 2011. Using memory-based reasoning for predicting default rates on consumer loans. *Review of Business Information Systems (RBIS)*, Vol. 11(1), pp. 1–16.

9

Multiobjective Routing in a Metropolitan City with Deterministic and Dynamic Travel and Waiting Times, and One-Way Traffic Regulation

Swaminathan Vignesh Raja, Chandrasekharan Rajendran, Ramaswamy Sivanandan, and Rainer Leisten

CONTENTS

9.1 Introduction

A transportation network is a network of roads and streets, which permits vehicles to move from one place to another. A directed or undirected graph can be used to represent a transportation network, where edges of the graph have a capacity and receive a flow. The vertices of the graph are called nodes and the edges of the graph are called arcs. In any transportation network, the vehicle starts from the source node and moves toward the sink node. There are several types of network problems/models available in litera-ture, for example, transportation problem, shortest path problem, minimum spanning tree problem, maximum flow problem, and minimum flow prob-lem. The shortest path problem has tremendous importance in network flow models because it is applicable to any type of transportation network seen in real life. One of the most important issues affecting the performance of a transportation network is routing. The goal of routing between two points in a network is to reach the destination as quickly as possible (shortest time path problem) or as cheaply as possible (minimum cost or distance path problem).

The shortest path problem is one of the most studied problems among net-work optimization problems (Bertesekas, 1991; Ahuja et al., 1993; Schrijver, 2003). Two kinds of labelling approaches, namely, label-setting algorithm and label-correcting algorithm, are in use to solve the shortest path problem. Label-setting algorithm is in use only for acyclic network with nonnegative costs, whereas label-correcting algorithm is applicable for acyclic network with negative and nonnegative costs.

The shortest path problem with time dependency is dealt as a single-objec-tive problem by several researchers, and one can classify the single-objective problem into minimum cost path and fastest path problem. The minimum cost path problem is solved to find the path having the minimum length with respect to the cost by considering the travel time, while in the fastest path prob-lem, the objective is to find the paths having the minimum length with respect to time-dependent travel time. Bellman's optimality principle (Bellman, 1958) is a modification of the single-objective shortest path problem where arc travel times are nonnegative integers for every time period, and achieve all-to-one fastest paths to the single destination node from all other nodes. Dreyfus (1969) proposed the modification of Dijkstra's static shortest path algorithm to obtain the fastest path between two vertices for a given departure time. The algo-rithm of Dreyfus (1969) is suitable to solve first-in-first-out network problems. Ziliaskopoulos and Mahmassani (1993) and Wardell and Ziliaskopoulos (2000) proposed a label-correcting algorithm to obtain all-to-one minimum cost paths and all-to-one fastest paths for all departure time. Chabini (1998) proposed a label-setting algorithm running backward in the set of time parameters to obtain all-to-one minimum cost and all-to-one fastest paths for all departure time without the first-in-first-out assumption.

The big data is characterized by volume, velocity, variety, and value of data. With the help of technological advancements, we are now able to collect the data about the transportation network of the entire city. The network representation of roads and streets allows us to explore the topological and geographical properties of the transportation network.

The multiple criteria decision making (MCDM) is an important area of research in Operations Research which deals with decision problems that involve multiple objectives. Real-life shortest path problem involves multi-objectives; for example, minimizing cost, time, distance, maximizing reliability, etc.

Typically, there does not exist a unique optimal solution for MCDM problems, and based on the decision maker's preference, a solution is chosen from the set of alternative solutions. The shortest path problem with multiobjectives is an nondeterministic polynomial time (NP)-hard problem (Hamacher et al., 2006; Mohamed et al., 2010). In multiobjective optimization problems, there may not exist a single solution that satisfies all the objectives simultaneously, and the objective functions in most of the cases are conflicting. A solution set is called nondominated set or Pareto set if none of the objective function values can improve without degrading any of the other objective function values. Ruzika and Wiecek (2005) presented a comprehensive survey of the literature (from 1975 to 2005) for the multiobjective shortest path problem. Ehrgott and Wiecek (2005) presented a survey on multiobjective integer programming, and discussed the scalarization techniques for general continuous multiobjective optimization.

The organization of this chapter is as follows: In Section 9.2, we review the literature; in Section 9.3, we define the multiobjective shortest path problem with time dependency of travel times along roads and of waiting times at junctions; we present our mathematical model and the multiobjective optimization approach in Section 9.4; we explain the model with a numerical illustration in Section 9.5; and in Section 9.6, we present the results with respect to the complete travel in the city of Chennai from a given origin to a given destination, followed by summary in Section 9.7.

9.2 Literature Review

In this section, we discuss the literature on multiobjective shortest path problems. Real-world shortest path problems are often time-dependent, with more than one set of time-dependent parameters (Müller-Hannemann and Schnee, 2007). Hamacher et al. (2006) studied the application of the shortest path problem for evacuation modeling where shortest paths represent evacuation routes. Several attributes associated with a route, such as its length or reliability, are of particular interest for an evacuee in case of emergency. A collection of possible evacuation routes are used to model a complete evacuation plan. They presented the interrelation of the time-dependent network optimization problem and evacuation modeling considering two objectives. They proposed a heuristic algorithm to solve the problem and compared its performance with respect to the existing algorithms. Sung et al. (2000) considered a flow speed model of shortest path problem with time-dependent networks. In their problem, as the interval changes, the flow speed on the link changes, but not the travel time. They modified the Dijkstra's label-setting algorithm and proved that the flow speed model is better than the link travel-time model. Dell'Amico et al. (2008) considered a shortest path problem, where the travel time on the arcs may vary with time, and allow for the waiting time at the nodes. Since the simple Dijkstra's algorithm adaptation may fail to solve the discontinuities that exist on the routes, they proposed a new Dijkstra-like algorithm to solve the problem. Mohamed et al. (2010) proposed a genetic algorithm to solve the shortest path problem with the two objectives, namely, minimizing cost and minimizing travel time. They compared the performance of their genetic algorithm with the algorithm of Brumbaugh-Smith and Shier (1989). Chitra and Subbaraj (2010) proposed an elitist multiobjective evolutionary algorithm on the basis of the nondominated sorting genetic algorithm, for solving the dynamic shortest path routing problem in computer networks. They addressed the problem by considering delay and cost as a weighted sum of objectives, and generated the Pareto-optimal solutions for the dynamic routing problem in a single run, and compared the result with the result of a single-objective weighting factor method.

Seshadri and Srinivasan (2010) proposed a new bound-based optimality criterion for the optimal reliability path problem. On the basis of the bounds, the authors proposed an algorithm to evaluate the path having the maximum travel-time reliability between given source and destination on the network with stochastic, normal, and correlated link travel times specified by the multivariate normal distribution. Seshadri and Srinivasan (2012) proposed an algorithm to compute the minimum robust cost path on the correlated and stochastic link travel-time network.

They transformed the robust cost objective to a link separation or sum of squares form. The level robust cost measure of the path is quantified by using a weighted combination of squared mean and variance, and the weights represent the importance of the travel-time variability of the user. Prakash and Srinivasan (2014) presented a sample-based algorithm for the minimum robust cost path on a network with link travel-time correlations, and formulated it as a separable multiobjective problem. The authors adopted a sample-based approach to represent the link travel-time distributions, implicitly capture the path correlations, and thus obviate the explicit estimation of the link travel-time correlation matrix.

The following gaps are identified from the review of literature.

- Several criteria exist in real-life situations. Hence, solving the problem as multiobjective considering two or more conflicting objectives takes the problem closer to reality. Our model includes two conflicting objectives, and provides the decision maker with a possible set of nondominated solutions allowing to choose her/his preferred solution, given ε_t.

- Many authors do not consider real-life situations such as one-way traffic in the rush period, time-dependent dynamic and deterministic travel times along the roads, and time-dependent dynamic and deterministic waiting times at junctions; we consider these real-life aspects in our mathematical model. It is possibly for the first time such a multiobjective mixed-integer linear programming (MILP) model is developed by the explicit consideration of time-dependent travel times and time-dependent waiting times, and time-dependent one-way traffic regulation in a major city road network.

9.3 Problem Definition: Multiobjective Shortest Path Problem with Time Dependency

In this study, we mainly focus on minimizing the distance, time, or cost to traverse between two nodes. The objective of this work is to find a shortest path with the minimum total travel time and distance between a given pair of origin and destination (O–D). The time taken to travel a particular link is called traverse time. Most of the authors consider the traverse time as deterministic and static; however, in reality, it is not static because it depends on various factors such as one-way traffic in rush periods, dynamic waiting time at signals, varying traffic conditions, etc. In this work, we consider multiobjective shortest path problem with time-dependent dynamic and deterministic traverse times and waiting times, and time-dependent one-way traffic regulation.

Let $G = (V,E)$ be a directed graph, where V is a set of nodes, and E is a set of edges. The departure time at the source node, source node s and destination node d are given as inputs. The solution to the time-dependent shortest path problem is to find an (s,d)-path that leaves a source node at a given time and minimizes the total travel distance as well as total travel time to reach the destination node which satisfies all the constraints. The objective of the model is to arrive at an optimal route to traverse over a given network with consideration of minimizing the total travel time and the total distance simultaneously.

In this work, we consider multiobjective shortest path problem with real-life constraints such as time-dependent dynamic and deterministic travel times, time-dependent dynamic and deterministic waiting times, and time-dependent one-way traffic, and we propose a mathematical model to solve the same. The biggest advantage of developing a mathematical model is the flexibility of the resulting model; many cost functions can be chosen, and many constraints can be added that otherwise would be difficult to satisfy with a Dijkstra-like approach. For example, we address the one-way traffic regulation in the proposed mathematical model. Dynamic programming becomes time consuming and is not very computationally efficient due to curse of dimensionality. Another motivation to go for MILP model is that the same model can be extended for developing a multiobjective optimization algorithm.

The technological advancements (Internet of Things) in big data enable us to collect the data about the entire transportation network of a city. We evaluate the proposed model with multiple objectives using the real-world (in Chennai city) network of major roads consisting of 1658 nodes and 4224 links, mostly undirected graph, except the roads involving one-way traffic. We consider the following aspects in our work that are taken into consideration while collecting data from the travel and incorporated into the MILP model with multiple objectives:

- Consider the source node as n' and the destination node as n'', and the arrival time at node n' be denoted by $A_{n'}$.
- As travelers do not wait at the origin, we assume the waiting time at the source node as 0.
- A day consists of a given number of travel-time intervals.
- Unit of the distance is 1 km and unit of the time is 1 min.
- Traffic corresponds to a given number of congestion levels, thereby influencing the travel time along a road and waiting time at a junction.
- Waiting time at a junction or node i depends on the time interval during which the actual arrival at node i takes place; waiting time is dynamic and deterministic.
- Travel time along the road (i,j) depends on the time interval in the day during which the actual travel takes place; travel time is dynamic and deterministic.
- Entry into the road (i,j) should be avoided inherently in the MILP model during the one-way traffic intervals. Most roads allow two-way traffic; however, there can be some roads that allow two-way traffic for most periods in a day, except for some periods when

the roads will allow traffic in one-way; for example, traffic flow is allowed along a given road (i,j) during 8 a.m. to 10 a.m., but no traffic along arc (j,i) during the same period, whereas traffic is allowed along the road (j,i) during 6 p.m. to 8 p.m., but no traffic is allowed along arc (i,j) during the same period.

- Two objectives, namely, the minimization of total travel time (including waiting times at junctions) and the minimization of total distance traveled are considered.

9.4 Mathematical Model for the Time-Dependent Shortest Path Problem When the Travel Times and Waiting Times, and One-Way Traffic Regulation Are Dynamic and Time-Dependent in a City Network

The mathematical model for the time-dependent shortest path problem is presented in this section. We explain the terminologies associated with the mathematical model in Section 9.4.1.

9.4.1 Terminology

Parameters	Description
n	Number of nodes in the network
n'	Origin node of travel
n''	Destination node of travel
i,j	A pair of nodes
d_{ij}	Distance from node i to node j /*note: It is not necessarily symmetric*/
k	Index for time interval
∇_i^w	Number of waiting-time intervals with respect to node i
φ_{ij}^t	Number of travel-time intervals with respect to road (i,j)
Ω_{ij}^o	Number of one-way and two-way traffic intervals with respect to road (i,j)
t_{ijk}	Travel time from node i to node j in the time interval k
W_{ik}	Waiting time at node i during the time interval k
LL_{ik}^w	Lower limit for the time interval k with respect to node i, to define the time-dependent waiting time
UL_{ik}^w	Upper limit for the time interval k with respect to node i, to define the time-dependent waiting time
LL_{ijk}^t	Lower limit for the time interval k with respect to road (i,j), to define the time-dependent travel time

(Continued)

Parameters	Description
UL^t_{ijk}	Upper limit for the time interval k with respect to road (i,j), to define the time-dependent travel time /*note: For a given road (i,j), the actual travel time can vary depending upon the actual time of arrival at the head of the road (i,j); for example, in a day of 1440 min and $\varphi^t_{ij} = 4$, if interval 1 corresponds to 7.30 a.m., 11 a.m., we set $LL^t_{ij1} = 0$, $UL^t_{ij1} = 210$, and $t_{ij1} = 10$; interval 2 corresponds to 11 a.m., 5 p.m., we have $LL^t_{ij2} = 211$, $UL^t_{ij2} = 570$, and $t_{ij2} = 8$; interval 3 corresponds to 5 p.m., 9 p.m., we have $LL^t_{ij3} = 571$, $UL^t_{ij3} = 810$, and $t_{ij3} = 9$; and interval 4 corresponds to 9 p.m., 7.30 a.m., we have $LL^t_{ij4} = 811$, $UL^t_{ij4} = 1440$, and $t_{ij4} = 7$ */
LL^o_{ijk}	Lower limit for the allowed traffic regulation during the time interval k with respect to road (i,j), to define the time-dependent one-way traffic time/regulation
UL^o_{ijk}	Upper limit for the allowed traffic regulation during the time interval k with respect to road (i,j), to define the time-dependent one-way traffic time/regulation /*note: For a given road (i,j) that has no traffic for interval between 7.30 a.m. and 11 a.m., two-way traffic between 11 a.m. and 5 p.m., one-way traffic between 5 p.m. and 9 p.m., and two-way traffic between 9 p.m. and 7.30 a.m., we have $\Omega^o_{ij} = 4$; $LL^o_{ij1} = 0$ (i.e., 7.30 a.m.), $UL^o_{ij1} = 210$ (i.e., 11 a.m.), and set $\Delta^o_{ij1} = 0$; we have $LL^o_{ij2} = 211$, $UL^o_{ij2} = 570$, and $\Delta^o_{ij2} \in \{0,1\}$; we have $LL^o_{ij3} = 571$, $UL^o_{ij3} = 810$, and set $\Delta^o_{ij3} = 0$; and we have $LL^o_{ij4} = 811$, $UL^o_{ij4} = 1440$, and $\Delta^o_{ij4} \in \{0,1\}$ */
$J(i)$	Set of nodes or junctions to which there exists a direct connectivity from/to node i
M	A large value; set to 10,000 in this study

Decision Variables	Description
Y_{ij}	A binary variable that takes a value 1 if the road (i,j) is chosen in the travel route; 0 otherwise /* note: If there exists no direct connectivity between nodes/junctions i and j, then we set $d_{ij} = \infty$ and/or set $Y_{ij} = 0$ */
δ_i	An indicator (binary variable) that takes value 1 if node i is visited in the travel route; 0 otherwise
Δ^w_{ik}	An indicator (binary variable) that takes value 1 if node i is reached during the interval k in the travel route; 0 otherwise
Δ^t_{ijk}	An indicator (binary variable) that takes value 1 if road (i,j) is traversed during the interval k in the travel route; 0 otherwise
Δ^o_{ijk}	An indicator (binary) that takes value 1 if the one-way is allowed during the interval k in the travel route; 0 otherwise
ω_i	Waiting time at node i that takes a value if $\delta_i = 1$. /*note: $\omega_{n'} = \omega_{n''} = 0$ */
A_i	Arrival time at node i that takes a value if $\delta_i = 1$./*note: $A_{n'}$ is given as input*/

(Continued)

Decision Variables	Description
τ_{ij}	Travel time along the road (i,j) that takes a value if $Y_{ij} = 1$.
A_{ik}^{w}	Arrival time at node i in the interval k, and it takes a value if $\delta_i = 1$ and other A_{ik}^{w}'s are equal to 0
A_{ijk}^{t}	Arrival time on the road (i,j) in the interval k, and it takes a value if $Y_{ij} = 1$, and other A_{ijk}^{t}'s are equal to 0
A_{ijk}^{o}	Arrival time on the road (i,j) (with the possible one-way traffic regulation) in the interval k, and it takes a value if $Y_{ij} = 1$, and other A_{ijk}^{o}'s are equal to 0

9.4.2 Mathematical Model

Objective function:

$$\text{Minimize}\, Z_1 = dist = \sum_{i=1}^{n} \sum_{j \in J(i)} d_{ij} Y_{ij}. \tag{9.1}$$

$$\text{Minimize}\, Z_2 = time = \sum_{i=1}^{n} \sum_{j \in J(i)} \tau_{ij} + \sum_{i=1}^{n} \omega_i. \tag{9.2}$$

subject to the following:
/* Constraints (9.3) through (9.7) represent the conditions for at most one inflow and one outflow */

$$\sum_{j \in J(i)} Y_{ij} \le 1 \qquad \forall\, i = 1,\dots,n,\, i \ne n' \text{ and } i \ne n''. \tag{9.3}$$

$$\sum_{j \in J(i)} Y_{ji} \le 1 \qquad \forall\, i = 1,\dots,n,\, i \ne n' \text{ and } i \ne n''. \tag{9.4}$$

$$\sum_{j \in J(i)} Y_{ij} = \sum_{j \in J(i)} Y_{ji} \qquad \forall\, i = 1,\dots,n,\, i \ne n' \text{ and } i \ne n''. \tag{9.5}$$

$$\sum_{j \in J(i)} Y_{ji} = \delta_i \qquad \forall\, i = 1,\dots,n,\, i \ne n' \text{ and } i \ne n''. \tag{9.6}$$

$$Y_{ij} + Y_{ji} \le 1 \qquad \forall\, i = 1,\dots,n, i \ne n',\, i \ne n'', \text{ and } \forall j \in J(i). \tag{9.7}$$

/* Constraints (9.8) through (9.12) ensure that there is only one outflow and no inflow for the source node n', and only one inflow and no outflow for the destination node n'' */

$$\sum_{j \in J(n')} Y_{n'j} = 1. \tag{9.8}$$

$$\sum_{j \in J(n'')} Y_{jn''} = 1. \tag{9.9}$$

$$\sum_{j \in J(n')} Y_{jn'} = 0. \tag{9.10}$$

$$\sum_{j \in J(n'')} Y_{n''j} = 0. \tag{9.11}$$

$$\delta_{n'} = \delta_{n''} = 1. \tag{9.12}$$

/* Constraints (9.13) through (9.21) capture the waiting time that is dependent on the arrival time at node i with respect to corresponding interval Δ_{ik}^w */

$$\omega_{n'} = \omega_{n''} = 0. \tag{9.13}$$

$$A_{n'} = arrival\_time \text{ /*given as an input to the model*/.} \tag{9.14}$$

$$A_i \leq M\delta_i \qquad \forall i = 1,\ldots,n. \tag{9.15}$$

$$A_j \leq A_i + \omega_i + \tau_{ij} + M(1 - Y_{ij}) \qquad \forall i = 1,\ldots,n, i \neq n'', \text{ and } \forall j \in J(i). \tag{9.16}$$

$$A_j \geq A_i + \omega_i + \tau_{ij} - M(1 - Y_{ij}) \qquad \forall i = 1,\ldots,n, i \neq n'', \text{ and } \forall j \in J(i). \tag{9.17}$$

$$LL_{ik}^w \Delta_{ik}^w \leq A_{ik}^w \leq UL_{ik}^w \Delta_{ik}^w \qquad \forall i = 1,\ldots,n, i \neq n' \text{ and } i \neq n'', \text{ and } k = 1,2,\ldots,\nabla_i^w. \tag{9.18}$$

$$\sum_{k=1}^{\nabla_i^w} \Delta_{ik}^w = \delta_i \qquad \forall i = 1,\ldots,n, i \neq n', \text{ and } i \neq n''. \tag{9.19}$$

$$\sum_{k=1}^{\nabla_i^w} A_{ik}^w = A_i \qquad \forall i = 1,\ldots,n,\, i \neq n',\, \text{and } i \neq n''. \tag{9.20}$$

$$\omega_i = \sum_{k=1}^{\nabla_i^w} (\Delta_{ik}^w W_{ik}) \qquad \forall i = 1,\ldots,n,\, i \neq n',\, \text{and } i \neq n''. \tag{9.21}$$

/* Constraints (9.22) through (9.29) capture the arrival time and travel time along the road (i,j) based on the departure time at node i, that is, the actual arrival time at the road (i,j), defined with respect to φ_{ij}^t time intervals*/

$$LL_{ijk}^t \Delta_{ijk}^t \leq A_{ijk}^t \leq UL_{ijk}^t \Delta_{ijk}^t \qquad \forall i = 1,\ldots,n,\, i \neq n'',\, \forall j \in J(i),\, \text{and } k = 1,\ldots,\varphi_{ij}^t. \tag{9.22}$$

$$\sum_{k=1}^{\varphi_{ij}^t} \Delta_{ijk}^t = Y_{ij} \qquad \forall i = 1,\ldots,n,\, i \neq n'',\, \text{and } \forall j \in J(i). \tag{9.23}$$

$$\sum_{k=1}^{\varphi_{ij}^t} A_{ijk}^t \leq A_i + \omega_i + M(1 - Y_{ij}) \qquad \forall i = 1,\ldots,n,\, i \neq n'',\, \text{and } \forall j \in J(i). \tag{9.24}$$

$$\sum_{k=1}^{\varphi_{ij}^t} A_{ijk}^t \geq A_i + \omega_i - M(1 - Y_{ij}) \qquad \forall i = 1,\ldots,n,\, i \neq n'',\, \text{and } \forall j \in J(i). \tag{9.25}$$

$$\sum_{k=1}^{\varphi_{ij}^t} A_{ijk}^t \leq MY_{ij} \qquad \forall i = 1,\ldots,n,\, i \neq n'',\, \text{and } \forall j \in J(i). \tag{9.26}$$

$$\tau_{ij} \leq MY_{ij} \qquad \forall i = 1,\ldots,n,\, \text{and } \forall j \in J(i). \tag{9.27}$$

$$\tau_{ij} \leq \sum_{k=1}^{\varphi_{ij}^t} \left(\Delta_{ijk}^t t_{ijk} \right) + M(1 - Y_{ij}) \qquad \forall i = 1,\ldots,n,\, i \neq n'',\, \text{and } \forall j \in J(i). \tag{9.28}$$

$$\tau_{ij} \geq \sum_{k=1}^{\varphi_{ij}^t} \left(\Delta_{ijk}^t t_{ijk} \right) - M(1 - Y_{ij}) \qquad \forall i = 1,\ldots,n,\, i \neq n'',\, \text{and } \forall j \in J(i). \tag{9.29}$$

/* Constraints (9.30) through (9.34) ensure the travel along the road (i,j) only during the allowed traffic-time intervals when we arrive at node i. The mathematical model should inherently avoid the entry into the roads having one-way traffic regulation in a particular interval (Section 9.4.1 for the settings of Δ_{ijk}^{o}).*/

$$LL_{ijk}^{o}\,\Delta_{ijk}^{o} \leq A_{ijk}^{o} \leq UL_{ijk}^{o}\,\Delta_{ijk}^{o} \qquad \forall i=1,\dots,n, i\neq n'', \forall j\in J(i), \text{ and } k=1,\dots,\Omega_{ij}^{o}. \tag{9.30}$$

$$\sum_{k=1}^{\Omega_{ij}^{o}}\Delta_{ijk}^{o} = Y_{ij} \qquad \forall i=1,\dots,n, i\neq n'', \text{ and } \forall j\in J(i). \tag{9.31}$$

$$\sum_{k=1}^{\Omega_{ij}^{o}}A_{ijk}^{o} \leq MY_{ij} \qquad \forall i=1,\dots,n, i\neq n'', \text{ and } \forall j\in J(i). \tag{9.32}$$

$$\sum_{k=1}^{\Omega_{ij}^{o}}A_{ijk}^{o} \leq A_i+\omega_i+M(1-Y_{ij}) \qquad \forall i=1,\dots,n, i\neq n'', \text{ and } \forall j\in J(i). \tag{9.33}$$

$$\sum_{k=1}^{\Omega_{ij}^{o}}A_{ijk}^{o} \geq A_i+\omega_i-M(1-Y_{ij}) \qquad \forall i=1,\dots,n, i\neq n'', \text{ and } \forall j\in J(i). \tag{9.34}$$

$Y_{ij}, \delta_i, \Delta_{ik}^{w}, \Delta_{ijk}^{t}$, and Δ_{ijk}^{o} are binary variables, and all other variables are nonnegative.

/* note: Ω_{ij}^{o} is the number of one-way and two-way traffic intervals operational along the road (i,j). If traffic is not allowed along the road (i,j) during interval k, then set $\Delta_{ijk}^{o}=0$. For example, one-way traffic along the road (13,11) in Figure 9.4 is being operational during interval 3, whereas (11,13) traffic is not allowed. Then we set $\Delta_{11,13,3}^{o}=0$ for the road (11,13) initially itself in the model*/

Equation 9.1 is the objective function to minimize the total distance, and Equation 9.2 is the objective function to minimize the total travel time. Constraint (9.3) ensures a maximum of one outflow from the node i, except the source node and the destination node. Constraint (9.4) ensures a maximum of one inflow to node i, except the source node and the destination node. Constraint (9.5) assures that the size of inflow equals the size of outflow at all nodes, except source node and destination node. Constraint (9.6) ensures every node in the path has an inflow. On the selection of an arc, Constraint (9.7) ensures unidirectional flow in the arc. Constraints (9.8) and (9.10) make sure that there is only outflow and no inflow at the source node. Constraints (9.9) and (9.11) ensure that there is only inflow and no outflow at the destination node. Constraint (9.12) ensures the selection of the source and destination nodes as part of the path. Constraints (9.15) through (9.17) determine the arrival time at node j from node i. Constraints (9.18) through (9.21) capture the time-dependent dynamic waiting time on the basis of the arrival time interval k at node i. Constraints (9.22) through (9.29) capture the time-dependent dynamic travel time along the arc (i,j) on the basis of the departure time from node i. Constraints (9.30) through (9.34) ensure that the travel takes place during the allowed traffic periods along road (i,j) during the day.

9.4.3 Multiobjective Optimization Algorithm to Obtain the Strictly Nondominated Solutions

Scalarization method or weighted sum method, ε-constraint method, goal programming, and multilevel programming are in use to solve multiobjective optimization problems. In this study, we use ε-constraint method or ε-approach to solve the bicriteria MILP model. In this method, the decision maker has to choose an objective out of two objectives, while the remainder of the objectives is constrained to satisfy the given target value (defined by ε_t). Below is the step-by-step procedure of the ε-constraint method applied to multiobjective routing in the metropolitan city (also see Tiwari, 2016 for a related work).

9.4.3.1 Terminology

Variables	Description
iter	Iteration number
$dist^1_{iter}$	Distance obtained in the iteration *iter*
$time^1_{iter}$	Minimum travel time obtained in the iteration *iter*, corresponding to $dist^1_{iter}$.
$time^*$	Optimal travel time
$\{\delta el^{iter}_{ij}\}$	Solution from the MILP, associated with $dist^1_{iter}$ and $time^1_{iter}$
γ^1_{iter}	Solution set with $\{\delta el^{iter}_{ij}\}$, $dist^1_{iter}$, and $time^1_{iter}$

/* note: Rest of the terminologies are introduced in Section 9.4.1*/

9.4.3.2 Step-by-Step Procedure to Generate a Set of Nondominated (Pareto-Optimal) Solutions, Given ε_t

/* Do Step 1 to get a solution with the minimum total distance and the corresponding minimum total travel time; also this step gives us in Step 1.4 the optimal total travel time */

Step 1: Set $iter = 1$.

Step 1.1: Execute the following MILP, called MILP-1:
Minimize

$$Z_1 = dist_{iter}^1 = \sum_{i=1}^{n} \sum_{j \in J(i)} d_{ij} Y_{ij}$$

with the MILP given in Section 9.4.2.
Step 1.2: Execute the following MILP, called MILP-2:
Minimize

$$Z_2 = time_{iter}^1 = \sum_{i=1}^{n} \sum_{j \in J(i)} \tau_{ij} + \sum_{i=1}^{n} \omega_i$$

and subject to all constraints in the MILP given in Section 9.4.2, and
with the following add-on constraint:

$$\sum_{i=1}^{n} \sum_{j \in J(i)} d_{ij} Y_{ij} = dist_{iter}^1. \qquad (9.35)$$

Denote the resultant solution for this MILP as $\left\{ \delta el_{ij}^{iter} \right\}$.

Step 1.3: Set $\gamma_{iter}^1 = \left\{ \delta el_{ij}^{iter} \right\}$, associated with $dist_{iter}^1$, $time_{iter}^1$, and their
respective Y_{ij}'s.

/* note: We have the Pareto-optimal solution with minimum total distance
and corresponding minimum total travel time */

Step 1.4: Execute the MILP given in Section 9.4.2 with the following
objective function:
Minimize

$$Z_2^* = time^* = \sum_{i=1}^{n} \sum_{j \in J(i)} \tau_{ij} + \sum_{i=1}^{n} \omega_i$$

subject to all constraints given in the MILP presented in Section 9.4.2.

Step 2:
/* Do this step to get further Pareto-optimal solutions, given ε_t (with
respect to time decrement)*/

/* note: Skip Step 2 if $time_1^1 = time^*$ */

Step 2.1: Set $iter = iter + 1$;

With respect to the original MILP given in Section 9.4.2, do the following:

Set:

Minimize

$$Z_1 = \sum_{i=1}^{n} \sum_{j \in J(i)} d_{ij} Y_{ij} = dist_{iter}^1$$

subject to all constraints given in the MILP presented in Section 9.4.2, and with the following additional constraint:

$$\text{if} \left(time_{iter-1}^1 - \varepsilon_t > time^* \right)$$

then

$$\text{add}: \sum_{i=1}^{n} \sum_{j \in J(i)} \tau_{ij} + \sum_{i=1}^{n} \omega_i \leq time_{iter-1}^1 - \varepsilon_t \qquad (9.36)$$

else

$$\text{add}: \sum_{i=1}^{n} \sum_{j \in J(i)} \tau_{ij} + \sum_{i=1}^{n} \omega_i = time^* \qquad (9.37)$$

and execute the resultant MILP.

Step 2.2: Execute the following MILP:

Minimize

$$Z_2 = time_{iter}^1 = \sum_{i=1}^{n} \sum_{j \in J(i)} \tau_{ij} + \sum_{i=1}^{n} \omega_i$$

and subject to all constraints in the MILP given in Section 9.4.2, and with the following additional constraint

$$\sum_{i=1}^{n} \sum_{j \in J(i)} d_{ij} Y_{ij} = dist_{iter}^1 . \qquad (9.38)$$

Denote the solution from the above MILP as $\left\{\delta el_{ij}^{iter}\right\}$.

Step 2.3: Set $\gamma_{iter}^1 = \left\{\delta el_{ij}^{iter}\right\}$, associated with $dist_{iter}^1$, $time_{iter}^1$, and their respective Y_{ij}'s.

Step 3: If

$$time_{iter}^1 = time^*$$

then proceed to Step 4,

else return to Step 2.

Step 4: STOP: the set of strictly Pareto-optimal solutions is obtained, denoted by $\left\{\gamma_{iter}^1, \forall iter\right\}$ with the corresponding $dist_{iter}^1$, $time_{iter}^1$, and their respective Y_{ij}'s, for the given ε_t.

9.5 Experimental Settings, Results, and Discussion

We now present the application of the proposed mathematical model to find the shortest path for the Chennai city network with the objectives of minimizing the total travel time and total distance travelled.

9.5.1 Data with Respect to Distance and Travel Times along the Roads and Waiting Times at Junctions

The data used in this study consists of two sets: Chennai city network data and a sample network topology of Chennai city network. City network data is obtained from the Centre of Excellence in Urban Transport at the Indian Institute of Technology (IIT) Madras. Data set includes the map of Chennai city with parameters—from node ID., to node ID., the corresponding distance and free-flow travel time (FFT) along the road. To illustrate the working of the mathematical model, we consider a sample network topology of 33 nodes as shown in Figure 9.1. We use this illustrative example to show the solutions with respect to different traffic conditions. In this example, node 1 is the source node and node 33 is the destination node. The distance and the FFT are given in Tables 9.1 and 9.2, respectively. Data for the other traffic levels (low-traffic and high-traffic travel times) are generated from FFT. For the purpose of illustrating the working of the proposed MILP model, we consider four travel-time intervals during a day: high-traffic interval 1 (7.30 a.m. to 11 a.m.), free-flow traffic or normal-traffic interval 2 (11 a.m. to 5 p.m.), high-traffic interval 3 (5 p.m. to

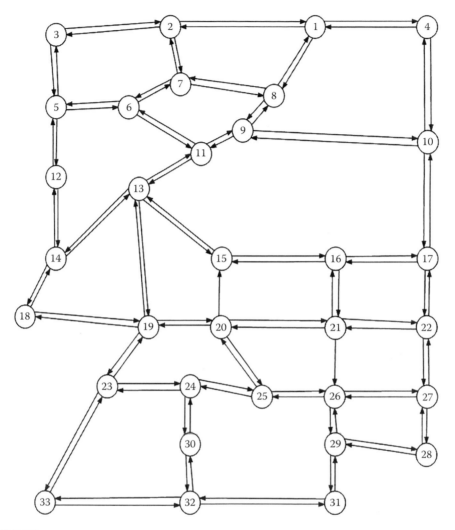

FIGURE 9.1
Chennai road network topology: a sample (not drawn to scale). Note: origin and destination nodes are denoted by node 1 and node 33, respectively.

9 p.m.), and low-traffic interval 4 (9 p.m. to 7.30 a.m.) in this study. We use 0.6 (i.e., 0.6 × FFT) as the multiplication factor for the low traffic and 1.8 (i.e., 1.8 × FFT) as the multiplication factor for the high traffic. The waiting times with respect to normal-traffic interval at nodes are tabulated in Table 9.3, and for other traffic levels, waiting times are generated using these multiplication factors. Details of one-way traffic are shown in Table 9.4. Interested readers can obtain the time data from the authors at the Indian Institute of Technology Madras.

TABLE 9.1

Distance between Nodes (in km): Chennai Road Network Topology (Sample)

To/From	1	2	3	4	5	6	7	8	9	10	11	12	13	14	15	16	17	18	19	20	21	22	23	24	25	26	27	28	29	30	31	32	33
1	–	1.6	–	–	–	–	–	1.6	–	–	–	–	–	–	–	–	–	–	–	–	–	–	–	–	–	–	–	–	–	–	–	–	–
2	1.7	–	0.08	0.28	–	–	–	–	–	–	–	–	–	–	–	–	–	–	–	–	–	–	–	–	–	–	–	–	–	–	–	–	–
3	–	0.1	–	0.24	–	–	–	–	–	–	–	–	–	–	–	–	–	–	–	–	–	–	–	–	–	–	–	–	–	–	–	–	–
4	4.3	–	0.85	–	–	–	–	–	–	0.3	–	–	–	–	–	–	–	–	–	–	–	–	–	–	–	–	–	–	–	–	–	–	–
5	–	–	–	–	–	0.39	0.25	–	–	–	–	0.62	–	–	–	–	–	–	–	–	–	–	–	–	–	–	–	–	–	–	–	–	–
6	–	0.26	–	–	–	–	0.26	–	–	–	0.14	–	–	–	–	–	–	–	–	–	–	–	–	–	–	–	–	–	–	–	–	–	–
7	–	–	–	–	–	0.09	–	0.09	0.11	–	–	–	–	–	–	–	–	–	–	–	–	–	–	–	–	–	–	–	–	–	–	–	–
8	1.7	–	–	–	–	–	0.26	–	0.26	–	–	–	–	–	–	–	–	–	–	–	–	–	–	–	–	–	–	–	–	–	–	–	–
9	–	–	–	–	–	–	–	0.26	–	0.28	0.39	–	–	–	–	–	–	–	–	–	–	–	–	–	–	–	–	–	–	–	–	–	–
10	–	–	–	2	–	0.14	–	–	0.28	–	–	–	–	–	–	–	4.3	–	–	–	–	–	–	–	–	–	–	–	–	–	–	–	–
11	–	–	–	–	0.85	–	–	–	0.25	–	–	–	0.91	–	–	–	–	–	–	–	–	–	–	–	–	–	–	–	–	–	–	–	–
12	–	–	–	–	–	–	–	–	–	–	–	–	–	0.62	–	–	–	–	–	–	–	–	–	–	–	–	–	–	–	–	–	–	–
13	–	–	–	–	–	–	–	–	–	–	1.9	–	–	–	1.1	–	–	0.68	0.68	–	–	–	–	–	–	–	–	–	–	–	–	–	–
14	–	–	–	–	–	–	–	–	–	–	–	2.2	–	–	–	4.8	–	0.49	–	–	–	–	–	–	–	–	–	–	–	–	–	–	–
15	–	–	–	–	–	–	–	–	–	–	–	–	4.3	–	–	3.7	1.1	–	–	–	–	–	–	–	–	–	–	–	–	–	–	–	–
16	–	–	–	–	–	–	–	–	–	–	–	–	–	–	3.7	–	4.4	–	–	–	3.7	–	–	–	–	–	–	–	–	–	–	–	–
17	–	–	–	–	–	–	–	–	–	4.3	–	–	–	–	–	4.4	–	–	–	–	–	4.4	–	–	–	–	–	–	–	–	–	–	–
18	–	–	–	–	–	–	–	–	–	–	–	–	4.8	0.73	–	–	–	0.16	0.96	0.55	–	–	–	–	–	–	–	–	–	–	–	–	–
19	–	–	–	–	–	–	–	–	–	–	–	–	1.6	–	6	–	–	–	–	–	0.91	–	1.9	–	–	–	–	–	–	–	–	–	–
20	–	–	–	–	–	–	–	–	–	–	–	–	–	–	–	–	–	–	1.9	–	–	–	–	–	0.11	–	–	–	–	–	–	–	–

(Continued)

TABLE 9.1 (Continued)

Distance between Nodes (in km): Chennai Road Network Topology (Sample)

To/From	1	2	3	4	5	6	7	8	9	10	11	12	13	14	15	16	17	18	19	20	21	22	23	24	25	26	27	28	29	30	31	32	33
21	–	–	–	–	–	–	–	–	–	–	–	–	–	–	–	6	–	–	–	0.95	–	2.4	–	–	–	2.4	–	–	–	–	–	–	–
22	–	–	–	–	–	–	–	–	–	–	–	–	–	–	–	–	2	–	–	–	1.6	–	–	–	–	–	0.32	–	–	–	–	–	–
23	–	–	–	–	–	–	–	–	–	–	–	–	–	–	–	–	–	–	0.96	–	–	–	–	0.73	–	–	–	–	–	–	–	–	0.69
24	–	–	–	–	–	–	–	–	–	–	–	–	–	–	–	–	–	–	–	–	–	–	0.21	–	0.49	–	–	–	–	0.69	–	–	–
25	–	–	–	–	–	–	–	–	–	–	–	–	–	–	–	–	–	–	–	0.55	–	–	–	0.17	–	0.31	–	–	–	–	–	–	–
26	–	–	–	–	–	–	–	–	–	–	–	–	–	–	–	–	–	–	–	–	–	–	–	–	0.35	–	0.67	–	0.88	–	–	–	–
27	–	–	–	–	–	–	–	–	–	–	–	–	–	–	–	–	–	–	–	–	–	–	1.5	–	–	1.5	–	1.5	–	–	–	–	–
28	–	–	–	–	–	–	–	–	–	–	–	–	–	–	–	–	–	–	–	–	–	1.5	–	–	–	–	1.5	–	0.62	–	–	–	–
29	–	–	–	–	–	–	–	–	–	–	–	–	–	–	–	–	–	–	–	–	–	–	–	–	–	1.5	–	2.2	–	–	0.95	–	–
30	–	–	–	–	–	–	–	–	–	–	–	–	–	–	–	–	–	–	–	–	–	–	–	0.88	–	–	–	–	–	–	–	0.21	–
31	–	–	–	–	–	–	–	–	–	–	–	–	–	–	–	–	–	–	–	–	–	–	–	–	–	–	–	–	0.35	–	–	0.67	–
32	–	–	–	–	–	–	–	–	–	–	–	–	–	–	–	–	–	–	–	–	–	–	–	–	–	–	–	–	–	0.31	0.3	–	3.3
33	–	–	–	–	–	–	–	–	–	–	–	–	–	–	–	–	–	–	–	–	–	–	1.9	–	–	–	–	–	–	–	–	3.3	–

TABLE 9.2

Free-Flow Travel Time (FFT in Minutes): Chennai Road Network Topology (Sample)

To/From	1	2	3	4	5	6	7	8	9	10	11	12	13	14	15	16	17	18	19	20	21	22	23	24	25	26	27	28	29	30	31	32	33
1	–	4.364	–	0.764	–	–	–	4.364	–	–	–	–	–	–	–	–	–	–	–	–	–	–	–	–	–	–	–	–	–	–	–	–	–
2	4.636	–	0.184	–	–	–	0.764	–	–	–	–	–	–	–	–	–	–	–	–	–	–	–	–	–	–	–	–	–	–	–	–	–	–
3	–	0.218	–	–	0.542	–	–	–	–	–	–	–	–	–	–	–	–	–	–	–	–	–	–	–	–	–	–	–	–	–	–	–	–
4	11.727	–	–	–	–	–	–	–	–	0.818	–	–	–	–	–	–	–	–	–	–	–	–	–	–	–	–	–	–	–	–	–	–	–
5	–	–	1.894	–	–	0.682	–	–	–	–	–	1.378	–	–	–	–	–	–	–	–	–	–	–	–	–	–	–	–	–	–	–	–	–
6	–	–	–	–	1.064	–	0.709	–	–	–	0.382	–	–	–	–	–	–	–	–	–	–	–	–	–	–	–	–	–	–	–	–	–	–
7	–	0.709	–	–	–	0.245	–	0.245	–	–	–	–	–	–	–	–	–	–	–	–	–	–	–	–	–	–	–	–	–	–	–	–	–
8	4.636	–	–	–	–	–	0.709	–	0.3	–	–	–	–	–	–	–	–	–	–	–	–	–	–	–	–	–	–	–	–	–	–	–	–
9	–	–	–	–	–	–	–	0.709	–	0.764	1.064	–	–	–	–	–	–	–	–	–	–	–	–	–	–	–	–	–	–	–	–	–	–
10	–	–	–	5.455	–	–	–	–	0.764	–	–	–	–	–	–	–	11.727	–	–	–	–	–	–	–	–	–	–	–	–	–	–	–	–
11	–	–	–	–	–	0.382	–	–	0.682	–	–	–	2.482	–	–	–	–	–	–	–	–	–	–	–	–	–	–	–	–	–	–	–	–
12	–	–	–	–	1.894	–	–	–	–	–	–	–	–	1.378	–	–	–	–	–	–	–	–	–	–	–	–	–	–	–	–	–	–	–
13	–	–	–	–	–	–	–	–	–	–	5.182	–	–	–	3	–	–	1.855	1.855	–	–	–	–	–	–	–	–	–	–	–	–	–	–
14	–	–	–	–	–	–	–	–	–	–	–	4.889	–	–	3	–	–	1.336	–	–	–	–	–	–	–	–	–	–	–	–	–	–	–
15	–	–	–	–	–	–	–	–	–	–	–	–	11.727	–	–	13.091	–	–	–	–	–	–	–	–	–	–	–	–	–	–	–	–	–
16	–	–	–	–	–	–	–	–	–	–	–	–	–	–	10.091	3	–	–	–	–	10.091	–	–	–	–	–	–	–	–	–	–	–	–
17	–	–	–	–	–	–	–	–	–	11.727	–	–	–	–	–	12	–	–	–	–	–	12	–	–	–	–	–	–	–	–	–	–	–
18	–	–	–	–	–	–	–	–	–	–	–	–	13.091	1.991	–	–	–	–	2.618	–	–	–	–	–	–	–	–	–	–	–	–	–	–
19	–	–	–	–	–	–	–	–	–	–	–	4.364	–	–	–	–	–	0.436	1.5	–	–	–	5.182	–	–	–	–	–	–	–	–	–	–

(Continued)

TABLE 9.2 (Continued)

Free-Flow Travel Time (FFT in Minutes): Chennai Road Network Topology (Sample)

To/From	1	2	3	4	5	6	7	8	9	10	11	12	13	14	15	16	17	18	19	20	21	22	23	24	25	26	27	28	29	30	31	32	33
20	–	–	–	–	–	–	–	–	–	–	–	–	–	–	16.364	–	–	–	5.182	–	2.482	–	–	–	–	–	–	–	–	–	–	–	–
21	–	–	–	–	–	–	–	–	–	–	–	–	–	–	–	16.364	–	–	–	2.591	–	6.545	–	–	0.3	6.545	–	–	–	–	–	–	–
22	–	–	–	–	–	–	–	–	–	–	–	–	–	–	–	–	5.455	–	–	–	4.364	–	–	–	–	–	–	–	–	–	–	–	–
23	–	–	–	–	–	–	–	–	–	–	–	–	–	–	–	–	–	–	2.618	–	–	–	–	1.991	–	–	0.873	–	–	–	–	–	–
24	–	–	–	–	–	–	–	–	–	–	–	–	–	–	–	–	–	–	–	–	–	–	0.573	–	1.336	–	–	–	–	–	–	–	1.882
25	–	–	–	–	–	–	–	–	–	–	–	–	–	–	–	–	–	–	–	1.5	–	–	–	0.464	–	0.845	–	–	–	1.882	–	–	–
26	–	–	–	–	–	–	–	–	–	–	–	–	–	–	–	–	–	–	–	–	–	–	–	–	0.955	–	1.827	–	2.4	–	–	–	–
27	–	–	–	–	–	–	–	–	–	–	–	–	–	–	–	–	–	–	–	–	–	4.091	–	–	–	4.091	–	4.091	–	–	–	–	–
28	–	–	–	–	–	–	–	–	–	–	–	–	–	–	–	–	–	–	–	–	–	–	–	–	–	–	0.873	–	1.378	–	–	–	–
29	–	–	–	–	–	–	–	–	–	–	–	–	–	–	–	–	–	–	–	–	–	–	–	–	–	4.091	–	4.889	–	–	2.591	–	–
30	–	–	–	–	–	–	–	–	–	–	–	–	–	–	–	–	–	–	–	–	–	–	–	2.4	–	–	–	–	–	–	–	–	–
31	–	–	–	–	–	–	–	–	–	–	–	–	–	–	–	–	–	–	–	–	–	–	–	–	–	–	–	–	0.955	–	–	0.573	–
32	–	–	–	–	–	–	–	–	–	–	–	–	–	–	–	–	–	–	–	–	–	–	–	–	–	–	–	–	–	0.845	0.818	–	9
33	–	–	–	–	–	–	–	–	–	–	–	–	–	–	–	–	–	–	–	–	–	–	5.182	–	–	–	–	–	–	–	1.827	9	–

TABLE 9.3

Waiting Time (at Nodes/Traffic Junctions) of Chennai Road Network
Topology

Node *i*	1	2	3	4	5	6	7	8	9
Waiting Time (min)	1.700	1.700	1.700	1.700	1.700	1.020	1.700	1.700	1.700

Node *i*	10	11	12	13	14	15	16	17	18
Waiting Time (min)	1.700	1.700	1.700	2.380	2.380	2.380	1.020	1.700	1.700

Node *i*	19	20	21	22	23	24	25	26	27
Waiting Time (min)	2.380	2.380	1.700	1.020	1.700	1.700	1.700	1.020	1.700

Node *i*	28	29	30	31	32	33
Waiting Time (min)	1.020	1.700	1.700	1.020	1.700	1.020

9.5.2 Sample Network Topology

We consider the following scenarios.

- *Scenario 1:* In this scenario, the entire travel is completed in one single traffic condition. For example, consider a traveler who wants to travel from node 1 to node 33 under normal traffic conditions. Suppose the traveler starts from node 1 at 4 p.m.; then the estimated arrival time at the destination and the total distance traversed, based on the optimization model, are 4.31 p.m. and 4.57 km (Table 9.5). Figure 9.2 gives an optimal route (1–4–10–9–11–13–19–20–25–24–23–33) with the minimum total distance and the corresponding total time taken.

TABLE 9.4

One-Way Traffic along the Road (*i,j*) of Chennai Road Network Topology

Road (*i,j*)	One-Way Traffic along the Road (*i,j*) (with No Traffic Allowed along the Road (*j,i*)) during the Period(s)
(2,3)	(7.30 a.m. to 11 a.m.); (11 a.m. to 5 p.m.); (5 p.m. to 9 p.m.).
(11,13)	(7.30 a.m. to 11 a.m.); (5 p.m. to 9 p.m.).
(17,22)	(7.30 a.m. to 11 a.m.); (5 p.m. to 9 p.m.).
(31,32)	(7.30 a.m. to 11 a.m.); (5 p.m. to 9 p.m.).

Legend: Road (*i,j*) is blocked in the particular traffic interval.

TABLE 9.5

Optimal Solution with Respect to Minimum Total Distance, and the Corresponding Travel Times, and Waiting Times; Travel Starts from Node 1 at 4 p.m.

Road (i,j)	Distance along the Road (km)	Time along the Road (min)	Waiting Time at the Junction (min)
(1,4)	0.28	0.764	1.020
(4,10)	0.30	0.818	1.700
(10,9)	0.28	0.764	1.700
(9,11)	0.39	1.064	1.700
(11,13)	0.91	2.482	2.380
(13,19)	0.68	1.855	2.380
(19,20)	0.55	1.500	2.380
(20,25)	0.11	0.300	1.700
(25,24)	0.17	0.464	1.700
(24,23)	0.21	0.573	1.700
(23,33)	0.69	1.882	0.000
Total distance = 4.57 km		**Total travel time = 30.826 min**	

- *Scenario 2:* In this scenario, let us assume that there is a change in traffic condition (say normal to heavy) during the travel from the source to the destination. Suppose a traveler decides to start the travel at 4.50 p.m. instead of 4 p.m., then based on our assumptions, the travel is in the normal level from 4.50 p.m. to 5 p.m. and at 5 p.m., it changes to heavy traffic. From Table 9.6, we see that the traffic condition changes from normal to heavy at node 11 resulting in a total travel time of 45.865 min (compared with 30.826 min in scenario 1) for the same route. Figure 9.3 displays the given route starting from node 1 at 4.50 p.m. and ending in node 33 at 5.36 p.m. It also displays an optimal route (1–4–10–9–11–13–19–20–25–24–23–33) with the minimum total distance and the corresponding total time taken.

- *Scenario 3:* In this scenario, we assume that there is one-way traffic in addition to changeover of traffic condition in the middle of the travel. Suppose the traveler encounters the unavailability of the arc from node 11 to node 13 due to one-way traffic, especially in high-traffic condition at 5 p.m. Here, our model considers that node 11 to node 13 is unavailable during this interval and finds the optimal route as (1–4–10–9–11–6–5–12–14–18–19–20–25–24–23–33) with a total travel time of 65.744 min and distance 6.20 km (Table 9.7). Figure 9.4 displays the given route starting from node 1 at 4.50 p.m. It also displays an optimal route with the minimum total distance and the corresponding total time taken.

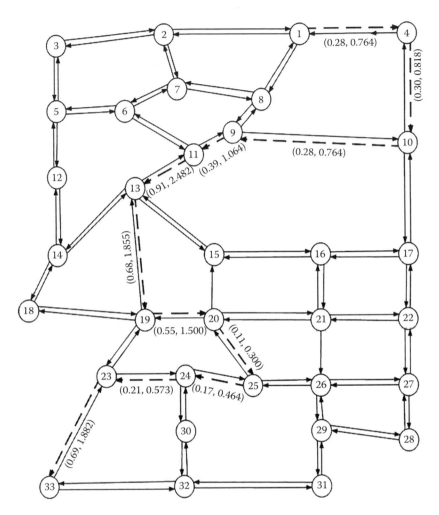

FIGURE 9.2
Travel starts from node 1 at 4 p.m. without changeover (with respect to traffic conditions) of travel time and waiting time (total distance = 4.57 km and total time covered = 30.826 min). *Note:* optimal route (with respect to minimum total distance and the corresponding total time covered/taken): {1–4–10–9–11–13–19–20–25–24–23–33}. Legend (distance, travel time): for example (0.28, 0.764) in the above figure corresponds to distance and travel time from node 1 to node 4, and rest of the values follow suit.

9.6 Implementation of the Proposed Multiobjective Model to the Complete Chennai Network

We implement the mathematical model for the shortest path problem with multiobjectives for the Chennai city network. The high-traffic travel time

TABLE 9.6

Travel Starts from Node 1 at 4.50 p.m.; Traffic Level Changes from 5 p.m. Onwards

Road (i,j)	Distance along the Road (km)	Time along the Road (min)	Waiting Time at the Junction (min)
(1,4)	0.28	0.764	1.020
(4,10)	0.30	0.818	1.700
(10,9)	0.28	0.764	1.700
(9,11)	0.39	1.064	1.700
(11,13)	0.91	2.482	4.280
(13,19)	0.68	3.339	4.280
(19,20)	0.55	2.700	4.280
(20,25)	0.11	0.540	3.060
(25,24)	0.17	0.835	3.060
(24,23)	0.21	1.031	3.060
(23,33)	0.69	3.388	0.000
Total distance = 4.57 km		**Total travel time = 45.865 min**	

is calculated and preprocessed using Bureau of Public Roads (BPR) volume-delay function (Ng and Waller, 2010) to make it closer to reality. In order to address real-life scenarios, we consider time-dependent dynamic and deterministic travel times, time-dependent dynamic and deterministic waiting times, and one-way traffic during high-traffic conditions, in our proposed mathematical model. In Section 9.5, we explain the scenarios such as changeover of traffic conditions and the one-way traffic during rush period with the help of a topology and the corresponding optimal solution. We also consider multiple objectives (total distance travelled and time taken to reach the destination) simultaneously in our mathematical model. When we implement the proposed approach to the city network, we have a set of Pareto-optimal solutions for the given ε_t which is assumed as 1 min. Results are presented by considering two levels of traffic. For the purpose of illustration, we choose two extreme solutions from the nondominated solution set. The routes are represented with the help of road map of Chennai city network. Two figures are shown here to illustrate the paths over the road map. An optimal solution with respect to total distance travelled (10.54 km) and an optimal solution with respect to the total time taken to reach the destination (97 min) during rush period are highlighted with thick line, respectively, in Figures 9.5 and 9.6. We tabulate the nondominated set of solutions with the total distance traversed and the total time taken for the travel from IIT Madras (Node ID: 746) to Chennai Central station (Node ID: 496), and present them in Table 9.8 and Figure 9.7. The Chennai city network data set with 1658 nodes and 4224 links as well as the results obtained from the model are available upon request.

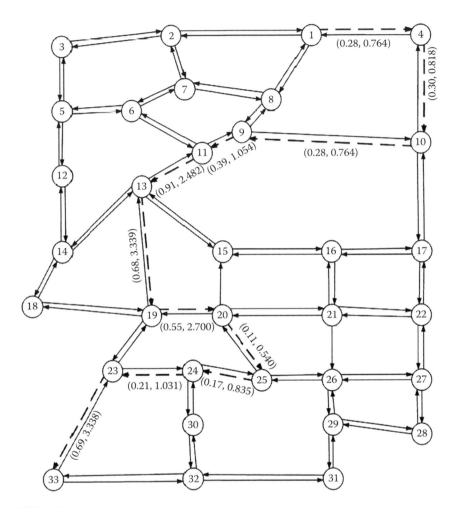

FIGURE 9.3
Travel starts from node 1 at 4.50 p.m. with changeover of travel time and waiting time from normal traffic period (up to 5 p.m.) to heavy traffic period (from 5 p.m. onwards) (total distance = 4.57 km and total time covered = 45.865 min). *Note:* optimal route (with respect to minimum total distance and the corresponding total time covered/taken): {1–4–10–9–11–13–19–20–25–24–23–33}. Legend (distance, travel time): For example (0.28, 0.764) in the above figure corresponds to distance and travel time from node 1 to node 4, and rest of the values follow suit.

9.7 Summary

We propose a mathematical model for the shortest path problem considering multiple conflicting objectives, namely, minimizing the total distance covered and total time taken, thereby considering the problem closer to reality. We also consider real-life aspects such as time-dependent dynamic and deterministic travel times, time-dependent dynamic and deterministic waiting times, and

TABLE 9.7

Travel Starts from Node 1 at 4.50 p.m., with the One-Way Traffic Regulation from 5 p.m. Onwards

Road (i,j)	Distance along the Road (km)	Time along the Road (min)	Waiting Time at the Junction (min)
(1,4)	0.28	0.764	1.020
(4,10)	0.30	0.818	1.700
(10,9)	0.28	0.764	1.700
(9,11)	0.39	1.064	1.700
(11,6)	0.14	0.688	1.840
(6,5)	0.39	1.915	3.060
(5,12)	0.62	2.480	3.060
(12,14)	0.62	2.480	4.280
(14,18)	0.73	2.405	3.060
(18,19)	0.96	4.712	4.280
(19,20)	0.55	2.700	4.280
(20,25)	0.11	0.540	3.060
(25,24)	0.17	0.835	3.060
(24,23)	0.21	1.031	3.060
(23,33)	0.69	3.388	0.000
Total distance = 6.20 km		**Total travel time = 65.744 min**	

Note: Due to the one-way traffic being operational from 5 p.m. onwards with respect to road (11,13), a new route starting from (11,6) is generated.

time-dependent one-way traffic regulation as constraints in our multiobjective mathematical model. A nondominated set of solutions are obtained using the ε-constraint method, thereby allowing the user to select the preferred solution. Our model inherently computes and determines the optimal route which accounts for time-dependent dynamic and deterministic travel times, time-dependent dynamic and deterministic waiting times, and time-dependent one-way traffic regulation along the roads. The model is validated using a sample topology for Chennai city network and also with a real-life large metropolitan city network (Chennai city network) consisting of 1658 nodes and 4224 arcs. Besides the quality of the solutions, an optimization model for Chennai city network aids the decision maker to compare and choose a preferable route from different alternatives. From the study, we observe that the MCDM models can be effectively applied to large transportation network to solve the multiple objectives that commonly arise in the shortest path problem. Our model has important real-life applications in intelligent transportation systems in relation to the route guidance and congestion mitigation on networks with time-dependent dynamic and deterministic travel times, and time-dependent dynamic and deterministic waiting times. As for future work, we propose to consider the stochastic travel times along the roads in the network which increase the available data size as well as the problem complexity.

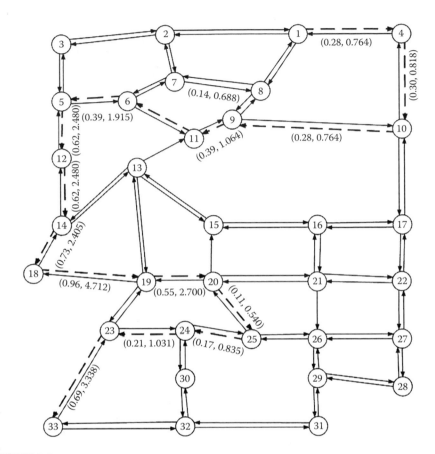

FIGURE 9.4
Travel starts from node 1 at 4.50 p.m. with changeover of travel time and waiting time from normal traffic to high-traffic period; one-way traffic along (13,11) being operational, but no traffic between (11,13) being operational between 5 p.m. and 9 p.m. (total distance = 6.20 km and total time covered = 65.744 min). *Note*: optimal route (with respect to minimum total distance and the corresponding total time covered/taken): {1–4–10–9–11–6–5–12–14–18–19–20–25–24–23–33}. *Legend* (distance, travel time): For example (0.28, 0.764) in the above figure corresponds to distance and travel time from node 1 to node 4, and rest of the values follow suit.

Acknowledgment

The authors gratefully acknowledge the support from the Centre of Excellence in Urban Transport at the Indian Institute of Technology Madras, which is funded by the Ministry of Urban Development, Government of India. The first author acknowledges the support from the DAAD for carrying out a portion of this work at the University of Duisburg-Essen. The authors are thankful to the reviewers and the editors for their valuable comments and suggestions to improve our chapter.

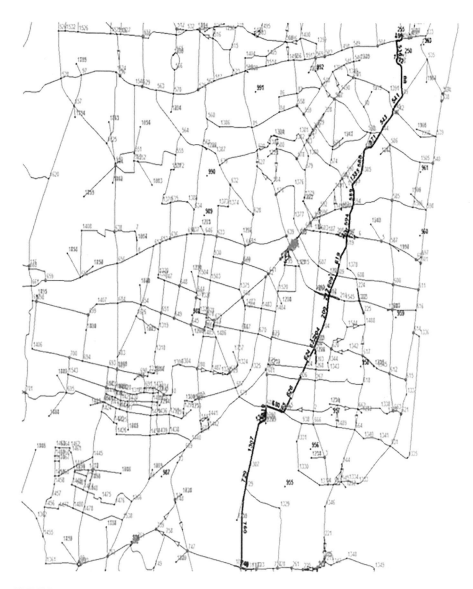

FIGURE 9.5
An optimal solution with respect to the objective function (minimizing total distance) for the travel from IIT Madras to Chennai Central Station.

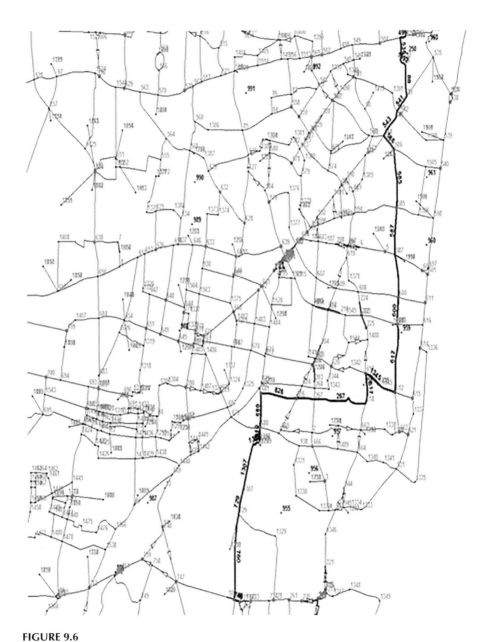

FIGURE 9.6
An optimal solution with respect to the objective function (minimizing total travel time) for the travel from IIT Madras to Chennai Central Station.

TABLE 9.8

A Set of Strictly Pareto-Optimal Solutions for Travel from IIT Madras to Chennai
Central Station, Given $\varepsilon_t = 1$ min

Distance (km)	10.54	10.77	11.42	11.72	11.95	12.60	12.83
Time (min)	114.6	111.4	107.6	105.0	101.8	100.2	97.0

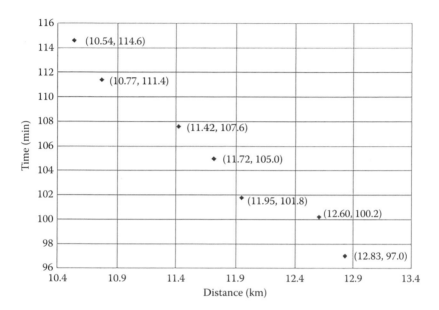

FIGURE 9.7
Nondominated solutions (heuristically) for the travel from IIT Madras to Chennai Central
Station, given $\varepsilon_t = 1$ min.

References

Ahuja, R., Magnanti, T. and Orlin, V. 1993. *Network Flows: Theory, Algorithms, and Applications.* Prentice Hall, Englewood Cliffs, NJ.

Bellman, R.E. 1958. On a routing problem. *Quarterly of Applied Mathematics* 16: 87–90.

Bertesekas, D. 1991. *Linear Network Optimization: Algorithms and Codes.* M.I.T. Press, Cambridge, MA.

Brumbaugh-Smith, J. and Shier, D. 1989. An empirical investigation of some bicriterion shortest path algorithms. *European Journal of Operational Research* 43: 216–224.

Chabini, I. 1998. Discrete dynamic shortest path problems in transportation applications. *Transportation Research Record* 1645: 8170–8175.

Chitra, C. and Subbaraj, P. 2010. A non-dominated sorting genetic algorithm for shortest path routing problem. *International Journal of Electrical and Computer Engineering* 5: 53–63.

Dell'Amico, M., Lori, M. and Pretolani, D. 2008. Shortest paths in piecewise continuous time-dependent networks. *Operations Research Letters* 36: 688–691.

Dreyfus, S. 1969. An appraisal of some shortest-path algorithms. *Operations Research* 17: 395–412.

Ehrgott, M. and Wiecek, M. 2005. Multiobjective programming. In: J. Figueira, S. Greco, and M. Ehrgott, editors, *Multicriteria Decision Analysis: State of the Art Survey*, Springer Verlag, Boston, Dordrecht, London: 667–722.

Hamacher, H.W., Ruzika, S. and Tjandra, S.A. 2006. Algorithms for time-dependent bicriteria shortest path problems. *Discrete Optimization* 3: 238–254.

Mohamed, C., Bassem, J. and Taicir, L. 2010. A genetic algorithm to solve the bicriteria Shortest Path Problem. *Electronic Notes in Discrete Mathematics* 36: 851–858.

Müller-Hannemann, M. and Schnee, M. 2007. Finding all attractive train connections by multi-criteria Pareto search. In: F. Geraets, L. Kroon, A. Schoebel, D. Wagner, and C. Zaroliagiis, editors, *Algorithmic Methods and Models for Railways Optimization*, Springer, Berlin, Heidelberg, 4359: 246–263.

Ng, M.W. and Waller, S.T. 2010. A computationally efficient methodology to characterize travel time reliability using the fast Fourier transform. *Transportation Research Part B* 44: 1202–1219.

Prakash, A. and Srinivasan, K.K. 2014. Sample-based algorithm to determine minimum Robust cost path with correlated link travel Times. *Transportation Research Record: Journal of the Transportation Research Board* 2467: 110–119.

Ruzika, S. and Wiecek, M.M. 2005. Approximation methods in multi-objective programming. *Journal of Optimization Theory and Applications* 126: 473–501.

Schrijver, A. 2003. *Combinatorial Optimization: Polyhedra and Efficiency*, Springer Verlag, Berlin.

Seshadri, R. and Srinivasan, K.K. 2010. Algorithm for determining most reliable travel time path on network with normally distributed and correlated link travel times. *Transportation Research Record: Journal of the Transportation Research Board* 2196: 83–92.

Seshadri, R. and Srinivasan, K.K. 2012. An algorithm for the minimum robust cost path on networks with random and correlated link travel times. In: D. M. Levinson, H. X. Liu, and M. Bell, editors, *Network Reliability in Practice*, Springer, New York: 171–208.

Sung, K., Bell, M.G.H., Seong, M. and Park, S. 2000. Theory and Methodology-Shortest paths in a network with time-dependent flow speeds. *European Journal of Operational Research* 121: 32–39.

Tiwari, A. 2016. Multi-objective convoy routing problem: Exact and heuristic methods, Unpublished MS thesis, IIT Madras, Chennai, India.

Wardell, W. and Ziliaskopoulos, A. 2000. A intermodal optimum path algorithm for dynamic multimodal networks. *European Journal of Operational Research* 125: 486–502.

Ziliaskopoulos, A. and Mahmassani, H. 1993. Time-dependent shortest path algorithm for real-time intelligent vehicle/highway system. *Transportation Research Record* 1408: 94–104.

10

Designing Resilient Global Supply Chain Networks over Multiple Time Periods within Complex International Environments

Rodolfo C. Portillo

CONTENTS

10.1 Introduction

With increased globalization, as stated by Friedman (2005), "In this world, a smart and fast global supply chain is becoming one of the most important ways for a company to distinguish itself from its competitors." As traditionally, the objective of every supply chain continues to be the maximization of the overall value by reducing procurement cost, increasing responsiveness to customers, and decreasing risk. The big change now is that global supply chain management involves a myriad of company's worldwide interests, customers, and suppliers rather than just a domestic perspective. Besides the financial aspects, companies now deal with a plethora of other factors when doing business abroad. Within this environment, as part of the company's strategy to manage its global supply chain, it must make decisions such as its overall sourcing plan, supplier selection, capacity and location of facilities, modes of transportation, etc. The emphasis of this research is on developing mathematical models to determine optimal supply chain designs that best support competitive strategies. Multicriteria mixed-integer linear programming models were developed to aid in a multiple echelon supply chain design. This work also includes the definition of a set of design selection criteria integrating financial, customer service, risk, and strategic factors.

A supply chain consists of (1) a series of physical entities (e.g., suppliers, plants, warehouses, and retailers) and (2) a coordinated set of activities concerned with the procurement of raw material and parts, production of intermediate and final products, and their distribution to the customers (Ravindran and Warsing 2013). The various decisions involved in managing a supply chain can be grouped into three types: strategic, tactical, and operational. Strategic decisions deal primarily with the design of the supply chain network, namely, the number and location of plants and warehouses and their respective capacities. They are made over a longer time horizon and have a significant impact with respect to the company's assets and resources, such as opening, expanding, closing, and downsizing facilities. Tactical decisions are primarily of a planning nature and made over a horizon of one or two years. They involve purchasing, aggregate production planning, inventory management, and distribution decisions. Finally, operational decisions are short term and made on a daily or weekly basis, such as setting customer delivery and weekly production schedules as well as inventory replenishment.

Optimal supply chain design needs to balance among multiple conflicting objectives, such as efficiency in terms of costs and profitability as well as speed to source, produce, and distribute products to customers. Resiliency is also an important objective and is measured in terms of the reliability of the supply chain network when there are disruptions to the supply chain. The case study presented in this chapter addresses strategic and tactical decisions in designing and managing an agile global supply chain considering the effect of multiple foreign currency exchange rates over multiple periods of time.

Decision makers often need to consider multiple criteria in order to determine the best course of action to solve a particular problem. The relationship among these decision criteria can be conflicting, which implies that trade-offs need to be considered and carefully evaluated. The search for an optimal solution for a multiobjective problem becomes a simultaneous process of optimizing two or more conflicting objectives. Refer to Ravindran (2008) and Ravindran et al. (2006) for further reading on multiple objective optimization methods.

As described by Masud and Ravindran (2008), a multicriteria decision-making problem in general can be represented as follows:

Maximize

$$C_1(x), C_2(x), \ldots, C_k(x),$$

where

$$x \in X$$

x is any specific alternative,

X is a set representing the feasible region or available alternatives, and

C_1 is the lth evaluation criterion.

According to the authors, multicriteria decision-making (MCDM) problems can be classified in two types: (1) the multicriteria mathematical programming problems (MCMP) with an infinite number of feasible alternatives implicitly determined by a finite number of explicitly stated constraints, and (2) the discrete multicriteria selection problems (MCSP) that consist of a finite number of alternatives stated explicitly. Refer to Chapter 2 for more details on MCDM models.

The material in this chapter is based on the doctoral dissertation of the author (Portillo 2009). With increased globalization, global supply chain management has become strategically important for many companies. The objectives of every supply chain continue to be maximizing the overall value generated by reducing the costs of procurement, increasing the responsiveness to customers, and decreasing the risks due to disruptions affecting the supply chain network. The big change now is that global supply chain management involves a company's worldwide interests: manufacturing facilities, distribution centers (DCs), customers, and suppliers located in several countries. Besides the conventional financial aspects, companies are now required to deal with a plethora of other factors for doing business abroad, such as duties, transfer prices, taxes, multiple exchange rates, and disruption risk. Transfer prices are charges among enterprise entities on goods and services. Disruption risk can be due to natural disasters, supplier quality issues, political tensions among countries, civil unrest, economic issues, government controls, and strikes from unions, among others. Within this environment, as part of their global supply chain management strategy, a company must make decisions on its overall outsourcing plan, supplier selection, the number of production plants and DCs that are needed, as well as the locations of those facilities, modes of transportation, and customer allocations to the DCs.

The emphasis of this chapter is on developing mathematical models to determine optimal supply chain design to support specific competitive strategies within the complex multinational environment across multiple periods of time. A multicriteria mixed-integer linear programming model has been developed to aid in a multiple echelon supply chain design, including manufacturing and distribution facilities' location/allocation, capacity and expansion requirements, production and distribution variables, international issues, exchange rates, lead times, transfer prices, and time periods. This work also considers a variety of semifinished goods of a health and hygiene consumer products company (e.g., tissue hard rolls, unpacked diapers, and oily soap solution) and finished products (e.g., facial tissues, toilet paper, toiletries, and gloves). Moreover, it includes the definition of a set of supply chain design criteria that integrates financial aspects, customer service, disruption risk, and strategic factors in the process. Strategic factors may include decisions to open new markets, increase market share, and to strengthen relationships with particular customers.

These methods for designing a resilient and responsive supply chain were applied to a leading global health and hygiene company listed in the Fortune 500 that operates in 37 countries; its global brands are sold in more than 150 countries and used by approximately 1.3 billion people, holding first or second market positions in the majority of markets. This case study focuses on the company's largest international division selling products across one continent with several offices, distribution and manufacturing facilities in 22 countries. As part of its competitive strategy, the firm serves from multinational chains to thousands of small "mom and pop" stores. A complex supply chain structure resulted after a series of mergers and acquisitions having 45 distribution facilities sourced by 21 plants spread out 10 countries within the division and other facilities located at different continents. In order to enhance the division's competitive strategy, a robust, flexible, and efficient global supply chain design was required to assure exceptional achievement of customer service and financial goals while considering related risk and strategic factors.

This application represented a major big data effort (refer to Chapter 3 on Big Data). An aggregated database consisting of approximately 75,000 data records, including marketing, sales, production, distribution, and purchasing information, was extracted from the company's enterprise resource planning (ERP) system, data warehouse, and other platforms by examining large sets containing structured and unstructured data of different types and sizes. Data quality assessment methods such as data profiling and standardization, cross tabulation, matching and linking, outlier detection, among others. In addition, query programming was critical to automate data extraction. Refer to an overview of the state of the art and focus on emerging trends to highlight the hardware, software, and application landscape of big data analytics (Kambatla et al. 2014).

10.2 Literature Review

Over 30 years ago, researchers recognized that systematic optimization should be used instead of common sense. Early models tended to treat only logistics aspects, while most developed thus far have focus primarily on single-criterion financial measures. A vast majority has addressed portions of the supply chain. Moreover, only a few have incorporated multinational and global criteria. Research for supply chain design started early on with a model developed for Hunt-Wesson Foods (Geoffrion and Graves 1974). After a decade, a system was implemented for Nabisco Foods, Inc. (Brown et al. 1987). Several applications followed for a petrochemical company (Van Roy and Wolsey 1985), Libbey–Owens–Ford (Martin et al. 1993), and Auli Foods (Pooley 1994). A comprehensive global supply chain model (GSCM) was applied to Digital Equipment Corporation (Arntzen et al. 1995). Later, a supply chain restructuring was supported at Procter & Gamble using mathematical optimization models (Camm et al. 1997). Other large-scale comprehensive models were implemented at Caterpillar (Rao et al. 2000), an agrochemicals company (Sousa et al. 2008), and to a global chemicals firm (Tsiakis and Papageorgiou 2008). All, except for the one at Digital Equipment Corporation (1995), have been established either for optimizing costs or profitability alone. This led to multiobjective approaches and surprisingly enough, despite the nature of the problem, very little work has been devoted using multicriteria techniques. The relationship among criteria can be conflicting, implying trade-offs, so the search for optimality becomes a simultaneous process. An early multicriteria approach was presented at Netherlands Car BV (Ashayeri and Rongen 1987). Analytic hierarchy process (AHP) and ranking methods for facility relocation were also proposed for solving supply chain design problems (Melachrinoudis and Min 1999). These authors later included two weighted objectives within a single nonpreemptive Goal Programing (GP) objective (Melachrinoudis et al. 2000). An AHP for optimizing the strategic importance of customers and their related risks was also proposed (Korpela et al. 2002). More recently, a bicriteria nonlinear stochastic optimization model was presented to determine the best supply chain (Gaur and Ravindran 2006). Afterwards, a multiobjective model was presented for solving a vendor selection problem (Wadhwa and Ravindran 2007). Arguably, most attention has been paid to methodologies that break the problem into pieces and simplify the inherent complexity of the supply chain.

10.3 Problem Description

In the last 15 years, a series of mergers and acquisitions across the continent led to a supply chain with a highly complex internal structure of many production and distribution facilities without standard assets and product technologies, together with redundancy of operational strategies and organizational structures. Most of the products sold in the region are supplied by these 21 manufacturing plants, and the rest are imported from other company facilities located all around the world. In addition, cross-sourcing activity within the continent has increased significantly in the last few years, by blending the advantages of single and multiple sourcing strategies. Today, more than 60% of the production facilities manufacture finished and semi-finished products that are distributed to different countries in addition to the local market. At least three facilities are continent-wide facilities sourcing all markets within the division, and they also export products to other company divisions worldwide. Until now, asset rationalization to improve operational efficiencies and structural reorganization efforts have focused on supply chain distribution designs for specific business units or division-wide designs considering manufacturing facilities for particular products only. In order to better support and enhance the division's competitive strategy, a robust, flexible, and efficient global supply chain design was required to ensure exceptional achievement of customer service levels and financial goals while considering related risk factors.

For a complete strategic and tactical optimization of the manufacturing and distribution network, the model needed to manage more than 100 customer zones or markets, as well as dozens of products manufactured in more than 250 production lines. Because of the diverse international nature of the problem, many global factors needed to be considered, such as domestic and international freights, transfer prices, taxes and duties, among others. Some products have a multistage production process and are processed in multiple echelons. In other words, the production process involves more than one stage before the final product is ready, where inventory may need to be managed and production technology and manufacturing lead times may differ. For example, in tissue production, you may have multiple processes to manufacture the paper itself in the form of semifinished hard rolls that are later converted to finished products such as napkins, facial tissue, toilet paper, and kitchen towels. Multiple production rates are considered for multiproduct machines and products. The costs in the optimization process include facility overheads, fixed and variable costs of production lines, and raw materials consumption costs.

10.4 Model Features

A base model for designing resilient global supply chain networks was presented by Portillo (2016). This chapter extends the application of this model to consider supply chain network design tactical decisions across multiple time periods as well as the impact of incorporating multiple foreign currency exchange rates given the global geographic scope of the analysis.

The multicriteria model integrates customer service levels, strategic factors, and disruption risk criteria along with the financial measure of performance.

Customer service level is measured using two factors: (1) demand fulfillment and (2) speed of delivery. Demand fulfillment is defined as the portion of the customer demand that is satisfied, namely the quantity that is effectively delivered to the customers. The ability to completely fulfill customer demand is modeled as a goal constraint by specifying demand fulfillment targets for all the combinations of products and customer zones. Speed of delivery is measured in terms of the lead time to deliver the products to the customers. This is also modeled as a goal, by minimizing the quantity weighted lead time, based on volume and the respective delivery lead times.* Weighted lead time targets, for each customer zone, are explicitly considered in the GP model. In addition, the multicriteria model considers the minimization of risk associated with supply chain disruptions. Different measures of risk for domestic and global sourcing are estimated for each manufacturing, converting, and distribution location. These measures incorporate facility- and country-specific risk factors. Facility-specific risk factors are determined based on assessments performed by the decision makers. Country-specific risk factors are obtained by considering the weighted average cost of capital rates for each country. The objective of minimizing the risk measure is also modeled as a goal constraint by setting the overall risk target value for the entire supply chain. Decisions related to supply chain network design may also require the modeler to consider strategic factors to open new markets, to increase market share, and to strengthen relationships with customers. This model includes measures for strategic factors for each facility, based on the ratings provided by the decision makers. A goal constraint is set to achieve the maximum possible overall strategic measure for the entire supply chain network.

Among other features, the model allows the evaluation of outsourcing decisions as well as the consideration of different product mix and corresponding productivity rates on different production lines and at different locations. The model supports both strategic and tactical decisions.

On the strategic side, the focus is on the design of the supply chain network, in which the optimization model determines the facilities that need to be opened and their locations, as well as the facilities that negatively affect profitability and therefore need to be closed. In the case where current network capacity (measured in product units) is not sufficient to fulfill customers' demand, the model provides for manufacturing and distribution decisions, evaluating where and how capacity should be expanded or outsourced. Note that, when combining products with significantly different specifications, a common standard unit of measure may be defined within the enterprise, such as weight (i.e., tons) and volume (i.e., cubic meters) for measuring capacity. Also, the ability to perform analysis at the production line level facilitates decisions associated with the transfer of equipment among facilities. Moreover, strategic decisions related to technological changes are

---

* Corresponding to each arc of the supply chain network that links a facility (plant/DC) to a customer zone.

supported by the model, such as what technologies are more convenient for the required expansions, or what specific equipment should be considered for write-off and replacement. The model also assists in tactical decisions, such as customer zone assignments to the DCs, the development of high-level production and distribution plans, product allocation to specific equipment, and cross-sourcing among production facilities. The objective then becomes the minimization of the deviations from the specified criteria targets: profit, demand fulfillment, lead time, disruption risk, and strategic factors.

A detailed description of the additions to the multicriteria base model (Portillo 2016) is included in the following sections, including some additional mathematical notation.

10.4.1 Multiperiod Model Features

Balancing supply and demand has been an art disguised as science, and the ability to precisely match production and sales volume has become more important than ever for highly globalized businesses—for both operational and financial efficiencies.

With this objective, the model presented in this chapter allows on one side the evaluation of production and sourcing capabilities compared with expected demand volume at different time periods. By optimizing multiple criteria, the model determines first the optimal configuration of manufacturing and distribution facilities for each period of time as well as the corresponding optimal production mix and cross-sourcing strategies to meet the desired objectives. It also provides for the evaluation of different supply chain network configuration alternatives over time, perhaps determining the optimal time period when a production line should be opened. In addition, different values for each time period can be set for most of the parameters included in the mathematical formulation, for example, allowing for the inclusion of price and cost trends, productivity learning curves and changes from continuous improvement, modifications in transfer prices and international commercial terms over time, among others. Moreover, the optimization model is flexible enough to handle different time horizons and periodicities, allowing both long-term strategic analyses as well as short-term tactical plans. Since the focus of this work is on strategic and some tactical types of decision, more operational decisions that involve the analysis of inventory levels have not been considered. However, including inventory variables and modifying the mathematical formulation allows this feature to be easily extended in the future.

10.4.1.1 *Notation*

10.4.1.1.1 *Index Sets*

Index m is included to represent time periods that can be defined depending on the level of analysis required as months, quarters, years, or others.

Index o is added to represent the starting period for specific design alternatives. By using the latter, it is possible, for example, to set different start-up periods for a production line for which the model will determine the optimal solution by considering the trade-offs from productivity learning curves, different sourcing options, and machine installation and operating costs. In this case, a particular production line under evaluation will have multiple production capacity and operating costs in a specific time period depending on when it started operations.

10.4.1.1.2 Parameters

This model considers a breakdown of customer demand as well as production and distribution capacities, balancing them at each period and optimizing the supply chain design and flows accordingly. Different values may be applied to sales prices, raw material and machine variable costs, transfer prices, duties, and freights over time, allowing for the analysis of the impact that changes with respect to these parameters in the network, creating cross sourcing and improving gross profit. These parameters are:

D_{kpm} = customer demand volume in market k for category p at time period m

P_{kpm} = sales price for category p in market k at time period m

MRC_{hpm} = raw material variable cost for category p in manufacturing plant h at time period m

MUC_{hpm} = CIF (cost, insurance, and freight) unit cost for category p for manufacturing facility h at period m

$I^{(2)}_{hipm}$ = cross-sourcing cost percentage for category p from manufacturing facility h to converting facility i at time period m

$T^{(2)}_{hipm}$ = freight in \$/ton for category p from manufacturing facility h to converting facility i at time period m

$TI^{(2)}_{hipm}$ = freight in \$/ton for category p from manufacturing facility h to entry port/customs when sending goods to converting facility i at time period m

$T^{(3)}_{hkpm}$ = freight in \$/ton for category p from manufacturing facility h to market k at time period m

CRC_{ipm} = raw material variable cost per ton for category p in converting plant i at time period m

CMC_{tm} = machine cost per hour for converting line t at time period m

CUC_{ipm} = CIF unit cost for category p for converting site i at time period m

$I^{(4)}_{ijpm}$ = cross-sourcing cost percentage for category p from converting facility i to distribution facility j at time period m

$T^{(4)}_{ijpm}$ = freight in \$/standard unit for category p from converting facility i to distribution center j at time period m

$TI^{(4)}_{ijpm}$ = transportation cost for product p from converting facility i to entry port/customs when sending goods to distribution facility j in \$/standard unit at m

$T^{(5)}_{ikpm}$ = freight in \$/standard unit for category p from converting facility i to market k at time period m

$T^{(7)}_{jkpm}$ = freight in \$/standard unit for category p from distribution facility j to market k at time period m

$T^{(6)}_{jlpm}$ = freight in \$/standard unit for category p from distribution facility j to distribution center l at time period m

$TI^{(6)}_{jlpm}$ = transportation cost for product p from DC j to entry port/customs when sending goods to DC l in \$ per standard unit at time period m

DUC_{jlm} = CIF unit cost for category p for distribution center j at time period m

$I^{(6)}_{jlpm}$ = cross-sourcing cost percentage for category p from distribution facility j to distribution facility l at time period m

VOC_j = variable operational cost of DC j in \$/cubic meters at time period m

In addition, parameters representing fixed operational costs for manufacturing, converting and distribution facilities and production lines, production and shipping capacities, and productivity rates consider not only m but also index o. Multiple values can be considered by the optimization model for one period of time depending on the start date when a given facility or production line started operations. These parameters are defined next:

FM_{hmo} = facility fixed costs of manufacturing plant h at time period m if opened at time period o

FMM_{htmo} = fixed costs of production line t in manufacturing site h at time period m if opened at time period o

MPT_{ptmo} = manufacturing time hours per ton for category p at production line t at time period m if opened at time period o

CPT_{ptmo} = conversion time hours per ton for category p at production line t at m for alternative o

MMC_{htmo} = machine cost per hour for manufacturing line t at time period m if opened at time period o

FC_{imo} = facility fixed costs of converting plant i at time period m if opened at o

FCM_{itmo} = fixed costs of production line t in converting site i at time period m if opened at o

CMC_{itmo} = machine cost per hour for production line t in converting facility i if opened at o

FD_{jmo} = fixed operational cost of DC j at time period m if opened at o

MC_{hmo} = manufacturing capacity of h at time period m if opened at o

CC_{imo} = converting capacity of i at time period m if opened at o

SC_{jmo} = shipping capacity of DC j at time period m if opened at o

10.4.1.2 Variables

Similarly, the variables corresponding to the manufacturing and converting supply chain echelons include the m and o indices indicating the corresponding flow between a pair of nodes at period m if the facility or production line is opened at period o.

$x^{(1)}_{hiptmo}$ = production of category p at production line t at manufacturing facility h sent to converting facility i at time period m if production line is opened at o

$x^{(2)}_{hkptmo}$ = production of category p at production line t at manufacturing facility h sent to market k at time period m if production line is opened at o

$y^{(1)}_{ijptmo}$ = production of category p at production line t at converting facility i sent to distribution facility j at time period m if production line is opened at o

$y^{(2)}_{ikptmo}$ = production of category p at production line t at converting facility i sent to market k at time period m if production line is opened at o

However, since it is assumed that the shipping capacity of a distribution center will not be significantly affected by when it started operations, for simplicity index o is not added to the distribution center outflow variables. These variables are:

$z^{(1)}_{jlpm}$ = distribution of category p from DC j to DC l at time period m

$z^{(2)}_{jkpm}$ = distribution of category p from DC j to market k at time period m

The binary variables only include index o, so multiple binary variables will exist for a particular facility or production line depending on different start-up period options. Note that index m is not included since the objective is to decide to open/close an asset or not and if so then determine when is the optimal time to perform the corresponding action.

$\delta^{(1)}_{ho}$ = 1 if manufacturing facility h is opened at time period o, 0 otherwise

$\gamma^{(1)}_{hto}$ = 1 if production line t in manufacturing facility h is opened at time period o, 0 otherwise

$\delta_{io}^{(2)} = 1$ if converting facility i is opened at time period o, 0 otherwise

$\gamma_{ito}^{(2)} = 1$ if production line t in converting facility i is opened at time period o, 0 otherwise

$\delta_{jo}^{(3)} = 1$ if distribution facility j is opened at time period o, 0 otherwise

The modified model formulation is presented in Equations 10.1 through 10.39.

10.4.2 Objective Function

In this model, the objective function includes the time period index m in the demand negative deviation variable term, which now adds the demand deviations for each of the market, product, and time period combinations. All other terms in the objective function are the same as in previous models. Similarly, the parameters w_g and P_g are used to incorporate the cardinal weights and ordinal priorities for a goal g, which are determined based on the decision maker preferences. Weights or preferences are used depending on the type of the objective function, nonpreemptive, or preemptive, respectively.

Note that in the model presented below the demand fulfillment deviation variable is indexed on the time period m as well, such that d_{5kpm}^-. This means that specific target values for demand fulfillment are specified for each product category p sold in a given market k at a particular time period m. The sum of all the corresponding deviation variables is then averaged arithmetically by the total number of market, product, and time period combinations with customer demand greater than zero, which is given by the product of the cardinality of the sets MKS, PRS, and PDS such that $D_{kp} > 0 \forall k, p$. This result is weighted then as a component of the goal-programming objective function. See below:

Nonpreemptive

$$z = w_1 d_1^- + w_2 \left(\frac{\sum_{k \in MKS} d_{2k}^+}{|MKS|} \right) + w_3 d_3^+ + w_4 d_4^- + w_5 \left(\frac{\sum_{m \in PDS} \sum_{k \in MKS} \sum_{p \in PRS} d_{5kpm}^-}{|MKS| * |PRS| * |PDS|} \right). \quad (10.1)$$

Preemptive

$$z = P_1 d_1^- + P_2 \left(\frac{\sum_{k \in MKS} d_{2k}^+}{|MKS|} \right) + P_3 d_3^+ + P_4 d_4^- + P_5 \left(\frac{\sum_{m \in PDS} \sum_{k \in MKS} \sum_{p \in PRS} d_{5kpm}^-}{|MKS| * |PRS| * |PDS|} \right). \quad (10.2)$$

10.4.3 Model Scaling

For the nonpreemptive version of the multicriteria objective function, scaling is necessary for proper optimization. Note that the goal constraints explained below can be given in different units of measure and can significantly vary in magnitude, for example, gross profit can be given in millions of currency units, lead time from single to no more than double-digit quantity of days, demand fulfillment as a percentage amount, and risk and strategy as single-digit measures. The formulation incorporates each goal target for scaling purposes assuring that the deviation variable values are such that $0 \leq d \leq 1$. In the nonpreemptive model, the objective function minimizes the weighted result of the scaled deviation variables value. For the preemptive version of the model, scaling is not necessary and consequently the goal constraint should incorporate the target values on the right hand side of the equation, allowing for the deviation variables to take a value according to the base unit of measure of each goal constraint (currency, days, percentage points, and risk and strategy units). In this case, the model will sequentially optimize each of the goals based on their priority.

10.4.4 Set of Goal Constraints

Similar to the objective function, some terms in the goal constraints include changes based on the modified parameters, including indices m and o. These are presented below:

f_1 = Max. gross profit

$$
\frac{1}{T\mu^{(1)}} \left(\sum_{k \in MKS} \sum_{p \in PRSK_k} \sum_{j \in DCSK_k} \sum_{m \in PDS} z^{(2)}_{jkpm} P_{kpm} \right.
$$

$$
+ \sum_{k \in MKS} \sum_{p \in PRSK_k} \sum_{i \in CFSK_k} \sum_{t \in CTSC_i} \sum_{m \in PDS} \sum_{o \in ODS} y^{(2)}_{ikptmo} P_{kpm} \tag{10.3}
$$

$$
\left. + \sum_{k \in MKS} \sum_{p \in PRSK_k} \sum_{h \in MFSK_k} \sum_{t \in CTSM_h} \sum_{m \in PDS} \sum_{o \in ODS} x^{(2)}_{hkptmo} P_{kpm} \right.
$$

$$
- \sum_{h \in MFS} \sum_{m \in PDS} \sum_{o \in ODS} FM_{hmo} \delta^{(1)}_{ho} - \sum_{i \in CFS} \sum_{m \in PDS} \sum_{o \in ODS} FC_{imo} \delta^{(2)}_{io} - \sum_{j \in DCS} \sum_{m \in PDS} \sum_{o \in ODS} FD_{jmo} \delta^{(3)}_{jo} \tag{10.4}
$$

$$
- \sum_{h \in MFS} \sum_{t \in CTSM_h} \sum_{m \in PDS} \sum_{o \in ODS} FMM_{htm} \gamma^{(1)}_{hto} - \sum_{i \in CFS} \sum_{t \in CTSC_i} \sum_m \sum_o FCM_{itm} \gamma^{(2)}_{ito} \tag{10.5}
$$

$$
- \sum_{m \in PDS} \sum_{h \in MFS} \sum_{p \in PRSM_h} \sum_{o \in ODS} MRC_{hpm} \left(\sum_{i \in CFSM_h} \sum_{t \in CTSM_h} x^{(1)}_{hiptmo} + \sum_{k \in MKSM_h} \sum_{t \in CTSM_h} x^{(2)}_{hkptmo} \right) \tag{10.6}
$$

$$
- \sum_{m \in PDS} \sum_{o \in ODS} \sum_{h \in MFS} \sum_{t \in CTSM_h} \sum_{p \in PRST_t} MMC_{tmo} MPT_{ptmo} \left(\sum_{i \in CFSM_h} x^{(1)}_{hiptmo} + \sum_{k \in MKSM_h} x^{(2)}_{hkptmo} \right) \tag{10.7}
$$

$$
- \sum_{i \in CFS} \sum_{p \in PRSC_i} \sum_{m \in PDS} CRC_{ipm} \sum_{t \in CTSC_i} \sum_{o \in ODS} \left(\sum_{j \in DCSC_i} \sum_{m \in PDS} y^{(1)}_{ijptmo} + \sum_{k \in MKSC_i} \sum_{m \in PDS} y^{(2)}_{ikptmo} \right) \tag{10.8}
$$

$$
- \sum_{m \in PDS} \sum_{o \in ODS} \sum_{i \in CFS} \sum_{t \in CTSC_i} \sum_{p \in PRST_t} CMC_{tmo} CPT_{ptmo} \left(\sum_{j \in DCSC_i} y^{(1)}_{ijptmo} + \sum_{k \in MKSC_i} y^{(2)}_{ikptmo} \right) \tag{10.9}
$$

$$- \sum_{m \in PDS} \sum_{j \in DCS} VOC_j \left(\sum_{l \in DCSC_j} \sum_{p \in PRSD_l} z_{jlpm}^{(1)} DF_{jkp}^{(2)} + \sum_{k \in MKSD_j} \sum_{p \in PRSK_k} z_{jkpm}^{(2)} DF_{jkp}^{(2)} \right) \tag{10.10}$$

$$- \sum_{m \in PDS} \sum_{h \in MFS} \sum_{i \in CFSM_h} \sum_{p \in PRSC_i} \sum_{t \in CTSC_h} \sum_{o \in ODS} T_{hipm}^{(2)} x_{hiptmo}^{(1)}$$
$$- \sum_{m \in PDS} \sum_{i \in CFS} \sum_{j \in DCSC_i} \sum_{p \in PRSD_j} \sum_{t \in CTSC_i} \sum_{o \in ODS} T_{ijpm}^{(4)} y_{ijptmo}^{(1)} \tag{10.11}$$
$$- \sum_{m \in PDS} \sum_{j \in DCS} \sum_{l \in DCSD_j} \sum_{p \in PRSD_l} T_{jlpm}^{(6)} z_{jlpm}^{(1)}$$

$$- \sum_{m \in PDS} \sum_{h \in MFS} \sum_{k \in MKSM_h} \sum_{p \in PRSK_k} \sum_{t \in CTSC_h} \sum_{o \in ODS} T_{hkpm}^{(3)} x_{hkptmo}^{(2)}$$
$$- \sum_{m \in PDS} \sum_{i \in CFS} \sum_{k \in MKSC_i} \sum_{p \in PRSK_k} \sum_{t \in CTSC_i} \sum_{o \in ODS} T_{ikpm}^{(5)} y_{ikptmo}^{(2)} \tag{10.12}$$
$$- \sum_{m \in PDS} \sum_{j \in DCS} \sum_{k \in MKSD_j} \sum_{p \in PRSK_k} T_{jjpm}^{(7)} z_{jkpm}^{(2)}$$

$$- \sum_{m \in PDS} \sum_{h \in MFS} \sum_{p \in PRSM_h} \sum_{i \in CFSM_h} \sum_{t \in CTSM_h} \sum_{o \in ODS} MUC_{hpm} I_{hipm}^{(2)} x_{hiptmo}^{(1)}$$
$$- \sum_{m \in PDS} \sum_{i \in CFS} \sum_{p \in PRSC_i} \sum_{j \in DCSD_i} \sum_{t \in CTSC_i} \sum_{o \in ODS} CUC_{ipm} I_{ijpm}^{(4)} y_{ijptmo}^{(1)} \tag{10.13}$$
$$- \sum_{m \in PDS} \sum_{j \in DCS} \sum_{p \in PRSD_j} \sum_{l \in DCSD_j} DUC_{jpm} I_{jlpm}^{(6)} z_{jlpm}^{(1)}$$

$$- \sum_{m \in PDS} \sum_{h \in MFS} \sum_{i \in CFSM_h} \sum_{p \in PRSC_i} \sum_{t \in CTSM_h} \sum_{o \in ODS} TI_{hipm}^{(2)} I_{hipm}^{(2)} x_{hiptmo}^{(1)}$$
$$- \sum_{m \in PDS} \sum_{i \in CFS} \sum_{j \in DCSC_i} \sum_{p \in PRSD_j} \sum_{t \in CTSC_i} \sum_{o \in ODS} TI_{ijpm}^{(4)} I_{ijpm}^{(4)} y_{ijptmo}^{(1)} \tag{10.14}$$

$$- \sum_{m \in PDS} \sum_{j \in DCS} \sum_{l \in DCSD_j} \sum_{p \in PRSD_l} TI_{jlpm}^{(6)} I_{jlpm}^{(6)} z_{jlpm}^{(1)} \Bigg) + d_1^- - d_1^+ = 1. \tag{10.15}$$

$f_2 =$ Min. lead time to markets

$$\frac{\sum_{i \in CFS} L_{ik}^{(5)} \sum_{o \in ODS} \sum_{m \in PDS} \sum_{p \in PRSK_k} \sum_{t \in CTSC_i} y_{ikptmo}^{(2)} + \sum_{j \in DCS} L_{jk}^{(7)} \sum_{p \in PRSK_k} z_{jkpm}^{(2)}}{\sum_{\substack{m \in PDS \\ D_{kpm} > 0}} \sum_{p \in PRSK_k} D_{kpm}}$$

$$+ d_{2k}^- - d_{2k}^+ = \frac{T\ell_k^{(2)} \left(\sum_{m \in PDS} \sum_{o \in ODS} \sum_{p \in PRSK_k} \sum_{t \in CTSC_i} y_{ikptmo}^{(2)} + \sum_{m \in PDS} \sum_{p \in PRSK_k} z_{jkpm}^{(2)} \right)}{\sum_{\substack{m \in PDS \\ D_{kpm} > 0}} \sum_{p \in PRSK_k} D_{kpm}} \quad \forall k \in MKS. \tag{10.16}$$

$f_3 =$ Min. risk of supply chain disruptions

$$\frac{1}{T\ell^{(3)}} \Bigg(\sum_{h \in MFS} \sum_{o \in ODS} R\ell_h^{(1)} \delta_{ho}^{(1)} + \sum_{i \in CFS} \sum_{o \in ODS} R\ell_i^{(2)} \delta_{io}^{(2)} + \sum_{j \in DCS} \sum_{o \in ODS} R\ell_j^{(3)} \delta_{jo}^{(3)}$$

$$+ \sum_{h \in MFS} \sum_{o \in ODS} Rc_h^{(1)} \left(1 - \delta_{ho}^{(1)} \right) + \sum_{i \in CFS} \sum_{o \in ODS} Rc_i^{(2)} \left(1 - \delta_{io}^{(2)} \right) + \sum_{j \in DCS} \sum_{o \in ODS} Rc_j^{(3)} \left(1 - \delta_{jo}^{(3)} \right) \Bigg) \tag{10.17}$$

$$+ d_3^- - d_3^+ = 1.$$

$f_4 =$ Max. strategic factors

$$\frac{\sum\limits_{h\in MFS}\sum\limits_{o\in ODS} St_h^{(1)}\delta_{ho}^{(1)} + \sum\limits_{i\in CFS}\sum\limits_{o\in ODS} St_i^{(2)}\delta_{io}^{(2)} + \sum\limits_{j\in DCS}\sum\limits_{o\in ODS} St_j^{(3)}\delta_{jo}^{(3)}}{T\mu^{(4)}} + d_4^- - d_4^+ = 1. \quad (10.18)$$

$f_5 =$ Max. demand fulfillment at markets

$$\frac{\sum\limits_{m\in PDS}\left(\sum\limits_{h\in MKSM_h}\sum\limits_{t\in CTSM_h}\sum\limits_{o\in ODS} x_{hkptmo}^{(2)} + \sum\limits_{i\in CFSK_k}\sum\limits_{o\in ODS} y_{ikpmo}^{(2)} + \sum\limits_{j\in DCSK_k} z_{jkpm}^{(2)}\right)}{D_{kpm}} \quad (10.19)$$

$$+ d_{5kpm}^- - d_{5kpm}^+ = 1 \quad \forall m \in PDS, k \in MKS, p \in PRSK_k.$$

10.4.5 Set of Constraints

The set of constraints containing parameters or variables with index m are now defined for each time period m. In addition, the constraints including index o are presented in one of two forms depending on the constraint type. On one side, in the capacity and binary constraints the variables are added based on index o, and on the other the balance constraints are defined for each alternative o, except the binary equation for the distribution to customers' echelon. Modifications to the constraints presented in the base model (Portillo 2016) are as follows:

Distribution capacity utilization

$$\sum\limits_{l\in DCSD_j}\left(\sum\limits_{p\in PRSD_l} z_{jlpm}^{(1)} DF_{jlp}^{(3)} + \sum\limits_{k\in MKSD_j}\sum\limits_{p\in PRSK_k} z_{jkpm}^{(2)} DF_{jkp}^{(2)}\right) \leq \sum\limits_{o\in ODS} SC_{jmo}\delta_{jo}^{(3)} \quad \forall m, j \in DCS. \quad (10.20)$$

Balance at 2-stage product manufacturing

$$\sum\limits_{h\in MFSP_p}\sum\limits_{i\in CFSC_i}\sum\limits_{t\in CTSM_h}\sum\limits_{o\in ODS} DF_{ip}^{(1)} x_{hiptmo}^{(1)} =$$

$$\sum\limits_{i\in CFSF_p}\sum\limits_{j\in DCSC_i}\sum\limits_{t\in CTSC_i}\sum\limits_{o\in ODS} y_{ijptmo}^{(1)} + \sum\limits_{i\in CFSP_p}\sum\limits_{k\in MKSC_p}\sum\limits_{t\in CTSC_j}\sum\limits_{o\in ODS} y_{ikptmo}^{(2)}, \quad \forall m, p \in PRSS. \quad (10.21)$$

Balance at 2-stage product conversion

$$DF_{ip}^{(1)}\left(\sum\limits_{h\in MFSC_i}\sum\limits_{t\in CTSM_h}\sum_o x_{hiptmo}^{(1)}\right) - \sum\limits_{j\in DCSC_i}\sum\limits_{t\in CTSC_i}\sum\limits_{o\in ODS} y_{ijptmo}^{(1)} - \sum\limits_{k\in MKSC_i}\sum\limits_{t\in CTSC_i}\sum\limits_{o\in ODS} y_{ikptmo}^{(2)} = 0$$
$$\forall m, i \in CFS, p \in PRSC_i. \quad (10.22)$$

Balance at single-stage production

$$\sum\limits_{i\in CFSP_p}\sum\limits_{j\in DCSC_i}\sum\limits_{t\in CTSC_i}\sum\limits_{o\in ODS} y_{ijptmo}^{(1)} = \sum\limits_{j\in DCSP_p}\sum\limits_{l\in DCSD_j} z_{jlpm}^{(2)}, \forall m, p \notin PRSS. \quad (10.23)$$

Balance at distribution

$$\sum_{i\in CFSD_j}\sum_{t\in CTSC_i}\sum_{o\in ODS} y^{(1)}_{iiptmo} + \sum_{l\in DCSA_j} z^{(1)}_{ljpm} = \sum_{l\in DCSD_j} z^{(1)}_{ljpm} + \sum_{k\in MKSD_j} z^{(2)}_{jkpm}, \quad \forall m,j,p\in PRSD_j \quad (10.24)$$

$$\sum_{l\in DCSA_j} z^{(1)}_{ljpm} = \sum_{l\in DCSD_j} z^{(1)}_{jlpm} + \sum_{k\in MKSD_j} z^{(2)}_{jkpm}, \quad \forall m,j\in DCSTOMARKETS, p\in PRSD_j. \quad (10.25)$$

Manufacturing capacity

$$\sum_{p\in PRST_t}\left(MPT_{hptmo}\left(\sum_{i\in CFSM_h} x^{(1)}_{hiptmo}\right) + \sum_{p\in PRST_t} MPT_{hptmo}\left(\sum_{k\in MKSM_h} x^{(2)}_{hkptmo}\right)\right) \le \sum_{o} MC_{htmo}\gamma^{(1)}_{hto}$$
$$\forall m,o,h\in MFS, t\in CTSM_h. \quad (10.26)$$

Conversion capacity

$$\left(\sum_{p\in PRST_t} CPT_{iptmo}\left(\sum_{j\in DCSC_i} y^{(1)}_{iiptmo}\right) + \sum_{p\in PRST_t} CPT_{iptmo}\left(\sum_{k\in MKSC_i} y^{(2)}_{ikptmo}\right)\right) \le \sum_{o\in ODS} CC_{itmo}\gamma^{(2)}_{ito}$$
$$\forall m,o,i\in CFS, t\in CTSC_i. \quad (10.27)$$

Manufacturing and conversion binary

Equations 10.28 and 10.29 make sure that at least one production line t is active in order to open a manufacturing facility h or converting facility i. Note that the binary variables are indexed based on the opening time period o, indicating that there may be multiple options to open production lines as well as manufacturing or converting facilities at different time periods. Consequently, the constraints below apply to each of the production network design options given by index o.

$$\gamma^{(1)}_{hto} \le \delta^{(1)}_{ho}, \forall o, h\in MFS, t\in CTSM_h \quad (10.28)$$

$$\gamma^{(2)}_{ito} \le \delta^{(2)}_{io}, \forall o, i\in CFS, t\in CTSC_i. \quad (10.29)$$

Distribution binary

Similarly, Equation 10.30 states a shipping capacity constraint in cubic meters for the flows out of the distribution facility j considering if it is open or closed. The capacity of a distribution facility may vary at different time periods m. As well, the formulation provides for the ability to evaluate opening a distribution center at different time periods, considering that its capacity may differ as well depending on the opening period o.

$$\sum_{l\in DCSD_j}\sum_{p\in PRSD_j} DF^{(3)}_{jlp} z^{(1)}_{jlpm} + \sum_{k\in MKSD_j}\sum_{p\in PRSK_k} DF^{(2)}_{jkp} z^{(2)}_{jkpm} - \sum_{o\in ODS} SC_{jmo}\delta^{(3)}_{jo} \le 0$$
$$\forall m,j\in DCS. \quad (10.30)$$

Production extensions binary

For each of the potential time periods when a production line can start operations, Equations 10.31 and 10.32 enforce that a new production line t' proposed to be installed at a production facility as a capacity expansion is only activated when all the existing machines of similar technology t are operating. If at least one production line of type t is idle at a given plant, then no capacity expansions can be done. In other words, a given capacity extension cannot open at any given period of time if the current installed equipment has idle capacity. Note that these constraints can be relaxed if the analyst wants to evaluate the replacement of equipment; in this case, new production lines could be opened even if it implies that existing equipment stays idle.

$$\gamma^{(1)}_{ht'o} \le \gamma^{(1)}_{hto}, \ \forall (h,o,t,t') \in MFSTT_{htt'} \tag{10.31}$$

$$\gamma^{(1)}_{it'o} \le \gamma^{(1)}_{ito}, \ \forall (i,o,t,t') \in CFSTT_{itt'}. \tag{10.32}$$

Plant/production line constraint binary

$$\sum_t \gamma^{(2)}_{ito} \ge \delta^{(2)}_{io} \qquad \forall i,o \tag{10.33}$$

$$\sum_t \gamma^{(1)}_{hto} \ge \delta^{(1)}_{ho} \qquad \forall h,o. \tag{10.34}$$

In a similar way, above equations make sure that at least one production line t is active in order to open a manufacturing facility h or converting facility i, applying this restriction to every possible opening period o for a production line or facility.

Time period binary

$$\sum_o \delta^{(1)}_{ho} \le 1 \qquad \forall h \tag{10.35}$$

$$\sum_o \delta^{(2)}_{io} \le 1 \qquad \forall i \tag{10.36}$$

$$\sum_o \delta^{(3)}_{jo} \le 1 \qquad \forall j \tag{10.37}$$

$$\sum_o \gamma^{(1)}_{hto} \le 1 \qquad \forall h \tag{10.38}$$

$$\sum_o \gamma^{(2)}_{ito} \le 1 \qquad \forall i. \tag{10.39}$$

The above equations are included to constrain that a given facility or production line that has different start-up options opens at only one of the time periods (i.e., can open only once).

All decision variables are continuous and nonnegative, except for $\delta(1)_{ho}$, $\gamma^{(1)}_{hto}$, $\delta(2)_{io}$, $\gamma^{(2)}_{ito}$, and $\delta(3)_{jo}$ that are binary.

10.4.6 Objective Function Considering Currency Exchange Rates

When designing supply chain structures to support commercial activities within international environments, country-specific monetary factors such as currency exchange rates may have an impact on the financial performance of multinational firms. Perhaps a particular business unit of a multinational firm may collect revenues in one currency and have costs in different or several currencies. The impact of currency exchange rates fluctuation on sales price, local operating costs, local and imported raw material costs, among others may be decisive to choose where to locate production and distribution facilities as well as to determine sourcing strategies. This has significant implications for companies with international supply chains and markets. Recognizing and considering these implications in strategy making activities now could mean the difference between success and failure in the future. Therefore, it is considered imperative to include the effect of exchange rate in this optimization model.

With this purpose, the parameter E_{cm} is added to the model to represent the currency exchange rate for country c at period m. This variable is defined to include a deterministic best estimate of the future behavior of the currency exchange rate for a given country. Index $c = (1, \ldots, n_c)$ relates to the set of countries (COS). This new set is broken down in subsets corresponding to manufacturing facilities (COS_h), converting plants (COS_i), DCs (COS_j), and markets (COS_k), establishing the relationship between these nodes and the country where they are located. This parameter is included in the gross profit goal constraint to convert financial parameters defined in multiple currencies from different countries to a single currency of comparison. The next chapter will deal with the variability of this parameter. For now, the modified deterministic goal constraint is presented next.

$f_1 = $ Max. gross profit

$$
\frac{1}{T\mu^{(1)}} \left(\sum_{c \in COSk} \sum_{k \in MKS} \sum_{p \in PRSK_k} \sum_{j \in DCSK_k} \sum_{m \in PDS} z^{(2)}_{jkpm} P_{kpm} E_{cm} \right.
$$

$$
+ \sum_{c \in COSk} \sum_{k \in MKS} \sum_{p \in PRSK_k} \sum_{i \in CFSK_k} \sum_{t \in CTSC_i} \sum_{m \in PDS} \sum_{o \in ODS} y^{(2)}_{ikptmo} P_{kpm} E_{cm} \tag{10.40}
$$

$$
+ \sum_{c \in COSk} \sum_{k \in MKS} \sum_{p \in PRSK_k} \sum_{h \in MFSK_k} \sum_{t \in CTSM_h} \sum_{m \in PDS} \sum_{o \in ODS} y^{(2)}_{hkptmo} P_{kpm} E_{cm}
$$

$$
- \sum_{c \in COSh} \sum_{h \in MFS} \sum_{m \in PDS} \sum_{o \in ODS} E_{cm} FM_{hmo} \delta^{(1)}_{ho}
$$

$$
- \sum_{c \in COSi} \sum_{i \in CFS} \sum_{m \in PDS} \sum_{o \in ODS} E_{cm} FC_{imo} \delta^{(2)}_{io} \tag{10.41}
$$

$$
- \sum_{c \in COSj} \sum_{j \in DCS} \sum_{m \in PDS} \sum_{o \in ODS} E_{cm} FD_{jmo} \delta^{(3)}_{jo}
$$

$$-\sum_{o\in COSh}\sum_{h\in MFS}\sum_{t\in CTSM_h}\sum_{m\in PDS}\sum_{o\in ODS}E_{cm}FMM_{htm}\gamma_{hto}^{(1)}-\sum_{c\in COSi}\sum_{i\in CFS}\sum_{t\in CTSC_i}\sum_{m}\sum_{o}E_{cm}FCM_{itm}\gamma_{ito}^{(2)} \quad (10.42)$$

$$-\sum_{o\in COSh}\sum_{m\in PDS}\sum_{h\in MFS}\sum_{p\in PRSM_h}\sum_{o\in ODS}E_{cm}MRC_{hpm}\left(\sum_{i\in CFSM_h}\sum_{t\in CTSM_h}x_{hiptmo}^{(1)}+\sum_{k\in MKSM_h}\sum_{t\in CTSM_h}x_{hkptmo}^{(2)}\right) \quad (10.43)$$

$$-\sum_{o\in COSh}\sum_{m\in PDS}\sum_{o\in ODS}\sum_{h\in MFS}\sum_{t\in CTSM_h}\sum_{p\in PRST_t}E_{cm}MMC_{tmo}MPT_{ptmo}\left(\sum_{i\in CFSM_h}x_{hiptmo}^{(1)}+\sum_{k\in MKSM_h}x_{hkptmo}^{(2)}\right) \quad (10.44)$$

$$-\sum_{c\in COSj}\sum_{i\in CFS}\sum_{p\in PRSC_i}\sum_{m\in PDS}E_{cm}CRC_{ipm}\sum_{t\in CTSC_i}\sum_{o\in ODS}\left(\sum_{j\in DCSC_i}\sum_{m\in PDS}y_{ijptmo}^{(1)}+\sum_{k\in MKSC_i}\sum_{m\in PDS}y_{ikptmo}^{(2)}\right) \quad (10.45)$$

$$-\sum_{c\in COSi}\sum_{m\in PDS}\sum_{o\in ODS}\sum_{i\in CFS}\sum_{t\in CTSC_i}\sum_{p\in PRST_t}E_{cm}CMC_{tmo}CPT_{ptmo}\left(\sum_{j\in DCSC_i}y_{ijptmo}^{(1)}+\sum_{k\in MKSC_i}y_{ikptmo}^{(2)}\right) \quad (10.46)$$

$$-\sum_{c\in COSj}\sum_{m\in PDS}\sum_{j\in DCS}E_{cm}VOC_j\left(\sum_{l\in DCSC_j}\sum_{p\in PRSD_l}z_{jlpm}^{(1)}DF_{jkp}^{(2)}+\sum_{k\in MKSD_j}\sum_{p\in PRSK_k}z_{jkpm}^{(2)}DF_{jkp}^{(2)}\right) \quad (10.47)$$

$$-\sum_{c\in COSh}\sum_{m\in PDS}\sum_{h\in MFS}\sum_{i\in CFSM_h}\sum_{p\in PRSC_i}\sum_{t\in CTSC_h}\sum_{o\in ODS}E_{cm}T_{hipm}^{(2)}x_{hiptmo}^{(1)}$$
$$-\sum_{c\in COSi}\sum_{m\in PDS}\sum_{i\in CFS}\sum_{j\in DCSC_i}\sum_{p\in PRSD_l}\sum_{t\in CTSC_i}\sum_{o\in ODS}E_{cm}T_{ijpm}^{(4)}y_{ijptmo}^{(1)} \quad (10.48)$$
$$-\sum_{c\in COSj}\sum_{m\in PDS}\sum_{j\in DCS}\sum_{l\in DCSD_j}\sum_{p\in PRSD_l}E_{cm}T_{jlpm}^{(6)}z_{jlpm}^{(1)}$$

$$-\sum_{c\in COSh}\sum_{m\in PDS}\sum_{h\in MFS}\sum_{k\in MKSM_h}\sum_{p\in PRSK_k}\sum_{t\in CTSC_h}\sum_{o\in ODS}E_{cm}T_{hkpm}^{(3)}x_{hkptmo}^{(2)}$$
$$-\sum_{c\in COSi}\sum_{m\in PDS}\sum_{i\in CFS}\sum_{k\in MKSC_i}\sum_{p\in PRSK_k}\sum_{t\in CTSC_i}\sum_{o\in ODS}E_{cm}T_{ikpm}^{(5)}y_{ikptmo}^{(2)} \quad (10.49)$$
$$-\sum_{c\in COSj}\sum_{m\in PDS}\sum_{j\in DCS}\sum_{k\in MKSD_j}\sum_{p\in PRSK_k}E_{cm}T_{jkpm}^{(7)}z_{jkpm}^{(2)}$$

$$-\sum_{c\in COSh}\sum_{m\in PDS}\sum_{h\in MFS}\sum_{p\in PRSM_h}\sum_{i\in CFSM_h}\sum_{t\in CTSM_h}\sum_{o\in ODS}E_{cm}MUC_{hpm}I_{hipm}^{(2)}x_{hiptmo}^{(1)}$$
$$-\sum_{c\in COSi}\sum_{m\in PDS}\sum_{i\in CFS}\sum_{p\in PRSC_i}\sum_{j\in DCSD_l}\sum_{t\in CTSC_i}\sum_{o\in ODS}E_{cm}CUC_{ipm}I_{ijpm}^{(4)}y_{ijptmo}^{(1)} \quad (10.50)$$
$$-\sum_{c\in COSj}\sum_{m\in PDS}\sum_{j\in DCS}\sum_{k\in PRSD_j}\sum_{l\in DCSD_j}E_{cm}DUC_{jpm}I_{jlpm}^{(6)}z_{jlpm}^{(1)}$$

$$-\sum_{c\in COSh}\sum_{m\in PDS}\sum_{h\in MFS}\sum_{i\in CFSM_h}\sum_{p\in PRSC_i}\sum_{t\in CTSM_h}\sum_{o\in ODS}E_{cm}TI_{hipm}^{(2)}I_{hipm}^{(2)}x_{hiptmo}^{(1)}$$
$$-\sum_{c\in COSi}\sum_{m\in PDS}\sum_{i\in CFS}\sum_{j\in DCSC_i}\sum_{p\in PRSD_l}\sum_{t\in CTSC_i}\sum_{o\in ODS}E_{cm}TI_{ijpm}^{(4)}I_{ijpm}^{(4)}y_{ijptmo}^{(1)} \quad (10.51)$$

$$-\sum_{c\in COSj}\sum_{m\in PDS}\sum_{j\in DCS}\sum_{l\in DCSD_j}\sum_{p\in PRSD_l}E_{cm}TI_{jlpm}^{(6)}I_{jlpm}^{(6)}z_{jlpm}^{(1)}\Bigg) \quad (10.52)$$

$$+d_1^--d_1^+=1. \quad (10.53)$$

10.5 Data Collection

Before conducting the analysis, significant effort was required to collect data of both types, historical and planned. Because of the large scale of the analyses, it is important to highlight the effort dedicated to build the databases for the different supply chain network scenarios. Each scenario analysis required extracting thousands of market-, finance-, and operations-related data records from the firm's business systems as well as obtaining information from external sources. At least three employees worked full time directly gathering or requesting information, as well as organizing it appropriately to run the optimization models. In addition, at least a dozen people were contacted to provide information. The historical data was obtained from the company's ERP system at very low levels of granularity to have the flexibility of aggregating it as required by the optimization model. Sales volume and prices were obtained at stock keeping unit (SKU) and customer levels and then aggregated to the product and customer zone level. Production volume and unit costs were extracted at the SKU and production line levels and then aggregated to a product for each production line. Plant cost information was generated at the financial account level by each cost center associated with each production line and then classified as plant overheads, production line fixed and variable costs, and raw material costs. Transportation and cross-sourcing cost was defined for each arc of the network at the SKU level and then aggregated to products. Also, mass conversion factors were determined to handle different volume units of measure used at the different echelons of the supply chain. Projected information was obtained from the company's most recent business plans, including forecasted demand volume, price, and cost projections. All this data was stored, analyzed, and processed using the company's internal web services enabled by a single worldwide data center in North America. The mathematical formulation was coded in ILOG and solved using a CPLEX solver. The optimization model consisted of approximately 7500 variables, from which 300 were binary and around 7000 were constraints. Three persons were involved in coding the models in ILOG and then using the CPLEX software to solve the problems. In general, optimal solutions were obtained very efficiently, taking less than three minutes for each scenario analysis to run.

10.6 Case Study

This section presents the case study results for a multiperiod scenario considering exchange rate. The objective is to expand capacity to strengthen the position of a line of products in the marketplace for a particular country.

In the last three years, the sales growth for this business has been 21% per year. Recent customer demand forecasts indicate that with the currently operating capacity plus a recent installation of one machine in a regional plant outside the country, this business unit will need to import from outside the region approximately 17% of the total volume over the next 5 years. The incremental cost of import ranges from $12 to 22/Standard Unit versus producing in the regional plant. The purpose is to determine the most efficient supply chain network configuration and sourcing plans that best support this expected growth. The analysis consists of comparing the status quo to three new different scenarios as results from a preliminary analysis suggest that the status quo scenario is not convenient for the long-term sustainability of the business because of the supply disruption risk of importing 17% of the demand over the next 5 years, besides incurring higher product costs.

The first scenario consists of adding one more machine in the regional plant leveraging centralization efficiencies. The regional plant has four machines currently running and a fifth machine that has been installed and will soon begin operating. The proposal is to add a sixth machine. The second scenario proposes installing a new machine at a local DC in a neighboring country. With this, relevant savings could be achieved from reductions in freights and duties; however, additional fixed operating costs may become necessary and operational efficiency in the local facility could be lower than in the regional plant. Results of these scenarios illustrate how installing a new machine, regardless of its location, significantly reduces the volume of imports from outside the region, having a positive impact on the cost of goods and reducing the overall risk of supply chain disruptions. On one side, installing the new machine in the regional plant instead of a newly opened plant locally increases the overall production volume due to a better productivity learning curve in the regional plant, reducing the imports out of the region. On the other hand, installing it in a new plant locally would reduce the imports within the region but slightly increase the imports outside the region. Note that increasing the production capacity in the regional plant implies that the local market would need to continue being sourced with imported goods from the regional plant, which has an increased risk of supply chain disruptions than producing locally.

The third scenario suggests installing the new machine in the local facility and transferring the recently installed machine from the regional plant. In this case, it is expected that the incremental fixed operating costs would get diluted in the higher throughput and that better operating efficiency could be reached. For Scenario 3, it is assumed that the new machines will have a faster learning curve and better productivity levels in the regional plant, leveraging the know-how and product mix efficiency because the high number of machines would allow to reduce the number of changes required to produce different product types. Installing the two machines in the local distribution center would still cost more than installing them in the currently operating regional plant. This is because existing overheads in the regional plant get diluted in much higher production volume even if new machines are not installed, compared with additional headcount and infrastructure required in a new production local facility. Moreover, it is important to determine the best time to install the new machine. This can happen immediately or the following year, balancing between machine utilization rate and the associated incremental fixed operating costs. Any of these options would allow sourcing within the region for all of the market requirements, so the increased costs from imports outside the region could be reduced and supply chain risk of disruptions minimized. The objective is to maximize gross profit and minimize risk. Results from the analysis indicate that installing two machines locally would significantly reduce the imports out of the region and almost eliminate the imports within the region, giving a very high autonomy to the local market and therefore strongly minimizing the risk of supply disruption.

In conclusion, the three new proposals generate savings compared with the current supply chain network design requiring only between 2% and 4% of imports from outside the region. The scenario of adding one new machine to the regional plant generates more than twice the savings obtained from the other alternatives. In addition, it provides at least half the invested capital payback than installing one or two assets locally. These results are driven by the fact that when installing assets locally the reductions in transportation and duties do not set off higher production costs. The unit cost of producing locally is higher due to lower productivity and increased operating costs. Therefore, neither opening a new production facility locally nor continuing importing from outside the region is beneficial considering profits and risk. The best solution to support the expected business growth locally is to continue with a centralized strategy in the regional plant moving forward with the newly installed machine plus adding a new one.

As a complementary analysis indicated, delaying 1 year on the installation of the new machine in the regional plant affects its production volume. Considering the high cost of importing goods out of the region, besides the operational convenience of this solution, the results indicate that it is economically favorable putting in operation and starting amortizing the new machine the first year.

The analysis above was conducted in dollars. A second round of analysis was performed in local currency. Given the expected monetary depreciation trend in the regional plant host country and a more stable currency in the local one, the results reinforced the previous recommendation. The currency exchange rate of the country, where the regional plant is located, presents a steep depreciation rate forecast reaching almost 16% accumulated in 4 years, while the exchange rate from the country proposed as the location for the new plant presents only a 2% depreciation projection within the same timeframe. The impact of such currency depreciation difference provides a competitive cost advantage for exports considering that a significant proportion of the product cost is associated with locally incurred operating expenses as well as some key raw materials sourced from local suppliers.

Following this exercise, similar analyses were performed using this model in other regions of Latin America, North America, and China, with the objective of optimizing manufacturing and distribution networks as a base of business strategic plans over multiple years and considering the impact of foreign currency exchange rates from multiple countries and their impact on business results.

10.7 Conclusions and Future Research

This research provides relevant insights from modeling to implementation designing resilient complex global supply chain systems, a contribution to academia and industry research. In this work, diverse techniques classified as multicriteria mixed-integer programming and discrete multicriteria selection methods were combined to develop a global supply chain design model. It provides a resilient solution for supporting supply chain strategic decisions related to footprint design and tactical plans within a highly complex global environment. Focus was on optimizing conflicting criteria such as financial, customer service, risk, and strategic measures. These models were able to handle the complexity of the system (i.e., multiple products sold in several markets in different countries that required dealing with different currencies and commercial practices across multiple periods of time). Decisions regarding the optimal location, relocation, and allocation of production and distribution facilities as well as specific assets were efficiently addressed.

In addition, decisions regarding production and distribution plans were defined balancing demand and available capacity along multiple periods of time. By incorporating multiple period, the models provided the ability to consider the impact of changes through time in demand, machine productivity, facility capacity, prices, costs, and other parameters. On the strategic side, the models allowed the evaluation of different alternatives for the modification of the supply chain design at different time periods. On the tactical side, the models allowed a better evaluation of the balance between supply and demand as well as more accurate cross-sourcing strategy definition.

Although the model presented above embodies a robust solution for supply chain optimization, extensions can focus on determining the efficient frontier of conflicting criteria, treating the uncertainty associated with key elements of the problem using more sophisticated methods such as stochastic programming, and on expanding the scope on multiple supply chain directions, from including external partners as customers and vendors to optimizing more operational aspects of supply chain as sales and operations planning, distribution networks optimization, as well as inventory management.

Future research can be done on big data mining, particularly on distributed mining as data sources increase complexity and become more diverse and decision makers need to act faster, finding user-friendly visualizations as data gets bigger, and implementing hidden big data techniques as large quantities of data get lost as it is untagged and unstructured, aiming to take these analysis to have more higher frequency and granularity as data becomes more diverse, larger, and of faster availability.

References

Arntzen, B. C., G. G. Brown, T. P. Harrison, and L. L. Trafton, 1995, Global supply chain management at digital equipment corporation, *Interfaces*, 25, 69–93.

Ashayeri, J. and J. M. J. Rongen, 1987, Central distribution in Europe: A multi-criteria approach to location selection, *The International Journal of Logistics Management*, 9(1), 97–106.

Brown, G. G., G. W. Graves, and M. D. Honczarenko, 1987, Design and operation of a multi-commodity production/distribution system using primal goal decomposition, *Management Science*, 33(11), 1469–1480.

Camm, J. D., T. E. Chorman, and F. A. Dull, 1997, Blending OR/MS judgment and GIS: Restructuring P&G's supply chain, *Interfaces*, 27, 128–142.

Friedman, T., 2005, *The World is Flat: A Brief History of the Twenty-First Century*, Farrar, Straus & Giroux, New York.

Gaur, S. and A. R. Ravindran, 2006, A Bi-Criteria model for the inventory aggregation problem under risk pooling, *Computers & Industrial Engineering*, 51, 482–501.

Geoffrion, A. and G. Graves, 1974, Multicommodity distribution system design by benders decomposition, *Management Science*, 29, 822–844.

Kambatla, K., G. Kollias, V. Kumar, and A. Grama, 2014, Trends in big data analytics, *Journal of Parallel and Distributed Computing*, 74(7), 2561–2573.

Korpela, J., K. Kyläheiko, and A. Lehmusvaara, 2002, An analytic approach to production capacity allocation and supply chain design, *International Journal of Production Economics*, 78, 187–195.

Martin, C. H., D. C. Dent, and J. C. Eckhart, 1993, Integrated production, distribution, and inventory planning at Libbey–Owens–Ford, *Interfaces*, 23, 68–78.

Masud, A. M. and A. R. Ravindran, 2008, Multiple criteria decision making. In *Operations Research and Management Science Handbook*, A. Ravi Ravindran (Editor). CRC Press, Boca Raton, FL, 2008.

Melachrinoudis, E. and H. Min, 1999, The dynamic relocation and phase-out of a hybrid, two-echelon plant/warehousing facility: A multiple objective approach, *European Journal of OR*, 123, 1–15.

Melachrinoudis, E., H. Min, and A. Messac, 2000, The relocation of a manufacturing/ distribution facility from supply chain perspectives: A physical programming approach, *Multi-criteria Applications*, 10, 15–39.

Pooley, J., 1994, Integrated production and distribution planning at Ault foods, *Interfaces*, 24, 113–121.

Portillo, R. C., 2009, *Resilient Global Supply Chain Network Design*. PhD dissertation, Pennsylvania State University, University Park, PA, 2009.

Portillo, R. C., 2016, Designing resilient global supply chain networks. In *Multiple Criteria Decision Making in Supply Chain Management*, A. Ravi Ravindran (Editor), CRC Press, Boca Raton, FL, 2016.

Rao, U., A. Scheller, and S. Tayur, 2000, Development of a rapid-response supply chain at caterpillar, *Operations Research*, 48(2), 189–204.

Ravindran, A. R., 2008, *Operations Research and Management Science Handbook*. CRC Press, Taylor and Francis Group, Boca Raton, FL.

Ravindran, A. R., K. M. Ragsdell, and G. V. Reklaitis, 2006, *Engineering Optimization: Methods and Applications*. Wiley, Hoboken, NJ.

Ravindran, A. R. and D. P. Warsing, Jr., 2013, *Supply Chain Engineering: Models and Applications*. CRC Press, Boca Raton, FL.

Sousa, R., N. Shah, and L. G. Papageorgiou, 2008, Supply chain design and multi-level planning—An industrial case, *Computers and Chemical Engineering*, 32, 2643–2663.

Tsiakis, P. and L. G. Papageorgiou, 2008, Optimal production allocation and distribution supply chain networks, *International Journal of Production Economics*, 111, 468–483.

Van Roy, T. J. and L. A. Wolsey, 1985, Valid inequalities and separation for uncapacitated fixed charge networks, *Operations Research*, 4, 105–112.

Wadhwa, V. and A. R. Ravindran, 2007, Vendor selection in outsourcing, *Computers & OR*, 34, 3725–3737.

11

MCDM-Based Modeling Framework for Continuous Performance Evaluation of Employees to Offer Reward and Recognition

S. S. Sreejith and Muthu Mathirajan

CONTENTS

11.1 Introduction

Periodic performance evaluations of employees are required in order to measure the contribution level of employee toward organizational objectives and also to calibrate individual performance. Performance evaluations of employees in organizations are customarily conducted via a formal performance appraisal system (PAS) (Gruenfeld and Weissenberg 1966, Rosen and Abraham 1966). In general, the PAS is administered on an annual basis to evaluate the performance of employee over a period of past 1 year (Bassett and Meyer 1968). Such annual PAS has been subject to criticism from the industry citing long frequency, among other drawbacks (Henderson 1980, Ilgen et al. 1981). Some industries such as Information Technology (IT) have tried to shorten this frequency by making it a half-yearly PAS. Nevertheless, the resentment with the existing PAS persists. Some of the most common issues with the existing PAS on annual or half-yearly basis are: recency error, bias, favoritism, subjectivity, selective memory, etc. (Bowman 1999, Facteau and Craig 2001, Gray 2002). In order to minimize some of these issues, it has been repetitively recommended that the performance evaluation of employees should be conducted in a continuous manner (Garafano and Salas 2005, Kondrasuk 2011).

Literature endorses the necessity of having a continuous performance evaluation of employees (CPEE), rather than periodic appraisals (Boice and Kleiner 1997, Schraeder et al. 2007, Palaiologos et al. 2011, Sreejith 2015). Despite the theoretical compulsion, the idea of CPEE has not known to be translated into action in organizations. One reason for such inaction could be due to a lack of model, which details the intricacies of having a system to successfully carry out CPEE. As CPEE is a frequent phenomenon (assumed to be weekly or fortnightly in this study), there has to be periodic tangible outputs so as to increase the acceptability of the CPEE process. One such output can be to identify the best-performing employee and offer reward and recognition (R&R).

R&R is regarded as a motivational method to indicate appreciation and to boost better performance (Cacioppe 1999). It can be considered that reward is a materialistic part of the appreciation while recognition is the spiritual part. Examples of reward could be certificates, mementos, souvenirs, etc., while recognition could be public appreciation, notifying the contribution of employees through organization-wide newsletter or a simple pat on the back of the employee. Although in principle, "reward" and "recognition" indicate different concepts (Hansen et al. 2002), in this chapter they are considered in unison and mutually inclusive and is referred to as R&R. In addition, in this study, the means of offering the R&R is assumed to be varied across organization.

It is obvious that, given a chance, every employee expects to get R&R for displaying superior performance (Brun and Dugas 2008). Moreover, it has been observed that performance of employee increases when deserving employees are provided proper and timely R&R (Bradler et al. 2012). In addition, in order to be effective, R&R should be offered to employees immediately after noticing the superior performance. However, there is no known formal system/process/framework whereby deserving employees receive proper and timely R&R. To offer proper and timely R&R in a continuous manner, employees need to be evaluated on a continuous basis and ranked according to their performance. The CPEE process could frequently generate a rank of well-performing employees, who can be offered R&R. Hence, there needs to be a CPEE system/framework to identify the eligible employee(s) to offer R&R. One such framework for CPEE to offer R&R is attempted in this study.

The proposed framework for CPEE to offer R&R is anticipated to work much better in a dynamic industry with younger workforce, such as IT industry. This is because IT industry is comprised of younger workforce (Arora et al. 2001), who anticipate immediate feedback regarding their performance (Smola and Sutton 2002), and due to the alarming rate of voluntary attrition

(Bhatnagar 2006, Economic Times 2015). It can be presumed that if the proposed framework for CPEE to offer R&R is properly implemented, it could address the issue of voluntary attrition to some extend and offer better motivation for employees to perform better. This study focuses on developing a framework for CPEE system to offer R&R for the IT industry where employees are represented by software engineers (SEs) and project managers (PMs).

Based on the analysis of the literature and based on the observation in the industries, the performance evaluation of employees is assessed based on a combination of multiple numerous variables (criteria). Accordingly, the CPEE to offer R&R should be based on multiple variables/multiple criteria. From the literature review on CPEE, particularly for SEs, it appears that there are no variables/criteria that focus on CPEE to offer R&R. Due to this, multiple variables/multiple criteria suggested in the literature for the traditional PAS (among all industries) are collected to understand the types of multiple criteria being considered in PAS. Based on the understanding from these, subsequently suitable explorative and descriptive research methods have been carried out to identify the required set of variables/criteria for CPEE to offer R&R, particularly for SEs. In order to effectively utilize these identified multiple variables/multiple criteria for evaluating the performance of employees, a suitable multicriteria decision-making (MCDM) method is necessary. Accordingly in this study, two MCDM methods, analytical hierarchy process (AHP) and modified Pugh matrix method (MPMM), are considered and appropriately implemented to develop the proposed framework for CPEE to offer R&R.

The rest of the chapter is organized as follows: Identification of a comprehensive list of variables/criteria from (a) literature review, (b) exploratory, and (c) descriptive research methods and determining the main criteria for CPEE to offer R&R are detailed in Section 11.2. The literature focusing on the analysis of the performance of employees using MCDM methods are highlighted in Section 11.3. The development of MCDM-based modeling framework for CPEE to offer R&R is elaborated in Section 11.4. A suitable numerical example is developed to demonstrate the workability of the proposed framework for CPEE to offer R&R in Section 11.5. Section 11.6 discusses the managerial implications of the proposed MCDM-based modeling framework for CPEE to offer R&R. The study concludes by highlighting the contributions, limitation, and further research of the study in Section 11.7.

11.2 Identification of Multiple Variables/Multiple Criteria for CPEE to Offer R&R

In general, performance evaluation of employees using PAS commences by setting forth a set of defined variables/criteria. As there is no known literature specifying the variables for CPEE to offer R&R, the literature dealing with performance evaluation of employees using the traditional PAS are reviewed. Based on this review, a list of 44 variables is identified from the literature and is shown in column 3 of Table 11.1. In addition, a set of six variables are intuitively proposed by the researcher for the purpose of performance evaluation of employees (Sreejith 2016) and these are also listed in column 4 of Table 11.1.

The first 51 variables listed (based on columns 3 and 4) in Table 11.1 are not exclusively identified for IT industry, so it cannot be confidently assumed that all these 51 variables are relevant for performance evaluation of SEs in the IT industry. In addition, in general, all the listed 44 variables in Table 11.1 (column 3) pertaining to the traditional PAS cannot be blindly assumed to hold good specifically for continuous evaluation of employee. Hence, it is necessary to conduct exploratory and/or descriptive research methods to identify the required variables/criteria for CPEE to offer R&R, directly from the employees of IT industry. Accordingly, a Caselet approach is carried out and seven SEs are interviewed with suitably prepared Caselet schedule to identify a set of variables/criteria and individual Caselets are developed. Due to the brevity of the chapter, the Caselets developed are not presented in this chapter. From the analysis of the seven Caselets developed, 27 unique variables/criteria are identified for CPEE of SEs.

As the inference and/or finding obtained from Caselet approach cannot be generalized, another phase of exploratory research based on semistructured interviews is conducted among 58 SEs by developing an appropriate interview schedule using the finding from the Caselet approach. At the end of this phase, 35 variables/criteria are identified based on the opinion from SEs for CPEE to offer R&R and they are presented in Table 11.1 (5th column).

The list of variables/criteria identified based on 58 SEs for CPEE to offer R&R is required to be cross-verified from the administrative employee's (i.e., PM's) perspective for the purpose of offering R&R. In order to cross-check and confirm the variables/criteria identified from SEs to offer R&R, 31 PMs are interviewed by developing a suitable interview schedule based on the 35 variables identified from SEs' perspectives. At the end of this stage, 29 variables are confirmed from the list of variables identified from SEs' perspective by the PMs. In addition, the PMs added four new variables/criteria from their perspectives to offer R&R. Accordingly, the list of 33 variables (i.e., 29 variables confirmed by PMs from the list of variables identified based on SEs' perspective and four variables exclusively included based on PM's perspective) is considered based on PMs' perspective and the same is presented in Table 11.1 (6th column).

By comparing the list of variables presented in column 2 to column 6 of the Table 11.1, in this study, 33 variables, which are exactly accepted by PMs, are considered to offer R&R. The 33 variables considered for offering R&R are listed with description in Table 11.2. Accordingly, in this study, the CPEE to offer R&R is considered as a function of all the 33 variables listed

TABLE 11.1

A Consolidated List of Variables Identified for CPEE to Offer R&R

No.	Name of the Variable/Criterion	Variables Considered from Literature	Variables/Criteria To Be Considered for CPEE to Offer R&R			Final Set of Variables/Criteria Considered for CPEE to Offer R&R
			Identified from Researcher's Perspective	Software Engineers' Perspective	Project Managers' Perspective	
1	Age	✓	–	✓	✓	✓
2	Gender	✓	–	–	–	–
3	Marital status	✓	–	–	–	–
4	Education	✓	–	✓	✓	✓
5	University/institution	✓	–	✓	✓	✓
6	Experience	✓	–	✓	✓	✓
7	Tenure	✓	–	✓	✓	✓
8	Personality	✓	–	–	–	–
9	Parent's education	✓	–	–	–	–
10	Child status	✓	–	–	–	–
11	Quantity of work	✓	–	✓	–	–
12	Timeline adherence	✓	–	✓	✓	✓
13	Customer interaction	✓	–	✓	✓	✓
14	Target achievement	✓	–	✓	–	–
15	Timely reporting	✓	–	✓	✓	✓
16	Documentation	✓	–	✓	✓	✓
17	Reviewing	✓	–	–	✓	✓
18	Analytical ability	✓	–	–	✓	✓
19	Work planning	✓	–	–	–	–

(Continued)

TABLE 11.1 (*Continued*)

A Consolidated List of Variables Identified for CPEE to Offer R&R

No.	Name of the Variable/ Criterion	Variables Considered from Literature	Variables/Criteria To Be Considered for CPEE to Offer R&R			Final Set of Variables/Criteria Considered for CPEE to Offer R&R
			Identified from Researcher's Perspective	Software Engineers' Perspective	Project Managers' Perspective	
20	Creativity	✓	–	✓	✓	✓
21	Communication skills	✓	–	✓	✓	✓
22	Knowledge updation	✓	–	✓	✓	✓
23	Initiative	✓	–	✓	✓	✓
24	Understanding big picture	✓	–	✓	✓	✓
25	Additional responsibilities	✓	–	✓	✓	✓
26	Presentation skills	✓	–	–	✓	✓
27	Negotiation skills	✓	–	✓	✓	✓
28	Ideas/suggestions	✓	–	–	✓	✓
29	Innovation	✓	–	–	–	–
30	Patents/publications	✓	–	–	–	–
31	Self-learning	✓	–	✓	✓	✓
32	Leadership	✓	–	✓	✓	✓
33	Team cooperation	✓	–	✓	✓	✓
34	Punctuality	✓	–	✓	✓	✓
35	Mentoring	✓	–	–	✓	✓
36	Perseverance	✓	–	–	–	–
37	Humor sense	✓	–	–	–	–
38	Critical thinking	✓	–	–	–	–

(*Continued*)

TABLE 11.1 (Continued)

A Consolidated List of Variables Identified for CPEE to Offer R&R

No.	Name of the Variable/Criterion	Variables Considered from Literature	Variables/Criteria To Be Considered for CPEE to Offer R&R			Final Set of Variables/Criteria Considered for CPEE to Offer R&R
			Identified from Researcher's Perspective	Software Engineers' Perspective	Project Managers' Perspective	
39	Passion	✓	–	–	–	–
40	Resilience	✓	–	–	–	–
41	Commitment	✓	–	✓	✓	✓
42	Knowledge sharing	✓	–	✓	✓	✓
43	Proactiveness	✓	–	–	–	–
44	Code of conduct	✓	–	✓	✓	✓
45	Social volunteering	–	✓	–	–	–
46	Agility	–	✓	✓	–	–
47	Corporate social responsibility	–	✓	–	–	–
48	Business domain knowledge	–	✓	✓	–	–
49	Multitasking	–	✓	✓	–	–
50	Parent's occupation	–	✓	✓	–	–
51	Parent's domicile	–	–	✓	–	–
52	Improving morale	–	–	✓	✓	✓
53	Quality of the job	–	–	✓	✓	✓
54	Process adherence	–	–	✓	✓	✓
55	Cost saving	–	–	✓	–	–
56	Cocurricular activities	–	–	–	–	–
57	Best practice	–	–	–	✓	✓
	Total number of variables/criteria	44	07	35	33	33

TABLE 11.2

Criteria Considered Based on the Perspectives of Both SEs and PMs for CPEE to Offer R&R

No.	Variables/Criteria Considered for CPEE to Offer R&R	Description
1	Age	Age of the software engineer (SE)
2	Education	Completed highest education level of SE
3	University	University/institution where the highest education was obtained
4	Tenure	Number of years spent in the current organization
5	Experience	Total years of relevant experience in a similar profile
6	Quality of the job	The output produced should conform to the requirements/expectation
7	Timeline adherence	The task should be completed on time or ahead of time
8	Process adherence	Standard process for the job execution should be adhered
9	Customer interaction	Ability to communicate with the client/customer to convey and elicit required information
10	Documentation	Creating, updating, and maintaining all documents relating to the job
11	Reviewing	Willingness and ability to review other documents, codes, etc.
12	Timely reporting	Reporting the progress or defects on time, so that corrective action can be taken with minimal loss
13	Analytical ability	Ability to think in a logical and analytical manner
14	Best practice	Display best practice in process and quality
15	Communication skills	Ability to convey ideas orally and verbally
16	Ideas and suggestions	Recommend valid and implementable improvement suggestions
17	Knowledge updation	Keeping oneself updated with the knowledge in the fields by certifications and other relevant qualifications
18	Negotiation skills	Ability to confer with another person/department in the team/organization in order to come to terms or reach an agreement
19	Cost saving	Demonstrate measures to save cost for the project and organization
20	Presentation skills	Ability to present the ideas to audience
21	Understanding big picture	Ability to comprehend and understand the big picture of the job assigned
22	Additional responsibilities	Willingness to take up additional administrative responsibilities that fall outside the normal scope of job (like interviewing, auditing, etc.)
23	Creativity	Ability to think and come up with some creative solutions/process
24	Initiative	Proactiveness, innovating processes, which influence the project

(Continued)

TABLE 11.2 (*Continued*)

Criteria Considered Based on the Perspectives of Both SEs and PMs for CPEE to Offer R&R

No.	Variables/Criteria Considered for CPEE to Offer R&R	Description
25	Self-learning	Willingness to learn required knowledge and skills required for the job on one's own, rather than depend on formal training
26	Knowledge sharing	Willingness to share knowledge with other team members
27	Team cooperation	Ability to be flexible, accommodative, and work in a team
28	Mentoring	Genuinely interested in and working toward developing the skills and abilities of a junior member
29	Commitment	The dedication, ownership, and accountability shown in completing the task
30	Code of conduct	Disciplined behavior; comply with the organization's code of conduct, and uphold organizational values
31	Punctuality	Attending the workplace and meetings on time, regularly
32	Leadership	Ability to step up and direct the team to the right course of action
33	Improving morale	Take effort to improve the team spirit

in Table 11.2. Upon carefully scrutinizing the variables listed in Table 11.2, it can be observed that the first five variables: age, education, university, tenure, and experience can be grouped to the *demographic characteristic of the SEs* and called DCSE in this study, while the remaining 28 variables can be grouped to *performance of SEs* and they are called PSE in this study.

Though the final list of 33 variables identified based on (a) analyzing the literature review, and (b) both perspectives' of SEs and PMs using exploratory research methods, the importance of these variables for CPEE to offer R&R is not ascertained based on large-scale opinion of SEs. Accordingly, this is done from the opinion sought a sample of 443 SEs from 12 different IT organizations by carrying out descriptive research.

For conducting the descriptive research, a questionnaire is designed incorporating the 33 variables and the respondents are asked to rate the importance of each of these variables for CPEE to offer R&R. The responses are sought on a 7-point Likert scale (7 being extremely important). Using the data obtained from 443 respondents, a bivariate analysis is conducted using *t*-test to identify the significance of five demographic variables which may influence the CPEE to offer R&R. The details on the bivariate analysis with *t*-test for the significance of demographic variables on the influence of CPEE of SEs are shown in Table 11.3. From Table 11.3, it is clear that only three demographic variables viz. *education (E)*, *university (U)*, and *experience (X)*

TABLE 11.3

Bivariate Analysis and the Significant Demographic Variables in DCSE

Demographic Variable	Bifurcation Characteristics	N	Proactive Mean (SD)	Prompt Mean (SD)	Resourceful Mean (SD)	Responsible Mean (SD)	Diagnostic Mean (SD)	Dynamic Mean (SD)
Age	≤23 years	218	22.9 (3.44)	23.763 (3.054)	24.645 (2.79)	34.503 (4.77)	28.813 (3.45)	18.504 (1.84)
	>23 years	225	23.802 (2.8)	23.114 (2.89)	24.870 (2.66)	34.24 (4.68)	28.41 (3.38)	18.395 (1.72)
			$t = 1.59$	$t = 2.23^*$	$t = 0.87$	$t = 0.59$	$t = 1.24$	$t = 0.64$
Education	UG	311	22.174 (3.401)	23.647 (2.69)	24.902 (2.97)	34.017 (4.23)	28.11 (3.74)	18.222 (1.9)
	PG	132	23.784 (2.78)	22.925 (3.07)	24.261 (2.4)	34.982 (4.95)	28.956 (3.56)	18.871 (1.82)
			$t = 4.79^{**}$	$t = 2.47^*$	$t = 2.19^*$	$t = 2.08^*$	$t = 2.21^*$	$t = 3.33^{**}$
University	IIX	27	23.781 (2.6)	24.48 (3.15)	25.63 (3.12)	35.288 (5.02)	30.04 (4.18)	19.104 (1.88)
	Non-IIX	416	23.047 (3.07)	23.072 (3.06)	24.477 (2.89)	33.47 (4.53)	28.205 (3.48)	18.377 (1.75)
			$t = 2.76^*$	$t = 2.31^*$	$t = 1.95$	$t = 2.00^*$	$t = 2.62^{**}$	$t = 2.08^*$
Tenure	≤3 years	262	23.115 (3.27)	23.440 (3.43)	24.512 (2.64)	34.63 (4.47)	28.422 (3.55)	18.51 (1.83)
	>3 years	181	22.94 (3.0)	23.911 (2.88)	24.808 (2.87)	34.895 (4.86)	28.83 (3.72)	18.79 (1.81)
			$t = 0.57$	$t = 1.51$	$t = 1.12$	$t = 0.59$	$t = 1.16$	$t = 1.56$
Experience	≤3 years	280	23.101 (3.1)	23.911 (2.84)	24.205 (2.44)	34.2.1 (4.56)	28.3 (3.29)	18.314 (1.86)
	>3 years	163	23.718 (2.55)	23.27 (3.1)	24.73 (2.74)	35.589 (4.8)	29.41 (3.86)	18.86 (1.78)
			$t = 2.15^*$	$t = 2.21^*$	$t = 2.09^*$	$t = 3.03^{**}$	$t = 3.21^{**}$	$t = 3.03^{**}$

*significant at $p < 0.05$, **significant at $p < 0.01$.

have significant influence in all the six main criteria. Accordingly, these three significant demographic variables (*E, U,* and *X*) are considered to constitute DCSE.

The importance of the 28 variables (i.e., variables numbered from 6 to 33 in Table 11.2), grouped under PSE, is subjected to statistical analysis using factor analysis and structural equation modeling (SEM). The factor analysis is performed so as to identify the latent structure of the 28 variables, if any, so as to group them under the same factor. The factor analysis yielded six factors and they are named as: *proactive, prompt, resourceful, responsible, diagnostic,* and *dynamic.* The corresponding variables being manifested by each of these six factors are shown in Table 11.4. These six factors are considered as the main criteria against which SEs will be evaluated to ascertain their performance.

Summarizing the analysis based on *t*-test and factor analysis, the three demographic variables (i.e., DCSE) and the six factors/main criteria: *proactive, prompt, resourceful, responsible, diagnostic,* and *dynamic* related to PSE for SEs are considered to be important for CPEE to offer R&R. Based on these, a framework is proposed for CPEE to offer R&R and is shown in Figure 11.1.

Furthermore, the following hypotheses are also proposed based on the framework shown in Figure 11.1:

H$_1$: DCSE is positively related to R&R.

H$_2$: DCSE is positively related to each of the six main factors/criteria in PSE.

H$_3$: Each of the six main factors/criteria in PSE is positively related to R&R.

In order to test the proposed hypotheses, a partial least square (PLS)-based SEM is performed using R software (version 3.2.3). The path coefficients and corresponding *t*-values (given in parentheses) are shown in Figure 11.2.

Upon analyzing the path diagram and the corresponding path coefficients linking each of the six factors related to PSE with that of DCSE, it is observed that DCSE is positively influencing all the main criteria in PSE, while it has a negative path coefficient with R&R. It indicates that the hypothesis H$_1$ is rejected and hypothesis H$_2$ is accepted. From this, it is inferred that (i) DCSE has no direct effect on R&R, and (ii) DCSE has positive and direct influence on PES. This is taken into consideration while developing the performance evaluation model. Further, it is also noticed that PSE and each of the main six factors/criteria in PSE have positive path coefficients relating to R&R. As the path coefficients confirm the proposed hypothesis H$_3$, it is concluded that each of the six main factors/criteria in PSE is positively related to R&R.

The result of the H$_2$ indicates that PSE is found to be influenced by DCSE. Although DCSE is not directly related to R&R, DCSE is considered to have a moderating effect on the relation between PSE and R&R from the point of view

TABLE 11.4

Main Factor/Criterion-Wise List of Variables Grouped under PSE by Factor Analysis

Factor 1: Proactive	Factor 2: Prompt	Factor 3: Resourceful	Factor 4: Responsible	Factor 5: Diagnostic	Factor 6: Dynamic
Knowledge updation	Timeline adherence	Understanding big picture	Additional responsibilities	Quality of the job	Customer interaction
Initiative	Timely reporting	Ideas and suggestions	Knowledge sharing	Documentation	Communication
Self-learning	Process adherence	Creativity	Commitment	Analytical ability	Negotiation
Leadership	Punctuality	Cost saving	Teamwork	Reviewing	–
–	–	–	Mentoring	Presentation	–
–	–	–	Improving morale	–	–

Note: The variables "Best Practice" and "Code of Conduct" listed under PES were not seen in the above table because these two variables are not considered for factor analysis owing to low KMO value.

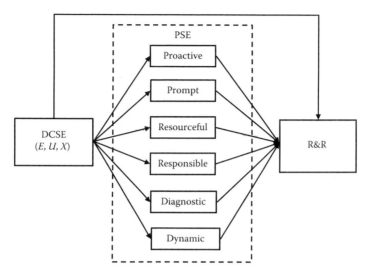

FIGURE 11.1
A proposed framework for CPEE to offer R&R.

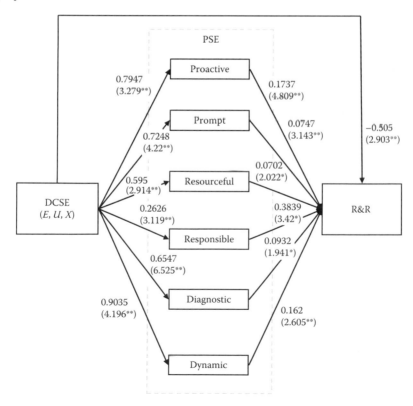

FIGURE 11.2
Proposed framework for CPEE to offer R&R with the path coefficients.

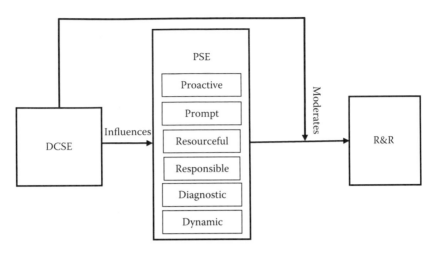

FIGURE 11.3
Final version of the framework for CPEE to offer R&R.

of the performance evaluation of an employee. That is, the DCSE influences the PSE as well as regulates the direct relation between PSE and R&R. Based on the result of these hypotheses, the framework shown in Figure 11.1 is modified and the final framework for CPEE to offer R&R is shown in Figure 11.3. From Figure 11.3, it is clear that multiple factors/criteria are needed for CPEE to offer R&R. Particularly these multiple criteria have to be jointly evaluated so as to determine the overall performance of the SE on these criteria.

As multiple criteria are involved for the CPEE, such evaluation can be effectively done using MCDM method(s). This necessitates the need of identifying suitable MCDM method(s), considering the factors/criteria related to PSE and the moderating factor related to DCSE, for CPEE to offer R&R. To identify and implement suitable MCDM method(s) for CPEE to offer R&R, various MCDM methods considered in the literature for performance evaluation of employee in general are reviewed and the same is discussed in the next section briefly.

11.3 Performance Evaluation of Employees Using MCDM Methods

Although the MCDM methods have gained prominence since 1970s (Zionts and Wallenius 1976), it appears that the application of such methods for performance evaluation of employees gained momentum only during the last decade. Though our aim in this study is not to give a complete and exhaustive list of MCDM methods considered in the literature for performance evaluation of employees, a sample of MCDM methods considered for performance evaluation of employees is presented to highlight the need of MCDM method(s) for CPEE of SEs to offer R&R.

A popular MCDM method called *analytical hierarchy process* (AHP) is introduced by Thomas L. Saaty during 1970s. The application of AHP for performance evaluation of employees through a hypothetical example is demonstrated in Saaty (1990). In the hypothetical example, the researcher evaluated three employees and created the hierarchy for evaluation and the relative priorities through pairwise comparison. The final scores are obtained for the three employees.

A real-time application of AHP is described by Taylor III et al. (1998) to evaluate candidates for the position of Dean at Texas A&M University at Kingsville. They have followed AHP with Saaty's scale to develop priorities and have compared a total of 33 employees who are considered for promotion. They faced difficulty in making pairwise comparison and divided the 33 employees considered for evaluation into three groups and the suitable candidate is chosen among a pool of prospective candidates. They have recommended that the number of employees being compared (i.e., the number of alternatives) should be limited to a maximum of eight.

Colson (2000) utilizes the outranking MCDM methods: *ELmination Et Choice Translating Reality* (ELECTRE) and *Preference Ranking Organization METHod for Enrichment Evaluations* (PROMETHEE) to evaluate the judgments made by decision makers on the selection of a suitable student to be considered for a scientific award. ARGOS (Aid to the Ranking to be made by a Group of decision makers using an Outranking Support), a software, is developed in order to validate the group decision. A model for performance evaluation of employees using AHP is proposed by Hemaida and Everett (2003). They developed the model in order to rank the employees for the purpose of salary allocation (i.e., the best-performing employee would receive a greater increase in pay). They have claimed that the director is able to use the proposed AHP to evaluate four employees and allocate the salary increase according to their performance. The authors have asserted that AHP provides a sense of objectivity with significantly different conclusion from the previous traditional evaluation methods used in the firm.

The application of AHP in a Malaysian service organization to evaluate the employees is detailed by Islam and Rasad (2006). The supervisors are provided training on using AHP and they are instructed to evaluate employees accordingly. Overall ranking of each employee is obtained, and they are able to identify the best-performing employees during the appraisal period. They have highlighted that the human resource department of the organization testified that the evaluation using AHP has generated a more reliable set of information. They recommended that performance evaluations should be an ongoing process. However, they do not provide any recommendation about how to use the AHP for CPEE.

A combination of two MCDM methods is attempted by Han and Ji (2009) to address the problem of performance evaluation of employees. They have developed a theoretical synthetic evaluation method based on AHP and fuzzy mathematics. By effectively combining the two techniques, the researchers confirm that a comprehensive performance evaluation of employees can be carried out for making the performance evaluation process more scientific and systematic.

Data envelopment analysis (DEA) is another MCDM method, which can be used to evaluate employee performance. Manoharan et al. (2009) detail one such process of performance evaluation using DEA to evaluate 18 employees of an Indian automotive organization. The factors considered by them for evaluation are classified into input and output factors. Based on these factors, they are able to rank the employees according to their performance. They underline that well-defined procedures and guidelines incorporated in DEA provide the performance appraisal and also reduce bias and subjectivity.

A fuzzy model for performance evaluation of employees using *Choquet Integral* (CI) and *Measuring Attractiveness by a Categorical Based Evaluation TecHnique* (MACBETH) is developed by Gurbuz (2010). This fuzzy model is validated in a medical equipment manufacturing organization, and a final ranking of five employees is produced. The author asserts that this model addressed the drawback of AHP (i.e., the assumption of independence among criteria/alternatives), by incorporating the interaction effect of criteria.

A decision support system is developed by Taufiq and Sugiharto (2011) using AHP for employee evaluation in a healthcare organization in Indonesia. The main objective of the performance evaluation is to select the employee for promotion purpose. The authors claim that the management is able to iden-tify the suitable employee for promotion. A popular outranking methods for evaluating and ranking the employees using MCDM technique—ELECTRE—is used to compare the performance of five employees in a telecom organiza-tion, based on seven criteria (Afshari et al. 2010). They have used ELECTRE method along with AHP technique to sort the employees according to their performance and rank them.

Pugh matrix method (PMM) is another MCDM method, which satisfies the pairwise comparison conditions of outranking methods put forward by Bouyssou (1996). Nada (2010) assesses an organizational innovation process and employee contribution toward organizational innovation. The author evaluates the employee participation in organizational innovation process using PMM. Specifically the author uses PMM method to outrank the ideas suggested by the employees. Although this is not widely used as a com-prehensive performance evaluation of employees method, PMM method is found to be simple compared with other outranking methods (Scharge 2010).

A conceptual framework based on MCDM methods, *fuzzy AHP* (FAHP), simple additive weighting (SAW), and *technique for order of preference by simi-larity to ideal solution* (TOPSIS) is proposed by Ardabili (2011) to evaluate the performance of bank staff. The author proposes that FAHP could be used for pairwise comparison of the criteria. Further, SAW and TOPSIS are proposed for evaluating the employees. The performance of vocational teachers is evaluated by Jati (2011) using MCDM methods: AHP and PROMETHEE II, particularly for evaluating three school teachers based on 10 criteria. AHP is used to determine the weights of the criteria using pairwise comparison. Further, the author uses PROMETHEE II to calculate preference functions for all pairs of alternatives, across the ten criteria considered in that study. The net outranking flow is calculated and based on which the performance of the three teachers is ranked.

Wu et al. (2012) uses a fuzzy MCDM approach to evaluate the employee performance of aircraft maintenance staff. They use FAHP and *Vise Kriterijumska Optimizacija I Kompromisno Resenje* (VIKOR) to develop a fuzzy MCDM method to evaluate 51 employees based on four performance dimensions. They observe that the ranking result obtained from the fuzzy MCDM method is differed from the traditional evaluation model. They also claim that their model brought in better fairness to the evaluation process.

Using DEA, a study is conducted to evaluate the performance efficiencies of banking professionals in Slovenia (Zbranek 2013) considering salary, working conditions, and benefits as input variables, and work motivation, job satisfaction, and organizational commitment as output factors. They observe that 12 employees among 60 are fully efficient and the remaining 48 are recommended for training. Islam et al. (2013) uses FAHP and TOPSIS to evaluate the performance of banking professionals based on five criteria. They use FAHP to calculate the weights of these criteria and subcriteria. Further, they evaluate the employees using TOPSIS by assigning triangular fuzzy numbers to the subcriteria. Based on these scores, the employee with a score closest to the ideal solution and farthest from the nonideal solution is identified for efficient one and the ones who need training.

Morte et al. (2013) conducted the performance evaluation of 31 road drivers in a Portuguese logistic company. As the number of employees (alternatives) is large, they do not use AHP so as to eliminate the pairwise comparison process. Instead they use PROMETHEE and *Methodologia Multicriterio para Apoio a Selecao de Sistemas de Informacao* (MMASSI) for evaluating two groups consisting of two and three decision makers, respectively, on 11 criteria and 31 employees. These evaluations are then compared with a set of self-evaluation made by the employee themselves to finally rank the employees according to their performance.

Gurbuz and Albayrak (2014) consider both *analytical network processing* (ANP) and CI to evaluate the employees of a pharmaceutical organization. They highlight that the traditional methods for performance evaluation of employees are highly subjective and do not provide a comprehensive information about the employee performance. They propose a framework based on ANP and CI based on three major criteria: sales-related performance criteria, customer-related performance criteria, and relations-related performance criteria. In addition, they consider organizational climate and demographic variables as moderating factor. Finally, they recommend that it is important to understand the interdependencies of the criteria and alternatives which influence the decision-making process.

A comparative study of various performance evaluation methods such as BARS, 360° method, assessment center, etc. as well as MCDM methods such as AHP, FAHP, fuzzy TOPSIS, etc. is done by Shaout and Yousif (2014). It appears that they have not provided an in-depth comparative analysis on the MCDM techniques considered for comparison and they have not suggested the conditions under which the suitable MCDM techniques can be adopted. Recently, Sahoo et al. (2016) used DEA method to identify the top researchers in various focus areas of management, based on six evaluation criteria and have identified the top 10 subject matter experts in eight academic areas.

A summary on the performance evaluation of employees using MCDM methods discussed in this section is presented in Table 11.5. It can be observed from Table 11.5 that AHP and its variants (FAHP) are widely used across industries for performance evaluation of employees. Further, from Table 11.5, it can be observed that there are few other methods, commonly called outranking methods, used for performance evaluation of employees. In general, the outranking methods build a preference function (called outranking relation) among the alternatives which are evaluated on several criteria (Bouyssou 1996, Bouyssou and Vincke 1997). This relation is built through a series of pairwise comparisons of alternatives. This would enable the decision maker to conclude that alternative A is better than B, if a majority of the evaluation criteria supports this conclusion (Bouyssou 2009). Some of the MCDM methods satisfying this property of building an outranking relation, which are already mentioned in Table 11.5, are: ELECTRE, PMM, TOPSIS, PROMETHEE, etc.

Inferring from Table 11.5, in general, the MCDM method(s) are mostly used to evaluate employees for the purpose of selection of a new employee, offering salary revision, offering promotion, etc. However, *there is no known literature for CPEE for choosing the best-performing employee to provide R&R, particularly for SEs.* In order to address this gap, an MCDM-based model for CPEE to offer R&R is attempted in this chapter. As the proposed CPEE system is a repetitive process, the choice of MCDM method(s) to offer R&R should satisfy the following characteristics:

1. The choice of the MCDM method(s) should complete the performance evaluation for offering R&R in the shortest possible time (*time-duration*).

2. The proposed MCDM-based modeling framework for CPEE to offer R&R should be easy to understand and easy to use by the PM (*easiness*).

3. Over a period of time, one or a few criteria considered would become irrelevant or not sufficient and due to this a few criteria have to be removed or have to be added to from the list of variables/criteria considered in the proposed MCDM-based modeling framework for CPEE to offer R&R. Thus, the selected MCDM method(s) should be flexible enough to add/remove criteria and/or alternatives (*flexibility*).

TABLE 11.5

Sample Applications of MCDM Methods in Performance Evaluation of Employees

No.	MCDM Method	Purpose	Industry	Reference
1	AHP	Selection and promotion	Service	Taylor III et al. (1998)
2	AHP	Salary revision	Service	Hemaida and Everett (2003)
3	AHP	Performance appraisal	Service	Islam and Rasad (2006)
4	AHP	Promotion	Service	Taufiq and Sugiharto (2011)
5	FAHP	Performance appraisal	Theoretical study	Han and Ji (2009)
6	DEA	Performance appraisal	Manufacturing	Manoharan et al. (2009)
7	DEA	Performance efficiency	Service	Zbranek (2013)
8	DEA	Subject expert identification	Service	Sahoo et al. (2016)
9	ELECTRE	Performance appraisal	Manufacturing	Afshari et al. (2010)
10	PMM	Evaluation of ideas from employees	Manufacturing	Nada (2010)
11	PROMETHEE	Performance prediction and selection	Service	Jati (2011)
12	FAHP and VIKOR	Performance appraisal	Service	Wu et al. (2012)
13	FAHP, TOPSIS	Performance appraisal	Service	Islam et al. (2013)
14	ELECTRE, PROMETHEE	Student award	Service	Colson (2000)
15	CI and MACHBETH	Performance appraisal	Manufacturing	Gurbuz (2010)
16	PROMETHEE and MMASSI	Performance appraisal	Service	Morte et al. (2013)
17	ANP and CI	Performance appraisal	Manufacturing	Gurbuz and Albayrak (2014)
18	FAHP, SAW, and TOPSIS	Performance appraisal	Service	Ardabili (2011)

Based on the expected characteristics of MCDM method(s), as stated above, for CPEE to offer R&R and frequent and successful implementation of AHP for performance evaluation of employees, in this study, we propose to develop a modeling framework based on AHP and an MPMM for CPEE to offer R&R. The MCDM, AHP, is considered for computing weights based on the importance of the six main factors/criteria: *proactive, prompt, resourceful, responsible, diagnostic,* and *dynamic,* and the moderating factor: DCSE considered for CPEE to offer R&R. Once the importance of the main factors/criteria along with the moderating factors is obtained in the form of weights, each SE (i.e., each alternative) is compared to select the best-performing SEs to offer R&R using the MPMM. The complete implementation details of AHP and MPMM for CPEE to offer R&R are discussed in the next section.

11.4 Development of an MCDM-Based Modeling Framework for CPEE to Offer R&R

In this section, the implementation details of the proposed MCDM-based modeling framework for CPEE to offer R&R are presented in the form of step-by-step procedure as follows:

Step 1: Determine the importance of the variables/criteria of DCSE using AHP

For k variables in DCSE, the relative importance or weights (w_j) can be determined using AHP, using Saaty's basic scale.

Step 2: Determine the normalized demographic score (NDS) for each SE

A decision matrix can be developed to assign uniform range of ordinal values (d_{ij}) for m SEs to each of the demographic variables, considered for DCSE. For each of the m SEs, the sum of products of weights (w_j) and ordinal value (d_{ij}) for three demographic variables would give the demographic score (DS) and it can be represented as:

$$DS_i = \sum_j^k w_j * d_{ij} \quad \text{for all } i = 1, 2, \ldots, m. \tag{11.1}$$

The normalized DS (NDS) of the m SEs is calculated as follows:

$$NDS_i = \frac{DS_i}{TDS} \quad \text{for all } i = 1, 2, \ldots, m, \tag{11.2}$$

where TDS (total demographic score) = $\sum_{i=1}^{m} DS_i$.

Step 3: Determine the importance of each of the main factors/criteria of PSE using AHP

For *each of the* main factors/criteria of PSE, the relative importance or weights (v_j) can be determined using AHP with Saaty's basic scale.

Step 4: Evaluate the SEs based on the main criteria of PSE using MPMM

In the PMM, one alternative is selected as a baseline (B), and all other alternatives are compared against B, with respect to each criterion (Pugh 1991). The comparison is denoted as −1 for worse score, 0 for equal score and +1 for better score. This evaluation results in a column vector with scores for all the alternatives other than the baseline (whose score will be zero). The alternative with the highest positive score would be chosen as the best alternative.

The PMM has been criticized highlighting three major limitations (Mullur et al. 2003): (i) the criteria weights are not incorporated, (ii) the rating scale is too small, and (iii) the selection of random baseline alternative could lead to bias. These limitations are addressed in this chapter by (i) using AHP to calculate the criteria weights, (ii) by increasing the evaluation range of the scale similar to a 5-point Likert scale, and (iii) by selecting all alternatives to be baseline and all other alternatives could be evaluated against the baseline. This implies a modification to the original PMM and referred as modified PMM (MPMM) in this chapter.

The evaluation scale for the MPMM for comparing the performance of two SEs A and B (where B is the baseline) would be by providing a qualitative measure of performance evaluation (g_{ij}) for a given criterion, *j* such that:

$$g_{ij} = \begin{cases} +2, \text{ if A performs much better than B in criterion } j \\ +1, \text{ if A performs slightly better than B in criterion } j \\ 0, \text{ if A performs as good as B in criterion } j \\ -1, \text{ if B performs slightly better than A in criterion } j \\ -2, \text{ if B performs much better than A in criterion } j \end{cases}$$

The g_{ij} is calculated for *m* SEs with respect to *each of the* main criteria considered for CPEE to offer R&R. Using the values of g_{ij} and v_j for *n* criteria, the performance score (PS) of SEs can be calculated as:

$$PS_i = \sum_j^n v_j * g_{ij} \quad \text{for all } i = 1, 2, \ldots, m; \text{ and } j = 1, 2, \ldots, n. \quad (11.3)$$

The normalized PS (NPS) for each of m SEs can be calculated as follows:

$$NPS_i = \frac{PS_i}{TPS} \quad \text{for all } i - 1, 2, ..., m, \tag{11.4}$$

where TPS (total performance score) = $\sum_{i=1}^{m} PS_i$.

In the original PMM, the NPS_i for m SEs with baseline SE as B_1 results in an $m \times 1$ column matrix (named as $\mathbf{C_1}$) with the first set of NPS_i for m SEs. In the MPMM, all m alternatives serve as baseline and hence it results in an $m \times m$ matrix such that:

$$\mathbf{A} = \begin{pmatrix} a_{11} & a_{12} & \cdots & a_{1m} \\ a_{21} & a_{22} & \cdots & a_{2m} \\ \cdots & \cdots & \cdots & \cdots \\ a_{m1} & a_{m2} & \cdots & a_{mm} \end{pmatrix},$$

where $a_{ij} = \mathbf{NPS}_{ij}$ which represents the NPS of ith SE when jth SE is the baseline.

Step 5: Determine the final performance score (FPS) of SEs

After executing Step 4 using the MPMM, it would result in m NPS score for every SE. The mean of NPS needs to be ascertained, such that

$$\overline{NPS_i} = \left[\sum_{j=1}^{m} \frac{NPS_{ij}}{m} \right] \quad \text{for all } i = 1, 2, ..., m.$$

For each of the m SEs, the final performance score (FPS_i) can be obtained by incorporating the moderating effect of DCSE with the six main factors relating to PSE. This can be computed in this study as follows:

$$FPS_i = \frac{\overline{NPS_i}}{NDS_i} \quad \text{for all } i = 1, 2, ..., m. \tag{11.5}$$

At the end of Step 5, the FPS_i of m SEs is considered for ranking the given set of SEs. Accordingly, the SE who has highest FPS can be offered R&R. If more than one has equal highest score, then everyone should be offered R&R. This completes one cycle (evaluation period) of CPEE of SEs.

The FPS_i of m SEs should be stored in an R&R database (RRDB), which could be accessed at a later point of time. Furthermore, at the end of every cycle (i.e., evaluation period) of CPEE, the FPS_i score of m SEs needs to be reset to zero. After initializing the FPS_i score of m SEs, a fresh evaluation needs to be carried out and the process needs to be repeated from Step 4.

The validation of the proposed MCDM-based modeling framework for CPEE of SEs for R&R presented here is illustrated with a numerical example in the following section.

11.5 Validation of the Proposed MCDM Modeling Framework for CPEE to Offer R&R

In order to validate the proposed MCDM-based modeling framework for CPEE to offer R&R, a hypothetical team of eight SEs with similar job profile is considered. Indicative demographic characteristics with respect to the variables: education (E), university (U) and experience (X) of the eight SEs (that is the required data for measuring the moderating factor/criterion: DCSE) are shown in Table 11.6. With these data, the performance evaluation process of eight SEs is detailed as follows:

Step 1: Determine the importance of the demographics variables: E, U, and X of DCSE using AHP

The initial evaluation process commences by determining the importance for the demographic variables: E, U, and X of DCSE. The weights are obtained for these demographic variables using AHP with Saaty's basic scale and the same is presented in Table 11.7. From Table 11.7, it can be observed that the demographic variable experience (X) demands more importance compared to the demographic variables: education (E) and university (U).

TABLE 11.6

Data on Demographic Characteristics of Eight Software Engineers for the Numerical Example

		Demographic Characteristics of Each Software Engineer		
No.	Name of the Software Engineer	Education (E)	University (U)	Experience (X) in Years
1	Anita	BTech	Cochin University	5
2	Benny	BE	Amrita University	2
3	Casper	BTech	NIT, Trichy	2
4	Deepak	MCA	Anna University	1
5	Esther	MTech	IIT Delhi	2
6	Faizel	BE	Manipal University	4
7	Gopi	ME	VTU	6
8	Harsha	BE	Anna University	1

TABLE 11.7

Computed Weights for Demographic Variables of DCSE Using AHP

Demographic Variables	E	U	X	Relative Importance of Demographic Variables by Weights
Education (E)	1	2	1/3	0.216
University (U)	1/2	1	1/7	0.103
Experience (X)	3	7	1	0.682

In this study, the relative importance (i.e., the normalized weights) obtained for the demographic variables: E, U, and X would remain unchanged for the PAS period unless a new demographic variable becomes significant for measuring DCSE or one of the demographic variables considered for measuring DCSE loses its significance for measuring DCSE. So, using these normalized weights for the demographic variables: E, U, and X of DCSE, the team of eight SEs can be evaluated so as to ascertain their demographic score (DS). This is detailed in the next step.

Step 2: Determine the demographic score (DS) for each SE

After ascertaining the weights for demographic variables, each of the eight SEs are rated on an ordinal scale, with respect to their demographic characteristics (Table 11.6). The ordinal ratings for each of the eight SEs are obtained according to the following scale:

$$Education, E = \begin{cases} 2, \text{ if Post Graduate (Masters)} \\ 1, \text{ if Under Graduate (Bachelors)} \end{cases}$$

$$University, U = \begin{cases} 2, \text{ if IIX(IIT/IISc/IIIT/NIT)} \\ 1, \text{ if Non-IIX} \end{cases}$$

$$Experience, X = \begin{cases} 2, \text{ if >3 years} \\ 1, \text{ if <= 3 years} \end{cases}.$$

Based on the scale defined for the demographic variables and the data given in Table 11.6, the ordinal ratings are obtained for each of the eight SEs and the same is given in Table 11.8. In addition, using Equations 11.1 and 11.2, both DS and NDS for each SE are calculated and presented in Table 11.8. The SE-wise NDS presented in Table 11.8 could be fixed for utmost 1 year (or for the PAS period), and hence the Step 2 need not to be repeated unless there is a change in the team of SEs considered for performance evaluation.

TABLE 11.8

Software Engineer-Wise Ordinal Ratings and the Computed Score on DS and NDS

Name of the Software Engineer	Weights for the Demographic Variable:			Computed Score	
	E: 0.216	U: 0.103	X: 0.682	DS	NDS
Anita	1	1	2	1.682	0.154
Benny	1	1	1	1.000	0.092
Casper	1	2	1	1.103	0.101
Deepak	2	1	1	1.216	0.112
Esther	2	2	1	1.318	0.121
Faizel	1	1	2	1.682	0.154
Gopi	2	1	2	1.897	0.174
Harsha	1	1	1	1.000	0.092

Step 3: Determine importance of the six factors/main criteria of PSE using AHP

The six main criteria defined to represent PSE need to be weighted for their relative importance. This is done using AHP utilizing Saaty's basic scale. The Saaty's basic scale for the six main criteria and the weight obtained from AHP toward the relative importance (i.e., normalized weights) of each of the six main criteria are shown in Table 11.9.

The computed weights for the six main criteria considered for CPEE to offer R&R, presented in Table 11.9, can be fixed for a relatively long duration (or for the PAS period), unless some criteria become irrelevant or some new criteria need to be added as part of PSE. Using these computed weights for the six main criteria, the performance of SEs can be evaluated using MPMM approach. This is detailed in the next step.

Step 4: Evaluate PSE based on MPMM to determine the PS of each SE

After computing the importance of weights for each of the six main criteria of PSE, the team of eight SEs is compared against each of the main criteria for evaluating their performance using MPMM. For proceeding with MPMM, the qualitative measure of performance evaluation (g_{ij}) for a given criterion j, as defined in Section 11.4, is followed to generate the data for each of the eight SEs against each of the criteria.

The pairwise comparison and the qualitative measure are presented in Table 11.10. Using the data given in Table 11.10, each of the SEs is considered as baseline employee and the MPMM process is applied to obtain the $m \times m$ matrix (i.e., 8×8 matrix for the numerical example) with NPS score of m (i.e., 8) SEs and is shown in Table 11.11 (columns 2–9).

TABLE 11.9

Computed Weights for the Six Factors/Main Criteria, Representing PSE Using AHP

Main Criteria	Main Criteria Representing PSE						Relative Importance of Main Criteria
	Proactive	Prompt	Resourceful	Responsible	Diagnostic	Dynamic	
Proactive	1	1/7	1/3	1	3	1/3	0.088
Prompt	7	1	5	3	4	3	0.406
Resourceful	3	1/5	1	3	1/2	1	0.151
Responsible	1	1/3	1/3	1	4	2	0.141
Diagnostic	1/3	1/4	2	1/4	1	1/2	0.086
Dynamic	3	1/3	1	1/2	2	1	0.128

TABLE 11.10

Performance Evaluation of SE Using MPMM

Name of the Software Engineer	Main Criteria Representing PSE with Its Weight (i.e., Importance) in Bracket					
	Proactive (0.088)	Prompt (0.406)	Resourceful (0.151)	Responsible (0.141)	Diagnostic (0.086)	Dynamic (0.128)
Anita	0	0	0	0	0	0
Benny	−1	0	1	2	0	−1
Casper	1	1	0	−1	0	1
Deepak	2	0	1	2	−1	−2
Esther	−2	2	0	1	1	−1
Faizel	1	0	0	−2	2	−1
Gopi	−1	−1	1	0	2	0
Harsha	−1	1	1	−1	0	1

Finally the SE-wise mean of NPS (i.e., $\overline{NPS}$) is computed and presented in Table 11.11 (column 10). From Table 11.11, particularly the column titled $\overline{NPS}$, it can be observed that Esther has the highest $\overline{NPS}$ and ranks first. Accordingly, Esther can be offered R&R, *if the demographic profile of SEs is not considered*. In order to be judicious and equitable in evaluation, it is important to consider the moderating effect of NDS on the $\overline{NPS}$. This is explained in the next step.

Step 5: Calculate the FPS by incorporating the moderating effect of NDS on $\overline{NPS}$

Each SE's $\overline{NPS}$ is moderated by respective NDS presented in Table 11.8 so as to obtain an unbiased FPS for each SE using Equation 11.5. Accordingly, the FPS for all the eight SEs are calculated and are given in Table 11.11 (last Column). From the last column of Table 11.11, it can be observed that after incorporating the moderating effect of DCSE, Harsha emerges as the top ranker. After Harsha is identified as the SE with the top FPS for offering R&R, one cycle/ evaluation period (could be weekly or fortnightly) of CPEE will be completed. The FPS_i of m (i.e., 8) SEs should be stored in RRDB, which could be accessed at a later point of time. Finally, as each cycle of CPEE to offer R&R is independent, the score on FPS at the commencement of each cycle needs to be reset to zero.

11.6 Managerial Implications

The emphasis in the CPEE to offer R&R is in the word "continuous." The proposed MCDM-based modeling framework in this study is very simple so any PM or the concerned administrator of an organization can conduct the performance evaluation using the proposed MCDM models and decide an

TABLE 11.11

The $\overline{NPS}$ and FPS for Each Software Engineer

Software Engineer	NPS Obtained Using MPMM Considering Baseline Employee as								$\overline{NPS}$	FPS
	Anita	Benny	Casper	Deepak	Esther	Faizel	Gopi	Harsha		
Anita	0.000	0.309	0.116	0.028	-0.075	0.406	0.278	-0.174	0.111	0.720
Benny	0.118	0.000	-0.129	0.340	-0.084	-0.024	0.000	0.205	0.053	0.582
Casper	0.263	0.317	0.000	-0.020	0.105	0.326	0.391	0.046	0.179	1.765
Deepak	0.146	0.328	0.014	0.000	0.411	-0.128	0.455	0.417	0.205	1.841
Esther	0.344	0.928	-0.046	0.283	0.000	0.234	0.014	0.000	0.220	1.816
Faizel	-0.081	-0.509	0.372	-0.021	0.340	0.000	-0.279	0.000	-0.022	-0.144
Gopi	-0.094	-0.165	0.055	0.188	-0.001	0.000	0.000	0.505	0.061	0.350
Harsha	0.249	-0.209	0.641	0.201	0.183	0.149	0.151	0.000	0.171	1.859

employee to offer R&R. The proposed R&R system loses its main objective if it is not offered on a timely and continuous basis. The frequency (i.e., evaluation frequency) of the CPEE may be recommended as weekly or fortnightly. One can also think of having the FPS as transparent to every employee of the team, involved in the evaluation system. When there is a continuous and transparent output from the system (i.e., timely R&R based on proper and transparent evaluation), the employee (i.e., SE) might feel energized and motivated to exhibit better performance. In addition, the PM has to decide an appropriate R&R to the deserving employee.

The numerical example illustrated in the previous section enables the PM to identify the best-performing SE(s) based on the relative maximum FPS after incorporating their respective demographic and performance-related factors. Instead of having relative FPS, the PM may specify a threshold FPS, beyond which all SEs can be offered R&R. Moreover, the proposed framework can also be used to identify the relatively worst-performing SEs and based on this the PM may recommend the specific SE(s) with the least FPS for special training to get performance improvement.

The FPS obtained during every performance evaluation cycle serves two purposes, as shown in Figure 11.4. The obvious purpose is to identify the best-performing SE(s) to offer immediate R&R. The second purpose is to cumulatively store the individual FPS of all SEs at every performance evaluation cycle in RRDB. This cumulative score in RRDB can be linked to organization's existing PAS and can be used appropriately during the organization's regular PAS. In addition to the cumulative FPS, data-indicate the best- and worst-performing SE(s) over a longer period of time.

Given that the performance evaluation cycle is related to a shorter period of time, the FPS of SEs during one performance evaluation cycle does not really convey comprehensive information about the actual performance of SEs. Hence, the CPEE system has to be a repetitive process with equal frequency. It is also important that the PM needs to conduct the CPEE in

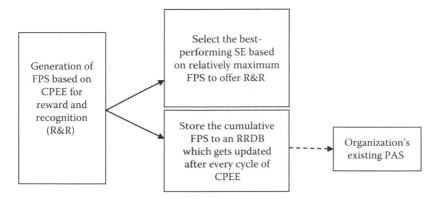

FIGURE 11.4
Purpose of the proposed MCDM-based modeling framework for CPEE to offer R&R.

an independent manner, that is, the FPS obtained during one cycle (performance evaluation cycle) should not influence the CPEE during the next cycle. To address this, after the top performing SE is offered R&R, at the end of one evaluation period, the FPS of all SEs in the team shall be set to zero and a fresh evaluation needs to be conducted for the next cycle. Such independent evaluation process when repeated in a continuous manner would produce a performance trend, which can easily be captured and interpreted using time series analysis.

The success of the proposed MCDM framework requires a commitment from the PM. The PM should tenaciously execute the CPEE process and identify the best-performing SE to offer R&R. Conducting the CPEE on a weekly or fortnightly frequency generates a large amount of data. The cumulative FPS data could serve as an objective input during the periodic appraisal process. Further, if a new PM assumes charge during the middle of an appraisal cycle, the data in the RRDB makes it easier for the new PM to understand the performance distribution in the team.

11.7 Conclusion

A new research problem on CPEE to offer R&R for an organization is attempted in this chapter, particularly focusing on IT organizations. A modeling framework for CPEE of SEs is proposed in order to offer R&R. In order to propose the modeling framework for CPEE to offer R&R, appropriate exploratory and descriptive research processes are carried out to identify suitable and adequate variables/criteria. Accordingly, 33 variables/criteria are identified. These variables/criteria are grouped into demographic variables (five numbers), which explain the DCSE and performance-related variables (28 numbers), which explain the direct PSE.

Based on the statistical test, out of five variables/criteria considered under DCSE, only three variables education (E), university (U), and experience (X) are statistically significant in representing the DCSE. Furthermore, the performance-related 28 variables/criteria are grouped into six main criteria/factors using factor analysis and in this study they are named as *proactive, prompt, resourceful, responsible, diagnostic,* and *dynamic* of PSE.

The AHP is used in this study to compute SE-wise demographic score, called normalized demographic score (NDS) and PSs, called normalized performance score (NPS). In addition, the proposed MPMM is appropriately implemented with the weighted scores: v_j obtained from AHP and finally obtained the mean NPS for each of the SEs. For introducing the moderating effect of DCSE and ranking the SEs, the data on NPS and NDS are used to obtain the score on FPS for each of the SEs. Finally, the SE who has the highest FPS will become the top performer for getting R&R.

This study makes the following contributions to the existing literature:

- A unique set of variables/criteria, related to demographic characteristics and performance evaluation characteristics of SEs, are identified for CPEE to offer R&R.
- A conceptual framework for CPEE to offer R&R is proposed, based on descriptive analysis and SEM analysis on the demographic and performance evaluation characteristics of SEs.
- After analysis on the existing literature on MCDM methods for performance evaluation of employees in general, MCDM methods, AHP and MPMM, have been implemented for demonstrating the proposed conceptual framework for CPEE to offer R&R.
- The proposed MCDM-based modeling framework for CPEE to offer R&R is demonstrated by developing a hypothetical data set.
- The score on cumulative FPS obtained from the MCDM-based modeling framework for CPEE to offer R&R is preserved for the possibility appropriately utilizing in the existing PAS.

Though the MCDM methods, AHP and MPMM, are successfully implemented for demonstrating the workability of the framework for CPEE of SEs to offer R&R, identifying different MCDM method(s) for its applicability toward CPEE to offer R&R and following by a systematic process for comparing various possible MCDM methods are the immediate research directions in this area. The development of interactive Excel-based software could be another possible extension for the research considered in this study.

_____

References

Afshari, A. R., M. Mojahed, R. M. Yusuff, T. S. Hong, and M. Y. Ismail. 2010. Personnel selection using ELECTRE. *Journal of Applied Sciences*. 10(23): 3068–3075.

Ardabili, F. S. 2011. New framework for modeling performance evaluation for bank staff departments. *Australian Journal of Basic and Applied Sciences*. 10: 1037–1043.

Arora, A., V. S. Arunachala, J. Asundi, and R. Fernandes. 2001. The Indian software services industry. *Research Policy*. 30: 1267–1287.

Bassett, G. A. and H. H. Meyer. 1968. Performance appraisal based on self-review. *Personnel Psychology*. 21(4): 421–430.

Bhatnagar, S. 2006. Indian software industry. V. Chandra (Ed.) 95–124. *Technology Adaptation and Exports: How Some Developing Countries Got It Right*. World Bank Publications: Washington, DC.

Boice, D. H. and B. H. Kleiner. 1997. Designing effective performance appraisal systems. *Work Study*. 46(6): 197–201.

Bouyssou, D. 1996. Outranking relations: Do they have special properties? *Journal of Multi-Criteria Decision Analysis.* 5: 99–111.

Bouyssou, D. 2009. Outranking methods. C. A. Floudas and P. M. Pardalos (Eds.) 2887–2983. *Encyclopedia of Optimization.* Springer: New York.

Bouyssou, D. and P. Vincke. 1997. Ranking alternatives on the basis of preference relations: A progress report with special emphasis on outranking relations. *Journal of Multi-Criteria Decision Analysis.* 6: 77–85.

Bowman, J. S. 1999. Performance appraisal: Verisimilitude trumps veracity. *Public Personnel Management.* 28(4): 557–576.

Bradler, C., R. Dur, S. Neckermann, and A. Non. 2012. Employee Recognition and Performance: A Field Experiment. CESIFO Working Paper No. 4164, Munich, Germany. https://www.cesifo-group.de/DocDL/cesifo1_wp4164.pdf (accessed February 22, 2014).

Brun, J. and N. Dugas. 2008. An analysis of employee recognition: Perspectives on human resource practices. *The International Journal of Human Resource Management.* 17(4): 716–730.

Burge, S. 2009. *The Systems Engineering Tool Box.* http://www.burgehugheswalsh. co.uk/uploaded/1/documents/pugh-matrix-v1.1.pdf (accessed March 12, 2015).

Cacioppe, R. 1999. Using team–individual reward and recognition strategies to drive organizational success. *Leadership & Organization Development Journal.* 20(6): 322–331.

Colson, G. 2000. The OR's prize winner and the software ARGOS: How a Multijudge and Multicriteria Ranking GDSS Helps a Jury to Attribute a Scientific Award. *Computers & Operations Research.* 27: 741–755.

Economic Times. 2015. Indian IT in FY15: What went wrong with TCS, Wipro & others as they struggled to grow revenues? http://articles.economictimes.indiatimes. com/2015-04-27/news/61578067_1_indian-it-industry-leader-tcs-anant-gupta/2 (accessed April 29, 2015).

Facteau, J. D. and S. B. Craig. 2001. Are performance appraisal ratings from different sources comparable? *Journal of Applied Psychology.* 86(2): 215–227.

Garafano, C. M. and E. Salas. 2005. What influences continuous employee development decisions. *Human Resource Management Review.* 15: 281–304.

Gray, G. 2002. Performance appraisals don't work. *Industrial Management.* 44(2): 15–17.

Gruenfeld, L. W. and P. Weissenberg. 1966. Supervisory characteristics and attitudes towards performance appraisals. *Personnel Psychology.* 19(2): 143–151.

Gurbuz, T. 2010. Multiple criteria human performance evaluation using Choquet integral. *International Journal of Computational Intelligence Systems.* 3(3): 290–300.

Gurbuz, T. and E. Y. Albayrak. 2014. An engineering approach to human resources performance evaluation: Hybrid MCDM application with interactions. *Applied Soft Computing.* 21: 365–375.

Han, N. and X. Ji. 2009. The study on performance evaluation based on AHP-fuzzy. *Proceedings of the Eighth International Conference on Machine Learning and Cybernetics,* Baoding, China: 2750–2753. http://ieeexplore.ieee.org/stamp/ stamp.jsp?tp=&arnumber=5212097 (accessed January 12, 2015).

Hansen, F., M. Michelle, and R. B. Hansen. 2002. Rewards and recognition in employee motivation. *Compensation and Benefits Review.* 34(4): 64–72.

Hemaida, R. and S. Everett. 2003. Performance evaluation of employees using the Analytic Hierarchy Process. *Journal of Management Information and Decision Sciences (Academy of Information and Management Sciences Journal).* 6(1–2): 67–76.

Henderson, R. I. 1980. *Performance Appraisal*. Prentice-Hall: Englewood Cliffs, NJ.

Ilgen, D. R., R. B. Peterson, B. A. Martin, and D. A. Boeschin. 1981. Supervisor and subordinate reactions to performance appraisal sessions. *Organizational Behavior and Human Performance*. 28: 311–330.

Islam, R. and S. M. Rasad. 2006. Performance evaluation of employees by the AHP: A case study. *Asia Pacific Management Review*. 11(3): 163–176.

Islam, S., G. Kabir, and T. Yesmin. 2013. Integrating analytic hierarchy process with TOPSIS method for performance appraisal of private banks under fuzzy environment. *Studies in System Science*. 1(4): 57–70.

Jati, H. 2011. Study on performance appraisal method of vocational education teachers using PROMETHEE II. Proceedings of International Conference on Vocational Education and Teaching (ICVET 2011), Yogyakarta, Indonesia: 139–145. http://eprints.uny.ac.id/5926/1/ICVET_paper.pdf (accessed November 3, 2014).

Kondrasuk, J. 2011. So what should an ideal performance appraisal look like? *Journal of Applied Business and Economics*. 12(1): 57–71.

Manoharan, T. R., C. Muralidharan, and S. G. Deshmukh. 2009. Employee performance appraisal using data envelopment analysis: A case study. *Research and Practice in Human Resource Management*. 17(1): 92–111.

Morte, R., T. Pereira, and D. B. M. M. Fontes. 2013. MCDA applied to performance analysis and evaluation of road drivers: A case study in the road transport company. *Proceedings of the Third International Conference on Business Sustainability*, Póvoa de Varzim, Portugal. https://repositorio-aberto.up.pt/bitstream/10216/70496/2/13674.pdf (accessed May 15, 2015).

Mullur, A., C. Mattson, and A. Messac. 2003. New decision matrix based approach for concept selection using linear physical programming. *Proceedings of 44th AIAA/ASME/ASCE/AHS Structures, Structural Dynamics, and Materials Conference*, Paper No. AIAA 2003-1446, Norfolk. www.ee.uidaho.edu/ee/classes/ECE482-F03/DM/LP-DM.pdf (accessed August 19, 2014).

Nada, N. 2010. A framework for systematic application and measurement of the innovation management process. *The Journal of Knowledge Economy and Knowledge Management*. 5: 57–69.

Palaiologos, A., P. Papazekos, and L. Panayotopoulou. 2011. Organizational justice and employee satisfaction in performance appraisal. *Journal of European Industrial Training*. 35(8): 826–840.

Pugh, S. 1991. *Total Design: Integrated Methods for Successful Product Engineering*. Addison Wesley: New York.

Rosen, A. and G. E. Abraham. 1966. Attitudes of nurses towards a performance appraisal system. *Nursing Research*. 15(4): 317–322.

Saaty, T. L. 1990. How to make a decision: The analytic hierarchy process. *European Journal of Operations Research*. 48: 9–26.

Sahoo, B. K., R. Singh, B. Mishra, and K. Sankaran. 2016. Research productivity in management schools of India during 1968–2015: A directional benefit-of-doubt model analysis. *Omega*. 66(A): 118–139, http://dx.doi.org/10.1016/j.omega.2016.02.004i

Scharge, D. P. 2010. *Product Lifecycle Engineering (PLE): An Application. Encyclopedia of Aerospace Engineering*. John Wiley & Sons Ltd: Atlanta.

Schraeder, M., B. J. Becton, and R. Portis. 2007. A critical examination of performance appraisals. *The Journal for Quality and Participation*. 30(1): 20–25.

Shaout, A. and M. K. Yousif. 2014. Performance evaluation: Methods and techniques survey. *International Journal of Computer and Information Technology*. 3(5): 966–979.

Smola, K. W. and C. D. Sutton. 2002. Generational differences: Revisiting generational work values for the new millennium. *Journal of Organizational Behaviour.* 23: 363–382.

Sreejith, S. S. 2015. Performance evaluation in IT needs to move from the manufacturing model. *Human Resource Management International Digest.* 23(1): 32–34.

Sreejith, S. S. 2016. Development of a Multi-Criteria Decision Making Model for Continuous Evaluation of Employee to Offer Reward and Recognition. Unpublished Ph.D. Thesis, Department of Management Studies, Indian Institute of Science, Bangalore, India.

Taufiq, R. and A. Sugiharto. 2011. The decision support system design of employee performance appraisal using analytical hierarchy process (AHP) method. *Proceedings of the 1st International Conference on Information Systems for Business Competitiveness (ICISBC)*, Semerang, Indonesia: 283–288.

Taylor, III, F. A., A. F. Ketcham, and D. Hoffman. 1998. Personnel evaluation with AHP. *Management Decision.* 36(10): 679–685.

Wu, H., J. Chen, and I. Chen. 2012. Performance evaluation of aircraft maintenance staff using a fuzzy MCDM approach. *International Journal of Innovative Computing, Information and Control.* 8(6): 3919–3937.

Zbranek, P. 2013. Data envelopment analysis as a tool for performance evaluation of employees. *Acta Oeconomica et Informatica.* 26(1): 12–21.

Zionts, S. and J. Wallenius. 1976. An interactive programming method for solving multiple criteria problem. *Management Science.* 22(6): 652–663.

12

Use of DEA for Studying the Link between Environmental and Manufacturing Performance

Ramakrishnan Ramanathan

CONTENTS

12.1 Introduction

We presently live in an era where data are being generated continuously and in several forms. These data have been called as the next big innovation (Gobble, 2013), and data analysts strive to make business sense of such data by analyzing using appropriate tools (Bose, 2009). It is important that appropriate tools that have the ability to use such large data and generate useful business insights are explored and made available to data scientists. In this regard, this book and this chapter focus on the use of multicriteria decision-making (MCDM) methods to help data scientists make sense of data. In this chapter, we illustrate specifically how data envelopment analysis (DEA), an MCDM tool, can be advantageously employed to help in economic and policy analysis. The specific problem we focus is a well-researched topic in the field of environmental policy. This problem focuses on the relationship between environmental expenditure and manufacturing performance. Since we are interested in using publicly available data, we do not focus on firms but instead focus on manufacturing sectors at aggregated level.

12.2 Literature Review

The literature review first describes DEA and provides arguments on why it should be considered as an MCDM tool. Then the literature on environmental expenditure and its relationship with performance are reviewed.

12.2.1 Data Envelopment Analysis

DEA is briefly discussed in this section. More elaborate discussions are available elsewhere (Charnes et al., 1994; Ramanathan, 2003). DEA has been successfully employed for assessing the relative performance of a set of firms, usually called as the decision-making units (DMUs), which use a variety of identical inputs to produce a variety of identical outputs. It has been recognized as a benchmarking tool (Charnes et al., 1994).

Assume that there are N DMUs, and that the DMUs under consideration convert I inputs to J outputs. In particular, let the mth DMU produce outputs y_{mj} using x_{mi} inputs. The objective of the DEA exercise is to identify the DMUs that produce the largest amount of outputs by consuming the least amounts of inputs, subject to the limits imposed by the performance of other similar DMUs. A DMU is deemed to be efficient if the ratio of weighted sum of outputs to the weighted sum of inputs is the highest. Hence, the DEA program maximizes the ratio of weighted outputs to weighted inputs for the DMU under consideration subject to the condition that the similar ratios for all DMUs be less than or equal to one. Thus, a model for calculating the efficiency of mth DMU (called the base DMU) is the following:

$$\max \frac{\sum_{j=1}^{J} v_{mj} y_{mj}}{\sum_{i=1}^{I} u_{mi} x_{mi}}$$

such that

$$0 \le \frac{\sum_{j=1}^{J} v_{mj} y_{nj}}{\sum_{i=1}^{I} u_{mi} x_{ni}} \le 1; \quad n = 1, 2, \ldots, N$$

$$v_{mj}, u_{mi} \ge 0; \quad i = 1, 2, \ldots, I; \ j = 1, 2, \ldots, J,$$

(12.1)

where the subscript i stands for inputs, j stands for outputs, and n stands for the DMUs. The variables v_{mj} and u_{mi} are the weights (also called multipliers) to be determined by the above mathematical program, and the subscript m indicates the base DMU. Soon after formulating Model (12.1), its authors suggested that the nonnegativity restrictions should be replaced by strict positivity constraints to ensure that all of the known inputs and outputs have positive weight values (Charnes et al., 1979). The optimal value of the objective function is the DEA efficiency score assigned to the mth DMU. If the efficiency score is 1 (or 100%), the mth DMU satisfies the necessary condition to be DEA efficient and is said to be located on efficiency frontier; otherwise, it is DEA inefficient. Note that the efficiency is relative to the performance of other DMUs under consideration.

It is difficult to solve the above program because of its fractional objective function. However, if either the denominator or numerator of the ratio is forced to be unity, then the objective function will become linear, and a linear programming problem can be obtained. For example, by setting the denominator of the ratio equal to unity, one can obtain the following *output maximization* linear programming problem. Note that by setting the numerator equal to unity, it is equally possible to produce *input minimization* linear programming problem.

$$\max \sum_{j=1}^{J} v_{mj} y_{mj}$$

such that

$$\sum_{i=1}^{I} u_{mi} x_{mi} = 1; \tag{12.2}$$

$$\sum_{j=1}^{J} v_{mj} y_{nj} - \sum_{i=1}^{I} u_{mi} x_{ni} \le 0; \quad n = 1, 2, \ldots, N$$

$$v_{mj}, u_{mi} \ge 0; \quad i = 1, 2, \ldots, I; j = 1, 2, \ldots, J.$$

Model (12.2) is called the *output maximizing multiplier version* in the DEA literature. A complete DEA model involves solving N such programs (Model 12.2), each for a base DMU ($m = 1, 2, \ldots, N$), to get the efficiency scores of all the DMUs. In each program, the objective function and the first constraint are changed while the remaining constraints are the same.

Computation of efficiency score is usually done with the dual of Model (12.2). The dual constructs a piecewise linear approximation to the true frontier by minimizing the quantities of the different inputs to meet the stated levels of the different outputs. The dual is given below:

$$\min \theta_m$$

such that

$$\sum_{n=1}^{N} y_{nj}\lambda_n \geq y_{mj}; \quad j = 1, 2, \ldots, J$$

$$\sum_{n=1}^{N} x_{ni}\lambda_n \leq \theta_m x_{mi}; \quad i = 1, 2, \ldots, I \tag{12.3}$$

$$\lambda_n \geq 0; \quad n = 1, 2, \ldots, N; \quad \theta_m \text{ free.}$$

Model (12.3) is usually called *input-oriented envelopment version* in the DEA literature. Analogously, an *output-oriented envelopment version* could be developed as the dual of *input minimization* linear programming problem mentioned earlier. Model (12.3) rates a particular DMU (*m*th DMU here). This DMU is relatively efficient if and only if the optimal values of its efficiency ratio, θ_m, equal unity.

Two different assumptions could be made while computing efficiency scores using DEA—constant returns to scale (CRS) and variable returns to scale (VRS). The assumption of CRS is said to prevail when an increase of all inputs by 1% leads to an increase of all outputs by 1% (Golany and Thore, 1997). Model (12.3) assumes CRS. However, VRS can be incorporated in it by appending the convexity constraint: $\sum_{n=1}^{N} \lambda_n = 1$ (Banker et al., 1984). The assumption of VRS is said to prevail when the CRS assumption is not satisfied. It has been proved that DEA efficiency scores computed with CRS assumption (usually called the CRS efficiency scores) are less than or equal to the corresponding VRS efficiency scores (Charnes et al., 1994) owing to the difference in scale size of DMUs. VRS efficiency of a DMU measures its pure technical efficiency, while CRS efficiency accounts for both technical efficiency and efficiency loss when the DMU does not operate in its most productive scale size (Charnes et al., 1994). The ratio of CRS to VRS efficiency scores is called the scale efficiency. Thus, scale efficiency of a DMU operating in its most productive scale size is 1.

DEA has the ability to give a single index of performance, usually called the efficiency score, synthesizing diverse characteristics of different DMUs. Due to this ability, DEA has found a number of applications as a benchmarking tool and for measuring comparative performance of organizations, industries, schools, banks, as well as nations (Emrouznejad et al., 2008).

12.2.2 DEA as an MCDM Tool

As highlighted in Chapter 2, DEA has now been accepted as an MCDM tool. It has been proved that DEA scores provide a ranking of MCDM alternatives when maximization criteria are considered as outputs in a DEA model, and minimization criteria are considered as inputs (Joro et al., 1998; Belton and

Stewart, 1999). In this study, we use this MCDM perspective of DEA to rank manufacturing sectors in terms of their ability to produce maximum outputs by consuming minimum inputs.

12.2.3 Relationship between Environmental Performance and Financial Performance

Firms spend money and efforts in order to ensure that they meet the requirements of environmental regulations formulated by governments all over the world. Traditionally, this is considered a burden on firms, particularly on manufacturing, since pollution abatement and restrictions on the use of certain materials raise the cost of operations, thereby reducing profitability and productivity (Christiansen and Haveman, 1981). However, in the early 1990s, following a new wave of environmental concern in the public and political sphere, Porter (1991), among others, suggested that environmental regulations might in fact be beneficial for businesses: if properly designed and received with a "dynamic mindset," regulations could prompt a move toward leaner manufacturing practices, more efficient energy and resource use, etc. This is based on the notion, drawn from industrial ecology, that pollution and discarded waste is a sign of inefficiency in production processes (Frosch 1982). More efficient use of energy, leaner and cleaner production processes, and the recycling and reuse of expired products will lower costs and reduce both the input of new resources (a cost) and the output of undesirable resources such as pollution (whose abatement represents a cost), thereby meeting the regulations and yielding a benefit for the firm. This is the so-called win–win argument or Porter hypothesis.

Porter's hypothesis has been tested extensively on firms with varying results. For example, a positive relationship between environmental performance and financial performance has been reported by Hart and Ahuja (1996), Waddock and Graves (1997), Russo and Fouts (1997), Margolis et al. (2007), Callan and Thomas (2009), and Peloza (2009). Support for a positive link has also been highlighted in case studies (Porter and van der Linde, 1995; Rugman and Verbeke, 2000) as well. A negative relationship was reported by Konar and Cohen (2001), Moore (2001), Sarkis and Cordeiro (2001), and Brammer et al. (2006).

A notable feature of previous studies is that most of these studies are based on primary data collected either from questionnaire surveys or interviews. There has been no attempt to use publicly available data sources. In line with the concept of big data, we attempt to collect publicly available data sources to test Porter's hypothesis in this study. In line with the focus on MCDM, we measure manufacturing performance using multiple criteria with the help of DEA.

Several previous studies have used environmental protection expenditure by firms as a proxy for environmental performance of firms

(Majumdar and Marcus, 2001; Ramanathan et al., 2010). Thus, based on Porter's hypothesis and previous literature, we propose the following hypothesis:

> Hypothesis 1. Environmental protection expenditure to meet environmental regulations is significantly positively related to performance.

12.3 Data and Analysis

12.3.1 Sample and Data Collection

A governmental regulation is applicable to all the firms in a particular sector. However, the influence of regulations will differ from sector to sector—more polluting sectors will face higher level of regulation. The question of how environmental regulations affected performance of firms in highly polluting and lowly polluting sectors is of much policy interest. Hence, we focus on sector level in this study.

We have considered 12 sectors in this study (Table 12.1). We have used data on these 12 sectors for a period of 5 years (2002–2006). Thus our sample size is 60. Though we have used data from multiple years, we consider them as cross-sectional data. This approach is similar to the ones used in the economics literature, for example, by Besley and Burgess (2004) mentioned above.

TABLE 12.1

Sectors Analyzed

SIC Code	Description	SIC Code	Description
10–14	Mining and quarrying	26	Manufacture of other nonmetallic mineral products
15–16	Manufacture of food, beverages, and tobacco products	27–28	Manufacture of basic metals and fabricated metal products
21–22	Manufacture of pulp, paper and paper products publishing, and printing	29	Manufacture of machinery and equipment not elsewhere classified
23	Manufacture of coke, refined petroleum products, and nuclear fuel	30–33	Manufacture of electrical and optical equipment
24	Manufacture of chemicals, chemical products, and man-made fibers	34–35	Manufacture of transport equipment
25	Manufacture of rubber and plastic products	40–41	Electricity, gas, and water supply

12.3.1.1 Manufacturing Performance (Manufacturing Efficiency Scores)

As highlighted earlier, we have used DEA to measure manufacturing performance of various sectors. This approach is similar to that of Majumdar and Marcus (2001). The inputs and outputs used in the calculation of these manufacturing efficiency scores are shown in Table 12.2. Similar to the approach by Majumdar and Marcus (2001), we have used constant-returns-to-scale input-minimization DEA for calculating the efficiency scores.

12.3.1.2 Environmental Expenditure

We obtained sector-level data from the Department for Environment, Food, and Rural Affairs (DEFRA) for both the operating and capital expenditure associated with pollution abatement in several different media. These data are based on the U.K. Environmental Protection Expenditure by Industry Survey 2006, which is a survey of a stratified random sample of 7850 companies belonging to various industrial sectors with 20.4% response rate (DEFRA, 2008). The survey found that gross spending on environmental protection in 2006 by the U.K. industry amounted to an estimated £4.2 billion, and that operating expenditure accounted for 71% of the total environmental protection expenditure. The primary spending sectors as per the survey were electricity and gas (37% of total spend), food, beverages, and tobacco products (12% of total spend), and basic metals and metal products (8% of total spend). The survey also found that the use of environmental management systems was more widespread in the larger companies.

In our study, we use two measures of pollution control expenditure. Operating expenditure (OPEX) covers in-house expenditure associated with the operation of pollution control abatement equipment and payments to external organizations for environmental services, including, labor costs, leasing payments, maintenance costs for equipment, and the treatment and disposal of waste. Capital expenditure (CAPEX) covers expenditure on

TABLE 12.2

DEA Inputs and Outputs for Measuring Manufacturing Efficiency and Descriptive Statistics

Inputs or Outputs	Mean	Std. Dev.	Min.	Max.
Inputs				
Compensation of employees (£ millions)	8704	4332	2048	14,870
Net capital stock (£ billions)	24.96	21.78	6.1	88.6
Intermediate consumption (£ millions)	25,035	12,110	6429	45,790
Outputs				
Gross value added (£ millions)	14,596	6597	2377	32,202
Gross fixed capital formation (£ millions)	4816	7221	0	20,917

end-of-pipe pollution control equipment and on integrated processes—new or modified production facilities that have been designed so that environmental protection is an integrated part of the process.

We have used data on pollution abatement expenditure for waste, water pollution, air pollution, and all other pollution. We have aggregated the operating and capital expenditure for a particular pollution (waste, water, air, and other) to get a single measure of pollution abatement for use in our analysis.

12.3.1.3 Control Variables

To control for the potential relationship between sector size and efficiency, we include the number of employees in each sector. We also incorporate R&D expenditure and energy consumption as control variables (Majumdar and Marcus, 2001). Table 12.3 shows the summary statistics for the variables of interest and the correlations between.

12.3.2 Regression Model

We verify our hypothesis using statistical regression. The dependent variable (manufacturing efficiency scores) is regressed on the various measures of environmental expenditure and the three control variables outlined above. Results are shown in Table 12.4. Note that the regression has a high R^2 value,

TABLE 12.3

Summary Statistics and Correlation Coefficients

Variable	Correlation Coefficients						
	1	2	3	4	5	6	7
1. Manufacturing efficiency	1						
2. Number of employees (hundred thousands)	0.38***	1					
3. Energy consumption (million tonnes of oil equivalent)	−0.55***	−0.09	1				
4. Other pollution abatement expenditure (£ hundred millions)	−0.49***	0.02	0.00	1			
5. Waste pollution abatement expenditure (£ hundred millions)	−0.13	0.40***	−0.03	0.18	1		
6. Air pollution abatement expenditure (£ hundred millions)	−0.48***	0.02	0.26**	0.43***	0.05	1	
7. Water pollution abatement expenditure (£ hundred millions)	−0.20	0.48***	0.18	0.41***	0.23*	0.37***	1
Mean	77.73	6.51	2.85	0.45	0.77	0.38	0.83
Std. Dev.	19.62	3.58	2.42	0.39	0.53	0.31	1.01
Min	32	1.22	0.38	0.06	0.03	0.04	0.13
Max	100	15.75	9.15	1.8	2.97	1.27	5.09

***$p < 0.01$; **$p < 0.05$; *$p < 0.1$.

TABLE 12.4

Results of the Simple Regression Model (Dependent
Variable: Manufacturing Efficiency)

Variables	Regression Coefficient
Controls	
Energy consumption	−3.63***
Other pollution expenditure	−17.08***
Number of employees	2.69***
Direct Effects	
Waste expenditure	−8.99***
Air expenditure	−11.55**
Water expenditure	−1.71
R^2	0.73
F	23.45***

***$p < 0.01$; **$p < 0.05$.

indicating that 73% of the variation in the dependent variable is explained
by the independent variables. The regression is statistically significant as
shown by the *F*-test. Expenditure incurred on waste and air is significant in
explaining performance but has negative values. The expenditure incurred
on water is not influencing performance significantly.

Our results thus fall short of validating Hypothesis 1 and further analysis
is needed to identify the reasons. We believe that the key lies in understanding the different regulation regimes followed in air/water/waste regulations
(Majumdar and Marcus, 2001). However, such more detailed analysis is not
attempted in this chapter since the focus of this chapter is to demonstrate
how DEA could be combined with statistics for environment policy analysis.
Obviously, this forms scope for further work.

12.4 Summary and Conclusions

We have illustrated in this study how DEA, an MCDM tool, can be usefully
combined with big data sources to perform further analysis for understanding an important issue of environmental policy. We measured manufacturing
efficiency of various sectors using DEA and hypothesized that environmental protection expenditure will be significantly related to efficiency. However,
our illustration has shown that the results do not support the hypothesis but
we have pointed out that more detailed analysis on the flexibility of environmental regulations in various media (air, water, and waste) may be required
to make more sense of available data.

Despite the fact that our hypothesis is not validated, we believe that this chapter shows how MCDM models can be used to make business sense of publicly available big data for policy analysis.

References

Banker, R.D., Charnes, A., and Cooper, W.W. 1984. Some models for estimating technical and scale efficiencies in data envelopment analysis. *Management Science*, 30(9), 1078–1092.

Belton, V. and Stewart, T.J. 1999. DEA and MCDA: Competing or complementary approaches? In: Meskens, N. and Roubens, M. (Eds.). *Advances in Decision Analysis*, Springer Science & Business Media, Dordrecht, Netherlands, 87–104.

Besley, T. and Burgess, R. 2004. Can labor regulation hinder economic performance? Evidence from India. *Quarterly Journal of Economics*, 119, 91–134.

Bose, R. 2009. Advanced analytics: Opportunities and challenges. *Industrial Management & Data Systems*, 109(2), 155–172.

Brammer, S., Brooks, C., and Pavelin, S. 2006. Corporate social performance and stock returns: UK evidence from disaggregate measures. *Financial Management*, 35(3), 97–116

Callan, S.J. and Thomas, J.M. 2009. Corporate financial performance and corporate social performance: An update and reinvestigation. *Corporate Social Responsibility and Environmental Management*, 16, 61–78

Charnes, A., Cooper, W.W., Lewin, A.Y., and Seiford, L.M. 1994. *Data Envelopment Analysis: Theory, Methodology and Applications*, Kluwer, Boston.

Charnes, A., Cooper, W.W., and Rhodes, E. 1979. Short communication: Measuring the efficiency of decision making units. *European Journal of Operational Research*, 3, 339–339.

Christainsen, G.B. and Haveman, R.H. 1981. The contribution of environmental regulations to the slowdown in productivity growth. *Journal of Environmental Economics and Management*, 8(4), 381–390.

DEFRA. 2008. UK environmental protection by industry survey. Available at: https://www.gov.uk/government/statistics/environmental-protection-expenditure-survey, Last accessed: September 29, 2015.

Emrouznejad, A., Parker, B.R., and Tavares, G. 2008. Evaluation of research in efficiency and productivity: A survey and analysis of the first 30 years of scholarly literature in DEA. *Socio-Economic Planning Sciences*, 42(3), 151–157.

Frosch, R.A. 1982. Industrial ecology: A philosophical introduction. *Proceedings of the National Academy of Sciences*, 89, 800–803.

Gobble, M.M. 2013. Big data: The next big thing in innovation. *Research Technology Management*, 56(1), 64–66.

Golany, B. and Thore, S. 1997. The economic and social performance of nations: Efficiency and returns to scale. *Socio-Economic Planning Sciences*, 31(3), 191–204.

Hart, S.L. and Ahuja, G. 1996. Does it pay to be green? An empirical examination of the relationship between emission reduction and firm performance. *Business Strategy and the Environment*, 5, 30–37.

Joro, T., Korhonen, P., and Wallenius, J. 1998. Structural comparison of data envelopment analysis and multiple objective linear programming. *Management Science,* 40, 962–970.

Konar, S. and Cohen, M.A. 2001. Does the market value environmental performance? *The Review of Economics and Statistics,* 83(2), 281–289.

Majumdar, S.K. and Marcus, A.A. 2001. Rules versus discretion: The productivity consequences of flexible regulation. *Academy of Management Journal,* 44, 170–179.

Margolis, J., Elfenbein, H.A., and Walsh, J. 2007. *Does It Pay To Be Good? A Meta-Analysis and Redirection of Research on the Relationship between Corporate Social and Financial Performance.* Mimeo, Harvard Business School, Boston, MA.

Moore, G. 2001. Corporate social and financial performance: An investigation of the UK supermarket industry. *Journal of Business Ethics,* 34, 299–315.

Peloza, J. 2009. The challenge of measuring financial impacts from investments in corporate social performance. *Journal of Management,* 35(6), 1518–1541.

Porter, M.E. 1991. America's green strategy. *Scientific American,* 264, 168.

Porter, M.E. and van der Linde, C. 1995. Toward a new conception of the environment-competitiveness relationship. *Journal of Economic Perspectives,* 9, 97–118.

Ramanathan, R. 2003. *An Introduction to Data Envelopment Analysis.* Sage, New Delhi.

Ramanathan, R., Black, A., Nath, P., and Muyldermans, L. 2010. Impact of environmental regulations on innovation and performance in the UK industrial sector, Special Topic Forum on Using Archival and Secondary Data Sources in Supply Chain Management Research. *Management Decision (Special issue on Daring to Care: A Basis for Responsible Management),* 48(10), 1493–1513.

Rugman, A.M. and Verbeke, A. 2000. Six cases of Corporate Strategic responses to environmental regulation. *European Management Journal,* 18(4), 377–385.

Russo, M.V. and Fouts, P.A. 1997. A resource-based perspective on corporate environmental performance and profitability. *Academy of Management Journal,* 40, 534–559.

Sarkis, J. and Cordeiro, J.J. 2001. An empirical evaluation of environmental efficiencies and firm performance: Pollution prevention technologies versus end-of-pipe practice. *European Journal of Operational Research,* 135(1), 102–113.

Waddock, S.A. and Graves, S.B. 1997. The corporate social performance-financial performance link. *Strategic Management Journal,* 18(4), 303–319.

13

An Integrated Multicriteria Decision-Making Model for New Product Portfolio Management

Pulipaka Kiranmayi and Muthu Mathirajan

CONTENTS

13.1 Introduction

> A successful new product does more good for an organization than any-
> thing else that can happen.
>
> **Crawford**

"Product is a multi-dimensional concept, so change in one or more dimensions or as a whole is considered to be a new product, depending upon the
extent of innovation into it" (Crawford and Di Benedetto 2008). In present
global competition, increasing customer requirements and rapidly changing
market requirements has made development of new product as an essential
task of any organization. The continuous development and market introduction of new products are important determinants of sustained organization
performance (Blundell et al. 1999). The ability to target the right customers in
the right way is a challenge. In addition to this, the complex characteristics of
the new product process (NPP) is one of the reasons for organizations to lose
touch with the reality of what consumers want and misspend huge amount
of investments on products that fail.

13.1.1 Characteristics of New Product Process (NPP)

New Product Process (NPP) generally is divided into five phases of development and they are (1) opportunity identification and selection, (2) concept
generation, (3) concept/project evaluation and selection (PES), (4) development, and (5) launch. Of all the five phases of NPP, the third phase: PES
plays a major key role in success of new product. The main purposes of the
PES phase are to (1) help an organization in deciding whether organization
should go forward with that particular project/product or not, (2) manage
the process in an optimized way as well as indicate better portfolio management by sorting the concepts and identifying the best ones, and (3) even
encourage cross-functional communication for new product success.

From implementation perspective, it appears that practitioners make two
types of errors in case of PES decisions: First, to consider a nonsuccessful project to be successful and risk the huge investments; second, to consider a successful project to be nonsuccessful and miss the opportunity of earning huge
profits. In case of traditional products, first error is considered to be more
serious compared with second. However, in the case of new product both the
errors would do equal damage to an organization, particularly the second
error can lead to the chances of wiping out the organization from the market

if competitor comes out with that particular project idea. In this scenario, effective and accurate decision making and managing of NPP is essential. In literature, the managing processes, decision-making perspective of new product development are referred to as new product management (NPM).

13.1.2 New Product Management

Ulrich and Eppinger (2004) defined NPM as "the set of activities beginning with the perception of a market opportunity and ending in the production, sale, and delivery of a product." Over the years this definition has been evolved and according to Loch and Kavadias (2002) "New Product Management (NPM) consists of the activities of the firm that lead to a stream of new or changed product market offerings overtime. This includes the generation of opportunities, their selection and transformation into artifacts and activities offered to customers and the institutionalization of improvements in new product development activities themselves."

NPM requires a static plan to sustain and win in a war for market territory, waging battles alongside alliance partners against competitors, and conquering market segments with products, services, and solutions. Efficient NPM outlines what market segment to target with what products, and what position to defend against competitors. NPM includes decisions about configuration and development of internal resources to build and defend the desired new products from the phase of idea selection to launch. NPM not only assists in making accurate decisions but also allows an organization to achieve strategic objectives and vision through development of new products. However, studies focusing on decision-making aspects of NPM are not as widespread (Yahaya and Abu-Bakar 2007). For all these specific purposes, an efficient management system for formulation of new product portfolio is essential for any organization to succeed. In the literature, the management system and decision making for formulation of portfolio is termed as new product portfolio management (NPPM). Often managers rate NPPM as the weakest NPM area and explicit evaluative dimensions are lacking (Cooper et al. 2001; McNally et al. 2009) in most of the reported portfolio framework. With these, the main objective of this study is to identify and study significance of different evaluative dimensions and subsequently build an appropriate and efficient decision-making model of PES for NPPM. Particularly, we intend to study different methodologies involved in literature, and identify limitation, and make an attempt to address the research gap. In addition, it is observed from literature that the integration of qualitative and quantitative methods is significant for future NPPM methods that seek practical implications. Accordingly, in this study, we propose to integrate balanced scorecard (BSC) and data envelopment analysis (DEA) methodologies for better NPPM performance.

The structural flow of this study is as follows: NPPM and the required evaluative dimensions for NPPM are discussed in Section 13.2. Literature on different methodologies implemented in NPPM is summarized in Section

13.3. In Section 13.4, a base BSC system along with proposed BSC index system for NPPM is first discussed and subsequently the proposed integrated DEA–BSC model is presented. The workability of the integrated decision-making model: DEA–BSC along with numerical example is presented in Section 13.5.

13.2 New Product Portfolio Management

In general, "Product Portfolio Management (PPM) is the centralized management of the processes, methods, and technologies used by project managers and project management offices (PMOs) to analyze and collectively manage current or proposed projects based on numerous key characteristics." PPM ascertains program and project managers with the capabilities needed to manage the time, resources, skills, and budgets necessary to accomplish all interrelated tasks.

NPPM in general is defined as "The dynamic decision process wherein the lists of active new products or R&D projects are constantly revised. In this process, new products or projects are evaluated, selected, and prioritized. Existing products or projects may be accelerated, killed, or deprioritized and resources are allocated to the active projects." In a competitive market along with goal of developing a new product, one has to keep in mind multiple goals such as reaching the market first, competitors' position, market sentiment, and so on. In this scenario, best mix of products or projects that ensures strategic alignment, balance of portfolio, and potential gain is compulsion. This can be achieved most efficiently through an NPPM that provides both high-level and detailed information on all products or projects for educated decision making—keeping the entire portfolio aligned with overall corporate strategies and objectives and securing the highest return on investments (ROIs). Based on the literature review and from practitioners' point of view, NPPM decisions are considered complex and significant. The reasons for the same are identified as follows:

- Decision maker may be interested in working toward multiple objectives but progress in one direction impedes progress in others. So in this case, different perspectives lead to different conclusions.
- New product development is characterized by a tremendous degree of complexity and uncertainty and involves choosing between different products competing for the same funding.
- Multiple nature of conflicting objectives (i.e., maximize return, maximize R&D productivity, minimize uncertainty, etc.) leads to considerations of multiple variables for taking a single decision.

- Decisions taken for the formulation of NPP determine the competitive position of business, product sales, and market share.
- Decisions implemented in NPPM play a major role in forging the link between product or project selection and organization strategy. As the portfolio is the expression of organizations' strategy, thus it is significantly important for NPP to align with organization strategy.
- Efficient NPPM achieves focus of not doing too many projects for the limited resources available but to allocate these resources to the efficient and profitable product or projects.
- Efficient NPPM achieves the right balance between long- and short-term product or projects, and high-risk and low-risk ones, consistent with the business goals.

Based on the reasons stated here on the complexity and the significance of NPPM, one needs to consider multiple evaluative dimensions for obtaining efficient NPPM. The reality and the literature revealed that product/project managers use different evaluation dimensions/criteria for NPPM. In addition, the ability of decision models for evaluating accurately the best set of products or projects varies depending on dimensions used and weights applied to these dimensions. So the identification of evaluative dimensions is inevitable. In this study, we made an attempt to identify and study the significance of different evaluative dimensions and its impact on efficiency of NPPM from the analysis of the literature. Accordingly, the evaluative dimensions identified through the related research studies on NPPM, R&D portfolio management, and PES for NPPM are discussed in detail in the following section.

13.2.1 Evaluative Dimensions for NPPM

There is a rich literature available for identification of evaluative factors or dimensions which affect NPPM performance. Carbonell-Foulquié et al. (2004) did a study on the evaluative dimensions and weights used in the formulation of NPP. The evaluative dimensions identified were named: technical feasibility, strategic fit (SF), customer acceptance, financial performance, and market opportunity. Cooper et al. (2001) report that managers use three broad dimensions: value maximization, portfolio and innovation balance (PIB), and SF to evaluate the organizations portfolio of new product projects.

Organizations that fail to manage their NPPM activities strategically are not only running their business from a position of disadvantage but are risking their future (Fitzsimmons et al. 1991). Thus, critical role of NPPM in the survival and success of organization and the need for managing it strategically is being increasingly recognized in both academic (Brown and Eisenhardt 1995; Griffin and Hauser 1996; Krishnan and Ulrich 2001) and practitioner (Gates 1999; Chesbrough and Teece 2002) point of views. Thus, the alignment between organization's missions/objectives and execution of NPP is a tactical challenge faced by every organization. Managers usually perceive that their new product projects are generally aligned with their organization's strategic objectives and goals (Cooper et al. 2004), which is important because this dimension correlates strongly with new product's performance.

Osawa and Murakami (2002) proposed a methodology to evaluate R&D projects in terms of SF and financial credibility. The implications of this study supported the inference: Adding SF as an important evaluative dimension improves NPPM performance, given in the study by Ronkainen (1985).

Balance is a critical NPPM dimension, as it is second most strongly correlated practice after value maximization with superior new product development performance (Cooper et al. 2004). Many researches (Cooper et al. 1997; Graves et al. 2000) have focused on the dimension of project portfolio balance, in terms of achieving balance among the set of projects that are selected in the portfolio. Oh et al. (2012) considered portfolio balance, but the limitation was that they considered balance between the set of projects which were yet to be selected for the portfolio and did not consider existing projects and the amount of innovation or type of innovation involved. However, in real-life scenario, there is a requirement to achieve balance among multiple directions (i.e., product innovation, process innovation, common product platforms, etc.). Though there is no strong empirical study which concludes that PIB increases the efficiency of NPPM, we strongly believe that PIB should be considered as one of the evaluative dimension for NPPM.

NPPM's critical task is allocation of resources between different innovation projects ranging from radical innovation to basic incremental innovation and each of these projects poses conflicting directions in terms of organizational strategy. Organizations prefer to develop more than one product in order to sustain the competition; as a result, interdependency between the projects increases. Simultaneously, resource allocation becomes complex when the number of projects and interdependency between them increases. Many researchers have developed different frameworks to increase effectiveness of NPPM with major concentration on resource allocation. For example, Loch and Kavadias (2002) focus on the optimal resource allocation across New Product Development (NPD) programs. But they do not consider how the types of the NPD investment or the investment horizon impact the allocation decision. Chao and Kavadias (2008) use concept of strategic buckets in order to allocate limited resources throughout project portfolio. Optimizing the available resources for the set of projects with the aim of increasing the efficiency of NPPM is very crucial for organization's success.

It is observed from the literature that cost–revenue or development cost evaluation or benefit–cost analysis is one of the evaluative dimensions, which is focused from past two to three decades and received major attention from academicians as well as practitioners. For example, Chiu and Park (1994) used the expected net present value (NPV) to evaluate the conditions for an R&D Project's success or failure. Mahmoodzadeh et al. (2007) reviewed four common methods: NPV, rate of return, benefit–cost analysis, and payback period (PBP) by using analytical hierarchy approaches (AHP). Lockett and Stratford (1987) and Ghasemzadeh and Archer (2000) focused on investments and other evaluative dimension: cost–revenue. Cooper et al. (2004) state, cost–revenue estimation (CRE) is a crucial dimension which distinguishes high-performing organizations in the area of NPD from low-performing organizations. The limitations of these research studies are that they did not consider multiple projects and overemphasized only on one particular evaluative dimension of cost–revenue/cost–benefit analysis.

Uncertainty management is an integral part of NPP and different approaches exist in literature to define and analyze uncertainty dimension. Recently, researchers have concentrated on minimizing risk and uncertainties involved in NPP (Kahraman et al. 2007; Mahmoodzadeh et al. 2007; Chiang and Che 2010). Kahraman et al. (2007) considered two types of risks, namely systematic risks (financial and technical) and unsystematic risks (managerial and personnel) in their hierarchical structure of decision criteria and employed heuristic multiattribute utility function.

Chiang and Che (2010) concentrated on risk and uncertainty-evaluative dimension, that is, expected revenue risk, manufacturability risk, and time-to-market risk, which were calculated using Bayesian belief network. Feyzioğlu and Büyüközkan (2006) highlighted that NPPM decision, especially necessary at early stages of the NPD, as it contains considerable amount of uncertainty-causing elements, which confuses decision maker to reach the target performance. At this point, decision maker has to make a lot of decisions based on inadequate information about the project, vagueness in challenging issues, no dependency on previous data leads to uncertainty and increase the risk involved in development of product. However, these studies did not consider other evaluative dimensions along with risks and uncertainty dimension for PES decision in NPPM. In this study, the wide spectrum of risks and uncertainties are categorized and considered in accordance with other dimensions.

Cooper (1994) stated that an organization which emphasizes only on a particular evaluative dimension for NPPM is linked with poorer performance. Accordingly, there are few studies concentrating on multiple dimensions. For example, Cooper et al. (1997) report that managers use three broad dimensions to evaluate the organizations portfolio of new product projects: value maximization, portfolio optimization, and SF. Ozer (2005) presented an integrated framework for understanding various factors which affect decision making in NPP. The study carried out by Mohanty et al. (2005) considered four evaluative dimensions: SF, cost–revenue, resource allocation, and risk and uncertainty for R&D project selection. Bhattacharyya et al. (2011) developed a fuzzy R&D portfolio selection model by considering three objectives: minimization of risk, minimization of project cost, and maximization of project outcome.

Based on the analysis of the literature review presented here, it is observed that there are five evaluative dimensions: SF, PIB, optimized resource allocation (ORA), CRE, and risk–uncertainty estimation (RUE) are considered in different sets and combination (Table 13.1).

TABLE 13.1

Summary of Closely Related Literature on Evaluative Dimensions of NPPM

Related Studies	Evaluative Dimensions of Study				
	Strategic Fit	Portfolio Balance	Risk–Uncertainty	Cost–Revenue	Resource Allocation
R&D Project Evaluation and Selection					
Osawa and Murakami (2002)	+			+	
Mohanty et al. (2005)	+		+	+	+
Eilat et al. (2008)		+	+	+	+
R&D Project Portfolio Formulation					
Eilat et al. (2006)			+	+	+
Wang and Hwang (2007)				+	+
Bhattacharyya et al. (2011)			+	+	+
Abbassi et al. (2014)	+		+		+
New Product Project Evaluation and Selection					
Thieme et al. (2000)				+	+
Feyzioğlu and Büyüközkan (2006)	+			+	+
Mahmoodzadeh et al. (2007)				+	+
Chiang and Che (2010)			+	+	+
New Product Portfolio Formulation					
Oh et al. (2012)	+	+	+	+	+

Table 13.1 clearly indicates that no one has considered the entire five evaluative dimensions together for PES decision in NPPM. In addition, it appears that there is only one study: Oh et al. (2012) considering PIB evaluative dimension, but not in the perspective of this study and particularly they considered only portfolio balance. Furthermore, there is no significant study which is carried out to analyze the interrelationship between these evaluative dimensions for the significance of NPPM. Finally, in addition to these research gaps, there exists a gap where these identified evaluative dimensions are not modeled for implementation in industry. In this study, we attempt to develop an explicit decision-making model which fulfills these research gaps. In the next section, methodologies used in different studies for development of decision models for NPPM are presented.

13.3 Methodologies/Models for NPPM

There are many theoretical and practical attempts to develop models for decision making in case of product or project selection for portfolio. Earlier

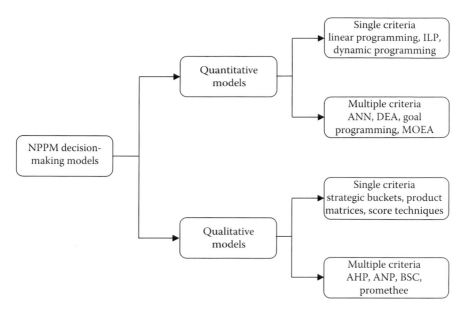

FIGURE 13.1
Classification of new product portfolio management models/methodology.

studies of portfolio management focused on constrained optimization models based on theoretical operation research. Brown and Eisenhardt (1995) classify NPD research depending on its methodological approach. Liao (2005) conducted a literature survey on NPPM methodologies and applications. Based on the analysis of the literature, models developed for NPPM can be classified based upon type of study as qualitative and quantitative, and further can be classified based on a number of criteria being considered into single criterion and multicriteria. Figure 13.1 depicts this classification and only recent literature which discusses these classified models is summarized in Table 13.2. In addition, a brief overview of the methods/models based on the classification given in Figure 13.1 is presented as follows.

13.3.1 Quantitative Models

Quantitative models consist of numeric data as input and adopt procedures such as mathematical algorithms, mathematical programming, or economic/quantitative indices for evaluating NPPM and produce a numerical output. These quantitative models again can be categorized based on a number of criteria considered for evaluation. Accordingly, these quantitative models are classified into single-criterion models and multiple-criteria models.

TABLE 13.2

Summary of Recent Related Literature on Models/Methodologies for NPPM

Approach	Number of Criteria	Methodology/Model	Reference
Quantitative	Single	Financial indices and methods	Patah and de Carvalho (2007), Ibbs et al. (2004)
		Linear programming[a]	Chien (2002)
		Integer programming	Mavrotas et al. (2006), Melachrinoudis and Kozanidis (2002)
		Dynamic programming[a]	Kyparisis et al. (1996)
	Multiple	Goal programming	Wey and Wu (2007), Lee and Kim (2000)
		Data envelopment analysis	Chang et al. (2014), Kumar et al. (2007)
		Multiobjective evolutionary algorithms	Metaxiotis et al. (2012), Gutjahr et al. (2010)
Qualitative	Single	Strategic buckets, score techniques	Chan and Ip (2010)
	Multiple	Multiattribute utility theory[a]	Duarte and Reis (2006)
		Promethee multicriteria[a]	Halouani et al. (2009)
		Balanced scorecard (BSC)	Asosheh et al. (2010), Eilat et al. (2008)
		Analytic hierarchy process/analytic network process	Ayağ and özdemr (2007), Wang and Hwang (2007), Yurdakul (2003)

[a] Details of these methods/models are not attempted to provide in this study.

13.3.1.1 Quantitative Models with Single Criterion

A major part of literature that concentrates on PES includes single-criterion quantitative models such as financial models and financial indices, linear programming, integer programming (IP), and dynamic programming.

13.3.1.1.1 Financial Models and Financial Indices

Financial models and financial indices are the most prominent and are used as evaluation models for decades by organizations. The existing literature on *Financial Models and Financial Indices* for evaluation can be categorized based on methodology used into two approaches. The first approach consists of cost–benefit analyses and the second approach consists of ROI-based analyses. The studies related to cost–benefit analyze the relationship between NPPM investments and the savings and the studies related to ROI analyze the relationship between NPPM maturity and project performance. Models that are frequently employed by practioners include: NPV, Internal Rate of Return (IRR), PBP, and ROI. Remera et al. (1993) conducted a survey on financial project evaluation techniques used in industries and mainly concluded that there is a shift from the use of IRR to NPV and a decrease in the use of the PBP.

13.3.1.1.2 *Integer Programming*

Most of the earlier reported studies highlighted that organizations develop and/or use IP models with different optimizing criteria in case of project selection. In literature, particularly theoretically, it is considered as the best method to achieve optimized solution where constraints (or) requirements can be represented in the form of linear relationships. However, in real-life scenario, there always exist complex constraints which cannot be modeled as linear relationships. Because of this, most of IP models developed did not come into use in practical scenarios.

13.3.1.2 **Quantitative Models with Multiple Criteria**

Multicriteria optimization is the problem of optimizing two or more objectives which may be conflicting in nature, subject to certain constraints simultaneously. It includes several approaches such as Goal Programming (GP), DEA, and multiobjective evolutionary algorithm (MOEA).

13.3.1.2.1 *Goal Programming*

GP can be considered as an extension or generalization of linear programming to handle multiple, normally conflicting objectives. Each of these objectives is considered as a goal and assigned a target value to be achieved. So while modeling, the unwanted deviations from this set of target values are then minimized in prime objective function. This objective function can be a vector or a weighted sum dependent on the GP variant used. A major strength of GP is its simplicity and ease of use. GP can handle relatively large number of variables, constraints, and objectives.

13.3.1.2.2 *Data Envelopment Analysis*

The basic concept of DEA is to measure the efficiency of a particular decision-making unit (DMU) against a projected point on an "efficiency frontier." It is a mathematical programming technique that calculates the relative efficiency of multiple DMUs on the basis of observed inputs and outputs, which may be expressed with different types of metrics. A DMU is considered efficient when no other DMU can produce more outputs using an equal or less amount of inputs. Additionally, DEA generalizes the usual efficiency measurement from a single-input single-output ratio to a multiple-input multiple-output ratio by using a ratio of weighted sum of outputs to weighted sum of inputs. Recent literature indicated that DEA is one of the efficient tools for evaluating projects (Eilat et al. 2008). In addition to evaluation of projects, the DEA also provides selection and ranking of projects for portfolio.

13.3.1.2.3 Multiobjective Evolutionary Algorithms

MOEA can be useful in complex problems, particularly NP-hard problem, for which no efficient deterministic algorithm exists. The first implementation of MOEA for project selection and evaluation dates back to the mid-1980s (Schaffer 1985). Over the years, researchers have developed several approaches for obtaining solution of multiobjective optimization problems with the use of *Evolutionary Algorithms*. One can refer to Metaxiotis and Liagkouras (2012) for detailed literature survey on different *Evolutionary Algorithms* existing and studies that implemented MOEAs.

13.3.2 Qualitative Models

Qualitative models/approaches such as scoring models, checklist, strategic buckets, and predictive analysis of expert opinions can be modeled for single or multiple criteria.

13.3.2.1 Strategic Approaches, Scoring Models, and Checklists

An organization in general tends to develop new product based on future needs. In this scenario, the organization needs to study unique needs and finally, evaluate them and try to come up with new products that can fulfill the most important ones. In such cases, scenario analysis is being used for identifying future needs and generating new product concepts before competition. In addition, there are other methodologies such as different checklists varying with organization goals, strategic buckets, score models, product complexity matrix, and so on developed by an organization as decision aid models. The dynamic model utilizes an extensive amount of input from the multimedia exercises (e.g., online search by consumers, dealer visits, word-of-mouth communication, magazine reviews), historical data of similar products, industry sales, managerial judgments, and production constraints. All these inputs are used to forecast and simulate the market environment. For detailed description of other approaches for qualitative models, one can refer to Ozer (1999).

13.3.2.2 Balanced Scorecard

Among other techniques, BSC technique is promoted as an efficient tool for strategic alignment of projects along with satisfying other objectives and measures for specific organizational unit (Eilat et al. 2008). The core idea of BSC is to display the organizational strategic trajectory through mutual-driven causal relationship between four perspectives and they are: financial, customer, internal processes, and learning and growth. Indeed, many organizations have adopted the BSC approach to (a) accomplish critical management processes, (b) clarify and translate their vision and strategy, (c) communicate and link strategic objectives and measures, (d) plan and align strategic initiatives, (e) enhance strategic feedback and learning, etc. The significance and implementation of BSC is trending exponentially in present competitive scenario (Eilat et al. 2006).

13.3.2.3 Analytical Hierarchy Approaches/Analytical Network Process

AHP/analytical network process (ANP) methods are special ones as both quantitative and qualitative and/or judgment criteria can be considered. In these methods, first the weights of different objectives are determined, then alternatives are compared on the basis of their contributions to these objectives, and finally a set of project benefit measures is computed. Once the alternatives have been arranged on a comparative scale, the decision maker(s) can proceed from the top of the list, selecting a subset of alternatives until the feasibility constraint is maintained. AHP has an advantage of allowing a set of complex issues that have an impact on an overall objective to be compared with the importance of each issue relative to its impact on the solution of the problem. However, the conventional AHP method assumes a unidirectional hierarchical relationship among decision levels and attributes; due to this, researches are using ANP, which is considered as a second generation of AHP. The ANP method has been designed to overcome the limitation and provide a solution for more complex decision problems with multidirectional relationships (Feyzioğlu and Büyüközkan 2006; Ayağ and özdemr 2007).

From the analysis of literature review and to the best of our knowledge, there are some specific shortcomings of the current methods, particularly for PES decision in NPPM and they are as follows:

- Current methods of PES in case of NPPM do not consider probabilities and risk dimensions (Cooper et al. 2001).
- Existing methods mostly depend on extensive financial and other quantitative data (Cooper et al. 2001).
- Major focuses of the current methods are given to financial and related criteria.
- No study considered both SF and PIB dimensions/criteria along with other evaluative dimensions.
- Inability to consider both qualitative and quantitative decisions while developing a model, particularly for PES problem.

Additionally, it is also implied that the integration of qualitative and quantitative methods is significant for future NPPM methods that seek practical implications.

13.4 Development of Integrated DEA–BSC Model for NPPM

This study considers both qualitative and quantitative data for PES decision in NPPM. The proposed methodology is based on relative evaluation of products or projects or portfolios. Particularly, this study proposes an evaluation model by appropriately integrating DEA and BSC models (this integrated model is called DEA–BSC model) for PES decision in NPPM. This integrated approach serves as an alternative to the conventional multidimensional knapsack approach which obtains an optimal portfolio with respect to a well-formulated objective function and multiple resource constraints (Eilat et al. 2006). The integrated approach proposed in this study has an additional feature of evaluating alternative new product projects in the presence of multiple objectives and possible interactions among the projects. Before detailing the proposed integrated DEA-BSC model, we discuss the base BSC and base DEA models in the following sections.

13.4.1 BSC for Achieving Strategic and Balanced NPP

Evaluating new product projects involves certain criteria for which managers cannot provide hard data, where decisions are to be made based on experience, intuition, and opinions. In order to inculcate this feature into decision-making tool, we propose to develop a BSC model for NPPM.

The BSC model is a management tool composed of a collection of evaluation indicators (such as customer trust, priority level, and supplier's satisfaction, etc.) arranged in groups (i.e., evaluation perspective) and denoted as cards. The BSC model is collection of nonfinancial and financial measures. The measures are related to four managerial perspectives, that is, marketing, strategic, operational, and financial, and are aimed to provide a comprehensive view to the top management of their NPP. A specific BSC model for projects was first proposed by Kaplan and Norton (1992).

The BSC model represents a translation of business strategy into a linked set of evaluation indicators that define both long-term and short-term objectives. It acts as a mechanism for achieving and obtaining feedback of the objectives (Kaplan and Norton 2001). Decision makers who rely on BSC model need not hang to just short-term financial measures as the sole indicators of project performance are for formulation of portfolio. The major advantage of using BSC model is that, it minimizes number of measures used, which in turn minimizes the information overloaded. This helps to summate seemingly disparate elements of the evaluation and finally provides with suboptimization of all the important measures.

The strategies and the lines of action that would enable the organization to achieve its strategic vision should be translated into each of the four perspectives. The organization's strategies that are formulated in alignment to the perspectives of learning and growth and in internal processes determine

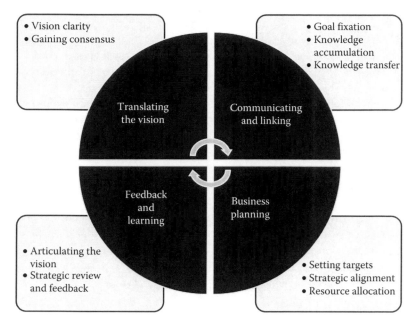

FIGURE 13.2
Managing strategy through balance scorecard.

the success of organization. These organization's strategies help one to meet organization's vision and objectives related to the satisfaction of its customers and shareholders. For linking strategy and transforming organization's visions into BSC model's perspectives, an efficient process is required. Accordingly, BSC model involves four processes which contribute to linking long-term strategic objectives to short-term actions. The four processes for managing strategy are: translating the vision, communicating and linking, business planning, and feedback and learning. Figure 13.2 depicts how these four processes help in formulation of BSCs.

13.4.2 DEA Approach for Ranking and Prioritization of Portfolios

All the quantitative models for NPPM capture only part of the scenario and they are often deemed as incomplete. For this reason, an increasing trend in the implementation and requirement of new performance measurement systems other than financial metrics is implied through many research studies (Andrews 1996; Banker et al. 2000, 2004).

DEA approach is a mathematical programming technique that calculates relative efficiency of multiple DMUs on the basis of observed inputs and outputs. Further, DEA approach is considered to be an efficient methodology for multiobjective decision making where qualitative and quantitative criteria are involved (Chames et al. 1978). The usefulness of DEA approach in evaluating multicriteria systems and providing targets for such system is reported in a large number of applications (Seiford 1996).

The basic DEA model estimates the relative efficiency of a DMU as the ratio of weighted outputs to weighted inputs, where the model takes weights for each DMU. As a result, model identifies its relative efficiency with respect to an "efficiency frontier," which is defined by assessing all the DMUs. However, in case of practical implementation perspective, the virtually unconstrained weights are not acceptable (Roll and Golany 1993). For this scenario, restricted DEA approaches were developed to allow control over the weights in the model. A general approach for controlling factor weights was developed by Chames et al. (1989). This weight-restricted method generalizes the original DEA model, by acquiring the values for input and output weights within the given closed cones. For the detailed weight restriction approaches, one can refer to Cook and Seiford (1978) and Roll and Golany (1993). The DEA model, proposed in this study, is based on Chames, Cooper, and Rhodes (CCR) model, which was first proposed by Chames et al. (1978).

13.4.2.1 Base CCR Model

For the development of base CCR model, we assume that there are k projects. Each project assumes to have varying amounts of l different inputs which produce m different outputs. The nomenclature involved in the development of the base CCR model is detailed as follows:

For Base CCR Model:

k—No. of projects

l—No. of different inputs consumed by a project

m—No. of different outputs produced by a project

Let P_j represent project j

$X_j = \{x_{ij}\}$ $i = \{1, ..., l\}$ input matrix

$Y_j = \{y_{nj}\}$ $n = \{1, ..., m\}$ output matrix

$\mu = \{u_i\}$ $i = \{1, ..., l\}$ input weight vector

$v = \{v_n\}$ $n = \{1, ..., m\}$ output weight vector

Consider, project $P_j(j = 1, ..., k)$ which consumes $X_j = \{x_{ij}\}$ of inputs ($i = 1, ..., l$) and produces $Y_j = \{y_{nj}\}$ of outputs ($n = 1, ..., m$). Further, when one considers k projects, we have $l \times k$ matrix of inputs is denoted by X and $n \times k$ matrix of output is represented by Y. Accordingly, the input and output weights are denoted by the vectors $\mu = \{u_i\}$ and $v = \{v_n\}$, respectively. The base CCR model defines the relative efficiency of a specific project P_0 as the ratio of sum of weighted outputs $\left(\Sigma_n v_n y_{n0}\right)$ by sum of weighted inputs $\left(\Sigma_i u_i x_{i0}\right)$. The objective function of this base CCR models is defined by Equation 13.1.

$$\max_{\mu, v} Z_0 = \frac{\sum_n v_n y_{n0}}{\sum_i u_i x_{i0}}, \tag{13.1}$$

where Z_0 represents the efficiency score of project P_0.

In order to bind the optimization problem presented in Equation 13.1, normalization constraint needs to be forced to the ratio of objective function as shown below:

$$\frac{\sum_n v_n y_{nj}}{\sum_i u_i x_{ij}} \leq 1, \quad \forall j \quad (\text{where } j = 1, \ldots, k). \tag{13.2}$$

The constrained optimization problem defined by Equations 13.1 and 13.2, including positivity constraints of weights, is included into base CCR model. Furthermore, the equivalent linear programming formulation for the base CCR model is presented as follows:

Base CCR model [A]

$$\max_{\mu, v} Z_0 = \sum_n v_n y_{n0}$$

such that,

$$\sum_i u_i x_{i1} = 1,$$

$$\sum_n v_n y_{nj} - \sum_i u_i x_{ij} \leq 0 \quad \forall j$$

$$v_n \geq \varepsilon, u_i \geq \varepsilon.$$

The constant ε is a small positive number that defines as a lower bound for the multipliers. By solving the Base CCR model (A) k times, we obtain k efficiency scores for all DMUs and these can be grouped into two categories. One group consists of the efficient ones that lay on efficient frontier and the other group consists of inefficient ones that fall below the frontier.

13.4.3 Proposed Integrated MCDM Model for NPPM

The proposed integrated DEA–BSC model helps to achieve the objectives of: (a) strategically aligned with organization goals and vision and (b) balance between innovation level, risk level, cost, and resources. The proposed integrated DEA–BSC model is based on DEA–BSC model developed by Eilat et al. (2006), who were the first to integrate DEA and BSC methodologies, for PES.

The objective of the proposed integrated DEA–BSC model is to relate the identified evaluative dimensions to the perspectives of BSC model. Once this is achieved, one needs to relate the input methodology of the BSC values to DEA model. Additionally, resource constraints, priorities of perspectives, and limits of each perspective also have to be included into DEA model.

In order to simplify the procedure and guide the development of the proposed integrated DEA–BSC model, we propose to develop a framework. Accordingly, the development of the proposed integrated DEA–BSC model is divided into the following three phases and more details are discussed in the following subsections:

> *Phase 1: Development of BSC index system*: In this phase, first we relate the identified evaluative dimensions to the perspectives of BSC. In this study, we proposed seven perspectives and these have to be measured. In addition, the required evaluation indicators to measure these perspectives also need to be determined. The details of these seven perspectives are described in detail in the next section.
>
> *Phase 2: Determination of balance constraints for DEA–BSC model*: In this phase, different perspective balance limits are determined. Subsequently, upper and lower bounds of each perspective are fixed and accordingly constraints are drawn which are implemented in base DEA model.
>
> *Phase 3: Development of integrated DEA–BSC model*: In this phase, priorities of each perspective to indicator level are determined. These priorities are considered as weights in DEA model. The balance constrains proposed are introduced and respective balance matrixes are developed and integrated into these balance constraints. Additionally, resource constraints and other feasibility constraints are introduced into base DEA model. With these three phases, the final proposed integrated DEA–BSC model is developed.

The proposed three phases involved in developing the integrated DEA–BSC model are semantically represented in the form of a framework and are presented in Figure 13.3. The proposed framework provides a comprehensive way to develop a holistic understanding of every perspective of BSC impact on NPPM decision by examining the interactions among the aforementioned dimensions. Furthermore, the framework assists in development of proposed integrated DEA–BSC model.

13.4.3.1 Phase 1: Development of BSC Evaluation Index System

In this study, the BSC evaluation index system is established using the principle of "SMART." According to this principle, the perspectives of BSC are determined and the respective evaluation indicators. Most of these evaluation indicators are determined based on subjective judgments. Table 13.3 presents a set of evaluation indicators that are considered in this study along with card labels and objectives for evaluation. All the evaluation indicators considered in the proposed BSC are identified from literature

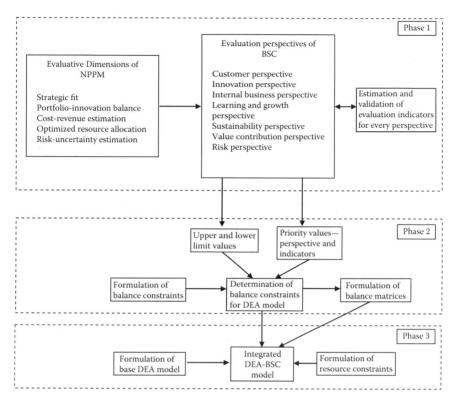

FIGURE 13.3
Proposed framework for integrated DEA–BSC model.

(Kiranmayi and Mathirajan 2013). There are in total seven cards considered in this study. For every card accounted, there exist certain measures that provide comprehensive evaluation of the NPPM performance.

The proposed BSC for NPPM considers seven perspectives. Out of seven perspectives, four are original perspectives: customer, internal business processes, learning and growth, financial, and three perspectives: innovation perspective, sustainability perspective, and uncertainty perspective are proposed explicitly for this study. The three proposed perspectives are expected to emphasize on the new product project features. In addition, in this study, we introduced an improved value contribution perspective rather than financial perspective where project cycle and cost are also taken into account. The details of these seven perspectives are presented as follows.

13.4.3.1.1 *Customer Perspective (O1)*

Customer perspective focuses on the responses of customers received through market surveys. As the required evaluation indicators for customer perspective are representative of new products/projects, the data/values of projects which are closely related or comparable to the project ideas are considered for

TABLE 13.3

Project Evaluation Index System of the Proposed BSC for NPP

Card Label	Evaluation Perspective	Evaluation Indicator	Evaluation Objective
O1	Customer perspective	Customer satisfaction Customer trust Degree of customer need met	Ensure that project meets and satisfy customer needs
O2	Internal business perspective	Employee satisfaction Supplier's satisfaction Project quality planning and tracking Internal communication Congruence Priority level	Ensure that the implementation of project plan, control, and other aspects, etc. optimize the organizational internal processes
O3	Learning and growth perspective	Propriety position Platform for growth Technical and market durability Team incentive Knowledge accumulation Project management maturity	Ensure that the implementation of the projects cultivates the organizational core technologies and competitiveness
O4	Innovation perspective	Technology newness Process newness Market newness	Ensure that project acquires the cutting edge to succeed
O5	Sustainability perspective	Ecological Social Economic	Ensure that project/product sustainability is achieved
O6	Value contribution perspective	Project profitability Project speed-up cycle Product sales	Ensure that the projects are completed in accordance with desired objectives and provide business value
O7	Uncertainty perspective	Operational–technical Organizational Financial Marketing	Ensure that projects are completed in accordance with time and specifications without any uncertainties and delay in development
I1	Resources	Investments Human resources Machinery and equipment	–

evaluation. Accordingly, customer satisfaction, customer trust, and degree of customer needs met are considered as the evaluation indicators of this perspective. Customer satisfaction addresses responsiveness, service satisfaction, product quality, market stability, and so on. Customer trust guarantees a part of demand volume, brand value, and customer relationship. Degree of customer needs met represents whether there's a necessity to change the product features, or add additional features, limitations of product, and so on.

13.4.3.1.2 *Internal Business Perspective (O2)*

Internal business perspective measures the degree of share that the new product/project aligns with the core competencies of the organization (i.e., missions and objectives). Accordingly, the evaluation indicators: employee satisfaction, supplier satisfaction, project quality planning and tracking, internal communication, congruence, and priority level are considered for internal business perspective. Employee and supplier satisfaction measures the commitment level and consistency for the development of new project. Project quality planning and tracking ensures the required actions and responsibilities of suppliers and employees are accurately and efficiently carried out and this accounts for smooth operation. Internal communication and congruence measure the understanding and level of commitment between top management and operational team members. Priority level of project decides the level of commitment required for completion of project.

13.4.3.1.3 *Learning and Growth Perspective (O3)*

Learning and growth perspective ensures that the development process does not suffer from knowledge asymmetry and transfer of knowledge. In present increasing global competition, an organization has to keep up with the advancement in technology. For this case, the learning ability of team and incentive of team are necessary to be accounted. For higher chances of organization success, a project should ensure higher chances of market and technology growth. To summarize, learning and growth perspective ensures that objectives of all the other perspectives are satisfied.

13.4.3.1.4 *Innovation Perspective (O4)*

In case of a new project there exist different levels and angles of newness brought by a project. An organization has to ensure the level of innovation that is incorporated by a project to a portfolio in order to achieve a balanced portfolio. This includes the measure of evaluation indicators such as technology newness, process newness, and market newness. The accounting for time and market newness ensures that the portfolios do not contain highly innovative projects and at the same time it should not be overwhelmed with just incremental innovation projects. Thus, innovation perspective brings the balance dimension to portfolio.

13.4.3.1.5 *Sustainability Perspective (O5)*

To the best of our knowledge, sustainability perspective is considered in BSC index system for the first time by this study. Increasing global competition and rapidly changing customer needs compete for the development of new product. In order to succeed in competition and keep up with trend development, sustainable products or technology are necessary.

Sustainability perspective includes estimation of evaluation indicators such as ecological, social, and economic. Though economical sustainability is ensured by many organizations, achieving the balance between all the evaluation indicators of sustainability is very important as this improves the efficiency of NPPM and increases the probability of success of new product.

13.4.3.1.6 Value Contribution Perspective (O6)

Financial perspective is a standard card in BSC index system. In this study, along with evaluation indicators of financial perspective, the performance indicator project cycle is introduced. In case of new product, first to market or time to market plays a crucial role in success of new product. Thus, estimation of project cycle as a whole ensures the time to market. Thus, we consider this perspective as value contribution perspective. In this perspective, along with project cycle, project cost and profitability are also measured. This financial objective serves as main focus for the objectives and measures in all the other scorecard perspectives.

13.4.3.1.7 Uncertainty Perspective (O7)

Uncertainty perspective ensures development process of new products with minimized uncertainty. It includes operational–technical, organizational, financial, and marketing uncertainties. The probabilities of these uncertainty evaluation indicators are estimated using Bayesian belief networks. Each and every evaluation indicator of uncertainty includes other measures which determine the total probability. This uncertainty perspective, though overlooked by many research studies, turned out to be the most prominent and essential perspective to be considered for PES decision of NPPM (Oh et al. 2012).

One of the key components of a BSC is establishment of baseline or benchmark against which the performance is measured. But the baseline or benchmark is hard to determine and can be misleading. Since DEA is based on relative analysis, the projects are evaluated against each other. In this study, BSC with DEA is integrated to overcome this constraint. Thus, the evaluation indicators are considered accordingly in formulation of DEA model.

13.4.3.2 Phase 2: Determination of Balance Constraints for DEA–BSC Model

To determine balance constraints for the proposed DEA–BSC model, we first brief how BSC evaluation index system is integrated into DEA model. Each card (i.e., the evaluation perspective) represents a major dimension of interest for the multiproject organization and the respective evaluation indicators are accordingly considered as inputs (x_i) and outputs (y_n) for DEA model. Accordingly, in the proposed BSC evaluation index system, there exist seven output cards (evaluation perspectives) and one input card. The base nomenclature of BSC index system is as follows:

For BSC index system:

S—No. of input cards

R—No. of output cards

$I_1...I_S$—Input of S cards

$O_1...O_R$—Output of R cards

$[L_{I_s}, U_{I_s}$—Lower and upper bounds for inputs

$[L_{O_r}, U_{O_r}$—Lower and upper bounds for output

To reflect the desired balance, we need to set limits regarded as upper and lower bounds on each cards. To formulate the proposed DEA–BSC model, we determine the balancing constraints and they are divided into two groups. They are lower and upper bound constraints for every output (O_R) and input (I_S) cards. The values are denoted as: $[L_{I_s}, U_{I_s}$—Lower and upper bounds for inputs; and $[L_{O_r}, U_{O_r}$—Lower and upper bounds for outputs.

With these, for a project $P_{j'}$ the constraints (13.3) and (13.4) are included to ensure the balance among the input and output cards, respectively. Furthermore, these balance constraints depicts the importance and variability of each card of the BSC index system.

$$L_{I_s} \leq \frac{\sum_{i \in I_s} u_i x_{ij}}{\sum_i u_i x_{ij}} \leq U_{I_s} \quad \forall j,... \tag{13.3}$$

$$L_{O_r} \leq \frac{\sum_{n \in O_r} v_n y_{nj}}{\sum_n v_n y_{nj}} \leq U_{O_r} \quad \forall j,... \tag{13.4}$$

13.4.3.3 Phase 3: Development of the Proposed Integrated DEA–BSC Model

For the proposed integrated DEA–BSC model, the set of balancing constraints (13.3) and (13.4) are added to the base CCR model (A). These constraints ensure that balance objective is achieved among the output and input cards. Lower bound constraints of output cards resemble the relative importance of that particular output card, while upper bound constraints ensure the variability balance. The formulation by default assumes nonnegativity constraints. With these, for example, for a given project P_1, the linear programming formulation of the proposed single-level integrated DEA–BSC model (B) is represented below:

The *proposed single-level integrated DEA–BSC model (B)*

$$\max_{u,v} Z_1 = \sum_n v_n y_{n1}$$

such that,

$$\sum_i u_i x_{i1} = 1,$$

$$\sum_n v_n y_{nj} - \sum_i u_i x_{ij} \leq 0,$$

$$L_{O_r} \sum_n v_n y_{n1} - \sum_{n \in O_r} v_n y_{n1} \leq 0,$$

$$\sum_{n \in O_r} v_n y_{n1} - U_{O_r} \sum_n v_n y_{n1} \leq 0,$$

$$L_{I_s} \sum_i u_i x_{i1} - \sum_{i \in I_s} u_i x_{i1} \leq 0,$$

$$\sum_{i \in I_s} u_i x_{i1} - U_{I_s} \sum_i u_i x_{i1} \leq 0.$$

13.5 Validation of the Proposed Integrated DEA–BSC Model for NPPM

In order to validate (that is illustrating the workability) the proposed integrated DEA–BSC model, a numerical example with 10 NPD projects is developed. Before getting into details, we first detail the preparatory steps for the implementation of proposed integrated DEA–BSC model:

Data:

1. The data required for the evaluation indicators of each of the BSC perspectives is suitably generated for 10 NPD projects. For generating values for each of the evaluation indicators (i.e., input and output values), we assume a range for every evaluation indicators and the same are presented in Table 13.4. Using the range defined in Table 13.4, randomly generated values for each of the evaluation indicators for the 10 NPD projects are given in Table 13.5.

2. In order to obtain weights for the evaluation perspectives and respective evaluation indicators, we used AHP. For this, an appropriate questionnaire was prepared and project/product managers were approached for their responses (in this study, we collected

TABLE 13.4

Range Considered for Generating Value for Each of the Evaluation Indicators

Card Label	Evaluation Perspective	Evaluation Indicator	Range Considered for Indicator's Value
O1	Customer perspective	Customer satisfaction (O11)	5–10
		Customers' trust (O12)	6–10
		Degree of customer need met (O13)	6–10
O2	Internal business perspective	Employee satisfaction (O21)	6–10
		Supplier's satisfaction (O22)	6–10
		Project quality planning and tracking (O23)	7–10
		Internal communication (O24)	5–10
		Congruence (O25)	4–10
		Priority level (O26)	5–10
O3	Learning and growth perspective	Propriety position (O31)	2–10
		Platform for growth (O32)	2–10
		Technical and market durability (O33)	4–10
		Team incentive (O34)	4–10
		Knowledge accumulation (O35)	6–10
		Project management maturity (O36)	4–10
O4	Innovation perspective	Technology newness (O41)	4–7
		Process newness (O42)	4–7
		Market newness (O43)	4–7
O5	Sustainability perspective	Ecological (O51)	4–10
		Social (O52)	4–10
		Economic (O53)	7–10
O6	Value contribution perspective	Project profitability(millions) (O61)	7–12
		Project speed-up cycle (years) (O62)	0.2–1
		Product sales (thousand units) (O63)	20–50
O7	Uncertainty perspective (probability of success)	Operational–technical (O71)	0.6–0.9
		Organizational (O72)	0.5–0.9
		Financial (O73)	0.7–0.9
		Marketing (O74)	0.7–0.9
I1	Resources	Investments (millions) (I11)	70–100
		Human resources (I12)	–
		Machinery and equipment (I13)	–

responses from 104 managers). The responses obtained are given as input to AHP and obtained the required weights. The AHP process on the given primary data is presented in Appendix A. The weights obtained for the evaluation perspectives and evaluation indicators are presented in Table 13.6.

TABLE 13.5

Project Wise, the Value of Evaluation Indicators

		Project No.									
Card Label		1	2	3	4	5	6	7	8	9	10
I11		73	82	96	87	75	78	96	89	83	91
O1	O11	6	5	7	7	8	6	8	7	9	8
	O12	8	6	7	7	6	6	8	7	8	6
	O13	8	6	7	8	9	8	7	8	9	7
O2	O21	7	7	6	6	8	7	8	6	8	7
	O22	7	8	9	8	7	7	6	6	8	7
	O23	7	8	9	9	8	8	7	7	7	9
	O24	9	6	7	7	7	5	8	5	8	8
	O25	4	6	8	10	5	6	7	4	4	8
	O26	5	6	5	8	8	8	7	6	6	8
O3	O31	3	4	5	8	4	8	8	2	6	5
	O32	3	5	5	8	9	9	10	5	2	6
	O33	9	6	7	7	7	5	8	5	8	8
	O34	4	6	8	10	5	6	7	4	4	8
	O35	9	8	9	5	8	8	7	8	7	9
	O36	7	4	5	8	4	9	8	4	6	5
O4	O41	4	6	7	6	5	6	5	5	5	6
	O42	5	6	7	7	5	7	5	7	7	6
	O43	5	6	7	5	5	6	7	6	5	7
O5	O51	7	6	5	9	5	8	8	7	8	8
	O52	7	7	5	5	8	4	9	8	4	8
	O53	7	7	8	7	8	8	8	7	7	8
O6	O61	8	9	11	10	12	11	10	12	8	9
	O62	0.9	0.9	0.5	0.6	0.5	0.4	1	0.8	0.6	0.5
	O63	25	24	29	38	45	35	40	42	35	44
O7	O71	0.7	0.6	0.9	0.8	0.8	0.6	0.7	0.7	0.9	0.9
	O72	0.7	0.8	0.5	0.6	0.5	0.8	0.9	0.7	0.8	0.8
	O73	0.7	0.8	0.8	0.7	0.9	0.7	0.8	0.9	0.7	0.8
	O74	0.8	0.9	0.9	0.8	0.9	0.8	0.8	0.7	0.9	0.8

3. In order to implement the proposed integrated DEA–BSC model, we need to determine the bounds for every evaluation perspective. However, the bounds allocated vary from organization to organization and depend on perspectives of the top management and project manager. One can consider maximal tolerance level (i.e., 100%) for limits of bounds, but in order to emphasize the importance of bounds, we assume the tolerance to be 60%. Accordingly, Table 13.7 presents the lower and upper bounds of every evaluation perspectives considered for the proposed integrated DEA–BSC model.

TABLE 13.6

Weights of the Evaluation Perspective and Evaluation Indictors

Card Label	Evaluation Perspective	First-Level Weights	Evaluation Indicator	Second-Level Weights
O1	Customer perspective	0.121	Customer satisfaction	0.0452
			Customer trust	0.0356
			Degree of customers' need met	0.0402
O2	Internal business perspective	0.101	Employee satisfaction	0.00887
			Supplier's satisfaction	0.00956
			Project quality planning and tracking	0.04756
			Internal communication	0.01548
			Congruence	0.00157
			Priority level	0.00896
O3	Learning and growth perspective	0.096	Propriety position	0.00945
			Platform for growth	0.00543
			Technical and market durability	0.04255
			Team incentive	0.00923
			Knowledge accumulation	0.02145
			Project management maturity	0.00789
O4	Innovation perspective	0.142	Technology newness	0.0756
			Process newness	0.0521
			Market newness	0.0143
O5	Sustainability perspective	0.068	Ecological	0.0149
			Social	0.0156
			Economic	0.0375
O6	Value contribution perspective	0.281	Project profitability	0.1457
			Project speed-up cycle	0.0478
			Product sales	0.0875
O7	Uncertainty perspective	0.191	Operational–technical	0.0145
			Organizational	0.0108
			Financial	0.0895
			Marketing	0.0762
I1	Resources	1	Investments	0.645
			Human resources	0.143
			Machinery and equipment	0.212

For generating the proposed integrated DEA–BSC model for any given data, a LINGO set code has been developed and presented in Appendix B. The data presented in Tables 13.5 through 13.7 can be given as input to the LINGO set code and the proposed integrated DEA–BSC model can be generated for every project and solved using LINGO. Finally, the efficiency scores of each of the projects along with relative ratings (rankings) are obtained and the same are presented in Table 13.8. From Table 13.8, one can select the set of projects for NPP based on the efficiency score.

TABLE 13.7

Lower and Upper Balance Bounds of Evaluation
Perspectives of DEA–BSC Model

Card Label	Evaluation Perspective	Lower Bound	Upper Bound
O1	Customer perspective	0.1	0.7
O2	Internal business perspective	0.2	0.8
O3	Learning and growth perspective	0.2	0.8
O4	Innovation perspective	0.1	0.7
O5	Sustainability perspective	0.1	0.7
O6	Value contribution perspective	0.3	0.9
O7	Uncertainty perspective	0.12	0.72

TABLE 13.8

Project wise the Efficiency Score and Relative
Rating Yielded by the Proposed DEA–BSC Model

Project	Efficiency Score	Rating
1	0.7432	7
2	0.9221	4
3	0.6043	8
4	1.0000	1
5	0.9910	2
6	1.0000	1
7	0.7496	6
8	0.9452	3
9	1.0000	1
10	0.8457	5

From the numerical example presented in this study, projects 4, 6, and 9 have almost same efficiency score; hence, they are having rating of "1" (Table 13.8). The third column (i.e., rating) in Table 13.8 represents the ranking of project. The projects with rating "1" have the maximum efficiency, so a decision maker tends to pick up these projects.

13.6 Summary

As significance for the development of new products is rapidly increasing, it becomes essential for an organization to have an effective and accurate decision-making process. In order to develop a successful new product, decision maker/project manager needs to identify right set of new projects/products and accordingly formulate a New Product Portfolio (NPP). Thus, the decisions taken at the phase of PES play a significant role in NPPM performance. There exist limited studies concentrating on (a) identifying factors/dimensions that influence decision making, (b) development of decision-making model for PES, and (c) improving the performance of NPPM. However, it is observed through literature that NPPM is the weakest research area and identification of explicit evaluative factors/dimensions is lacking (Cooper et al. 2001; McNally et al. 2009). In this study, we address these gaps.

Accordingly, in this study, we have identified five different evaluative dimensions (i.e., SF, PIB, Resource Allocation, CRE, and RUE) that are essential in the case of PES for NPPM. It is evident from literature that there exist very limited number of studies that concentrate on development of MCDM for PES in case of NPP. We make an attempt to employ all the evaluative dimensions in the development of the MCDM model for NPPM. To the best of our knowledge, this is the first study that considers all the five evaluative dimensions simultaneously in the development of MCDM in this area.

We further briefed different methodologies employed for MCDM model formulation in case of PES studies. From the discussion, it is identified that there exist certain limitations in the existing methodologies. In present NPPM scenario observed, it is essential to accommodate subjective data of a decision maker into the developed methodology. This probed us to the development of a methodology for NPPM in which qualitative and quantitative data can be considered. Thus, we proposed to develop an integrated DEA–BSC model.

The methodology is based on relative evaluation of entities (projects or portfolios), which is inspired by an integrated DEA–BSC model that was first presented by Eilat et al. (2008). For this, the identified evaluative dimensions are respectively reorganized into seven evaluation perspectives of BSC index system. Accordingly, each evaluation indicators are measured using different scales and metrics. The output of BSC is considered as input for DEA, along with certain other inputs such as weights or priorities of perspectives, obtained from AHP. Then, the proposed integrated DEA–BSC model estimates the efficiency score of each and every project and ranks them accordingly. Thus, the model proposed in this study provides clarity and accuracy regarding the subjective data associated to PES decisions.

In future work, we intend to extend the model for hierarchical level of BSC. We also intend to introduce an accumulation function that takes care of interactions between resources, benefit functions, and output functions. Finally, we propose to extend this model by introducing dynamic nature into the problem.

Appendix A: An Analytical Hierarchical Process for BSC

The AHP is a popular MCDM methodology which has been applied vastly in various fields. It was developed by Thomas Saaty (1960) for including qualitative variables for multicriteria decision-making model. AHP allows decision makers to model a complex problem in a hierarchical structure. In this method, a simple hierarchical model consists of a goal, criteria, and alternatives. In order to carry out an AHP analysis, the following iterative steps are involved:

1. The first and foremost activity in AHP is to analyze the problem, identify criteria/indicators/alternatives. In our case, evaluation perspectives and evaluation indicators are considered as criteria and subcriteria according to AHP terminology. Once these are identified, evaluation perspectives and indicators are represented in hierarchical level network. The network diagram of AHP of this study is presented in Exhibit 13.1.

EXHIBIT 13.1

Hierarchical Network Diagram of BSC

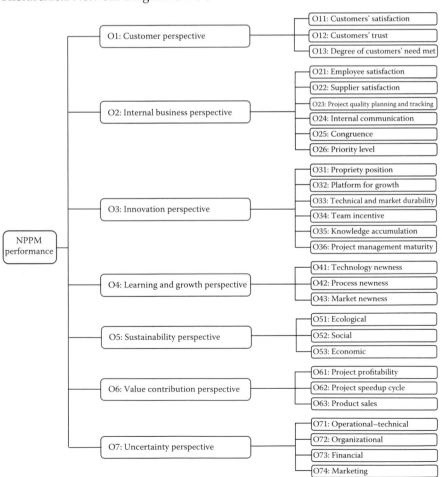

2. Once the evaluation indictors are identified, one needs to obtain pairwise comparison values for these evaluation indicators. A questionnaire is prepared in order to obtain relative weights for all the evaluation perspectives as well as evaluation indicators. These values lead to dominance matrices which are called pairwise comparison matrices. Ratio scales are derived in the form of principle eigenvectors from these matrices. The pairwise comparison matrices of evaluation perspective with respect to goal and evaluation indicators with respect to evaluation perspectives are presented in Exhibits 13.2 and 13.3, respectively.

3. The expert opinion results are either aggregated or considered independently to obtain scores of pairwise comparison. Through pairwise comparison matrices, one can obtain the priorities of evaluation perspectives and indicators by normalizing the matrices. The priority vectors along with pairwise normalized matrices are presented in Exhibits 13.4 and 13.5.

4. The overall priority vector for evaluation indicators is obtained by vector multiplication of the previously calculated perspective and indicator priority vectors. The resultant overall priority vector for indicator is presented in Exhibit 13.6.

EXHIBIT 13.2

Pairwise Comparison Matrix of Evaluation Perspectives with Respect to NPPM Performance

	O1	O2	O3	O4	O5	O6	O7
O1	1	1/4	6	4	1	1/3	1/5
O2	4	1	1/4	1/3	2	1/3	3
O3	1/6	4	1	2	5	1/3	1
O4	1/4	3	1/2	1	1	1/5	1/3
O5	1	1/2	1/5	1	1	1/5	1/3
O6	3	3	3	5	5	1	1/3
O7	5	1/3	1	3	3	3	1

EXHIBIT 13.3

Pairwise Comparison Matrix of Evaluation Indicators with Respect to Uncertainty Perspective

	O61	O62	O63	O64
O61	1	2	1	1
O62	1/2	1	1/2	1/4
O63	1	2	1	2
O64	1	4	1/2	1

EXHIBIT 13.4

Pairwise Normalized Matrix of Evaluation Perspectives with Respect to NPPM Performance

	O1	O2	O3	O4	O5	O6	O7	Priorities
O1	0.069	0.021	0.502	0.245	0.056	0.062	0.032	0.141
O2	0.277	0.083	0.021	0.020	0.111	0.062	0.484	0.151
O3	0.012	0.331	0.084	0.122	0.278	0.062	0.161	0.150
O4	0.017	0.248	0.042	0.061	0.056	0.037	0.054	0.074
O5	0.069	0.041	0.017	0.061	0.056	0.037	0.054	0.048
O6	0.208	0.248	0.251	0.306	0.278	0.185	0.054	0.219
O7	0.347	0.028	0.084	0.184	0.167	0.556	0.161	0.218

EXHIBIT 13.5

Pairwise Normalized Matrix of Evaluation Indicators with Respect to Uncertainty Perspective

	O61	O62	O63	O64	Priorities
O61	0.2	0.13	0.17	0.31	0.2
O62	0.2	0.13	0.17	0.08	0.14
O63	0.4	0.25	0.33	0.31	0.32
O64	0.2	0.5	0.33	0.31	0.34

EXHIBIT 13.6

Overall Priorities of Uncertainty Evaluation Indicators with Respect to Uncertainty Perspective

	Perspective Priority	Priorities	Overall Priorities
O61	0.219	0.2	0.0436
O62		0.14	0.0305
O63		0.32	0.0698
O64		0.34	0.0741

5. The matrices are validated by measuring consistency index and consistency ratio. This validation is considered to remove the response bias associated with pairwise comparisons. For this, one needs to calculate the maximum eigenvalue, λ_{max}, and the eigenvector, w, for the matrix. The consistency is performed by calculating the consistency index (CI) and consistency ratio (CR):

$$CI = \frac{\lambda_{max} - n}{n-1}$$

$$CR = \frac{CI}{RI},$$

where n is the number of items being compared in the matrix, and RI is random index. If CR is less than 0.1, the experts' judgments are consistent. If the consistency test is not passed, the part of the questionnaire must be done again.

Appendix B

A LINGO Set Code developed for generating the proposed integrated DEA–BSC model:

```
MODEL:
! DEA-BSC MODEL FOR NPPM ;
! PLEASE NOTE: THIS CODE ISFOR ILLUSTRATION PURPOSE;
! Part of the sample data is presented in this code
SETS:
    DMU/P1 P2 P3 P4 P5 P6 P7 P8 P9 P10/: ! 10 NEW PRODUCT PROJECTS;
        SCORE; ! Each decision making unit in this case new
product project has a score to be computed;
    FACTOR/I1 O11 O12 O13 O21 O22 O23 O24 O25 O26 O31 O32 O33 O34
O35 O36 O41 O42 O43 O51 O52 O53 O61 O62 O63 O71 O72 O73 O74/;
! ALL THE EVALUATION INDICATORS ARE CONSIDERED AS OUTPUT AND
INVESTMENTS AS INPUT;
DXF( DMU, FACTOR):   F; ! F( I, J) = Jth factor of DMU I;
ENDSETS
DATA:
    NINPUTS = 1;    ! The first NINPUTS factors are inputs;
!         The inputs,    the outputs;
F   =    73              6    8    8    --------    0.7    0.8
         82              5    6    6    --------    0.8    0.9
         96              7    7    7    --------    0.8    0.9
         87              7    7    8    --------    0.7    0.8
         75              8    6    9    --------    0.9    0.9
         78              6    6    8    --------    0.7    0.8
         96              8    8    7    --------    0.8    0.8
         89              7    7    8    --------    0.9    0.7
         83              9    8    9    --------    0.7    0.9
         91              8    6    7    --------    0.8    0.8;
ENDDATA
!-------------------------------------------------------------;
SETS:
! Weights used to compute DMU I's score;
DXFXD(DMU,FACTOR)  : W;
ENDSETS
DATA
W =   0.645
```

```
      0.0356
      0.0402
      0.00887
      |
      |
      |
      0.0108
      0.0895
      0.0762;
ENDDATA
!------------------------------------------;
SETS:
! Lower Bounds used to compute DMU I's score;
DXFXA(DMU,FACTOR) : LB;
ENDSETS
DATA
LB=    0
       0.1
       0.1
       0.1
       0.2
       |
       |
       |
       0.12
       0.12
       0.12
       0.12;
ENDDATA
!------------------------------------------;
SETS:
! Upper Bounds used to compute DMU I's score;
DXFXB(DMU,FACTOR) : UB;
ENDSETS
DATA
UB=    56
       0.7
       0.7
       0.7
       0.8
       |
       |
       |
       0.72
       0.72
       0.72
       0.72;
ENDDATA
! The Model;
! Try to make everyone's score as high as possible;
```

```
MAX = @SUM( DMU: SCORE);
! The LP for each DMU to get its score;
@FOR( DMU( I):
SCORE( I) = @SUM( FACTOR(J)|J #GT# NINPUTS:
F(I, J)* W(I, J));
! Sum of inputs(denominator) = 1;
@SUM( FACTOR( J)| J #LE# NINPUTS:
F( I, J)* W( I, J)) = 1;
! Using DMU I's weights, no DMU can score better than 1;
@FOR( DMU( K):
@SUM( FACTOR( J)| J #GT# NINPUTS: F( K, J) * W( I, J))
<= @SUM( FACTOR( J)| J #LE# NINPUTS:
F( K, J) * W( I, J)));
! Balance Constraints for outputs;
    (LB(I,J)* @SUM( FACTOR( J)| J #GT# NINPUTS: F( I, J) * W( I, J)))
    <= @SUM( FACTOR( J)| J #GT# NINPUTS: F( I, J) * W( I, J));
    @SUM( FACTOR( J)| J #GT# NINPUTS: F( I, J) * W( I, J))
    <= (UB(I,J)* @SUM( FACTOR( J)| J #GT# NINPUTS: F( I, J) *
    W( I, J)));
! Similarly Balance Constraint for input can be introduced;
    );
END
```

References

Abbassi, M., Ashrafi, M., and Tashnizi, E.S. 2014. Selecting balanced portfolios of R&D projects with interdependencies: A cross-entropy based methodology. *Technovation* 34 (1): 54–63.

Andrews, K.Z. 1996. Two kinds of performance measures. *Harvard Business Review* 74 (1): 8–9.

Asosheh, A., Nalchigar, S., and Jamporazmey, M. 2010. Information technology project evaluation: An integrated data envelopment analysis and balanced scorecard approach. *Expert Systems with Applications* 37 (8): 5931–5938.

Ayağ, Z. and Özdemr, R.G. 2007. An analytic network process-based approach to concept evaluation in a new product development environment. *Journal of Engineering Design* 18 (3): 209–226.

Banker, R.D., Chang, H., and Pizzini, M.J. 2004. The balanced scorecard: Judgmental effects of performance measures linked to strategy. *Accounting Review* 79 (1): 1–23.

Banker, R.D., Potter, G., and Srinivasan, D. 2000. An empirical investigation of an incentive plan that includes non-financial performance measures. *Accounting Review* 75 (1): 65–92.

Bhattacharyya, R., Kumar, P., and Kar, S. 2011. Fuzzy R&D portfolio selection of interdependent projects. *Computers & Mathematics with Applications* 62 (10): 3857–3870.

Blundell, R., Griffith, R., and Van Reenen, J. 1999. Market share, market value and innovation in a panel of British manufacturing firms. *The Review of Economic Studies* 66 (3): 529–554.

Brown, S.L. and Eisenhardt, K.M. 1995. Product development: Past research, present findings, and future directions. *Academy of Management Review* 20 (2): 343–378.

Carbonell-Foulquié, P., Munuera-Alemán, J.L., and Rodriguez-Escudero, A.I. 2004. Criteria employed for Go/no-Go decisions when developing successful highly innovative products. *Industrial Marketing Management* 33 (4): 307–316.

Chames, A., Cooper, W.W., and Rhodes, E. 1978. Measuring the efficiency of decision making units. *European Journal of Operational Research* 2: 429–444.

Chames, A., Cooper, W.W., and Rhodes, E. 1989. Cone ratio data envelopment analysis and multi-objective programming. *International Journal of Systems Science* 20 (7): 1099–1118.

Chan, S. and Ip, W. 2010. A scorecard-Markov model for new product screening decisions. *Industrial Management & Data Systems* 110 (7): 971–992.

Chang, Y.T., Park, H.S., Jeong, J.B., and Lee, J.W. 2014. Evaluating economic and environmental efficiency of global airlines: A SBM-DEA approach. *Transportation Research Part D: Transport and Environment* 27: 46–50.

Chao, R.O. and Kavadias, S. 2008. A theoretical framework for managing the new product development portfolio: When and how to use strategic buckets. *Management Science* 54 (5): 907–921.

Chesbrough, H.W. and Teece, D.J. 2002. *Organizing for Innovation: When Is Virtual Virtuous?* Harvard Business School Publishings, MA.

Chiang, T.-A. and Che, Z.H. 2010. A fuzzy robust evaluation model for selecting and ranking NPD projects using Bayesian belief network and weight-restricted DEA. *Expert Systems with Applications* 37 (11): 7408–7418.

Chien, C.-F. 2002. A portfolio-evaluation framework for selecting R&D projects. *R&D Management* 32: 359–368.

Chiu, C.-Y. and Park, C.S. 1994. Fuzzy cash flow analysis using present worth criterion. *The Engineering Economist* 39 (2): 113–138.

Cook, W.D. and Seiford, L.M. 1978. R&D project selection in a multi-dimensional environment: A practical approach. *Journal of the Operational Research Society* 21: 29–37.

Cooper, R.G. 1994. Perspective third-generation new product processes. *Journal of Product Innovation Management* 11 (1): 3–14.

Cooper, R.G., Edgett, S.J., and Kleinschmidt, E.J. 1997. Portfolio management in new product development: Lessons from the Leaders-II. *Research-Technology Management* 40 (6): 43.

Cooper, R.G., Edgett, S., and Kleinschmidt, E. 2001. Portfolio management for new product development: Results of an industry practices study. *R&D Management* 31 (4): 361–380.

Cooper, R.G., Edgett, S.J., and Kleinschmidt, E.J. 2004. Benchmarking best NPD practices-1. *Research-Technology Management* 47 (1): 31–43.

Crawford, C.M. and Di Benedetto, C.A. 2008. *New Products Management.* Tata McGraw-Hill Education, New York, NY.

Duarte, B.P. and Reis, A. 2006. Developing a projects evaluation system based on multiple attribute value theory. *Computers & Operations Research* 33 (5): 1488–1504.

Eilat, H., Golany, B., and Shtub, A. 2006. Constructing and evaluating balanced portfolios of R&D projects with interactions: A DEA based methodology. *European Journal of Operational Research* 172 (3): 1018–1039.

Eilat, H., Golany, B., and Shtub, A. 2008. R&D project evaluation: An integrated DEA and balanced scorecard approach. *Omega* 36 (5): 895–912.

Feyzioğlu, O. and Büyüközkan, G. 2006. Evaluation of new product development projects using artificial intelligence and fuzzy logic. *International Conference on Knowledge Mining and Computer Science* 11: 183–189.

Fitzsimmons, J.A., Kouvelis, P., and Mallick, D.N. 1991. Design strategy and its interface with manufacturing and marketing: A conceptual framework. *Journal of Operations Management* 10 (3): 398–415.

Gates, W. 1999. *Business @ the Speed of Sound*. Warner Books, New York, NY.

Ghasemzadeh, F. and Archer, N.P. 2000. Project portfolio selection through decision support. *Decision Support Systems* 29 (1): 73–88.

Graves, S.B., Ringuest, J.L., and Case, R.H. 2000. Formulating optimal R&D portfolios. *Research-Technology Management* 43 (3): 47–51.

Griffin, A. and Hauser, J.R. 1996. Integrating R&D and marketing: A review and analysis of the literature. *Journal of Product Innovation Management* 13 (3): 191–215.

Gutjahr, W.J., Katzensteiner, S., Reiter, P., Stummer, C., and Denk, M. 2010. Multi-objective decision analysis for competence-oriented project portfolio selection. *European Journal of Operational Research* 205 (3): 670–679.

Halouani, N., Chabchoub, H., and Martel, J.-M. 2009. Promethee-md-2t method for project selection. *European Journal of Operational Research* 195 (3): 841–849.

Ibbs, C.W., Reginato, J., and Kwak, Y.H. 2004. Developing project management capability: Benchmarking, maturity, modeling, gap analyses and ROI studies. In: *The Wiley Guide to Managing Projects*, John Wiley & Sons, Inc., Hoboken, NJ, pp. 1214–1233.

Kahraman, C., Büyüközkan, G., and Ateş, N.Y. 2007. A two phase multi-attribute decision-making approach for new product introduction. *Information Sciences* 177 (7): 1567–1582.

Kaplan, R.S. and Norton, D.P. 1992. The balance scorecard—Measures that drive performance. *Harvard Business Review* 70 (1): 71–79.

Kaplan, R.S. and Norton, D.P. 2001. Transforming the balanced scorecard from performance measurement to strategic management: Part I. *Accounting Horizons* 15 (1): 87–104.

Kiranmayi, P. and Mathirajan, M. 2013. A theoretical framework for project evaluation and selection in new product management. *Proceedings of International Conference on Sustainable Innovation and Successful Product Development for a Turbulent Global Market*, Indian Institute of Technology-Madras, Chennai, India.

Krishnan, V. and Ulrich, K.T. 2001. Product development decisions: A review of the literature. *Management Science* 47 (1): 1–21.

Kumar, D.U., Saranga, H., Ramírez-Márquez, J.E., and Nowicki, D. 2007. Six sigma project selection using data envelopment analysis. *The TQM Magazine* 19 (5): 419–441.

Kyparisis, G.J., Gupta, S. K., and Ip, C.-M. 1996. Project selection with discounted returns and multiple constraints. *European Journal of Operational Research* 94 (1): 87–96.

Lee J.W. and Kim H.S. 2000. Using analytic network process and goal programming for interdependent information system project selection. *Computers & Operations Research* 27: 367–382.

Liao, S.-H. 2005. Expert system methodologies and applications—A decade review from 1995 to 2004. *Expert Systems with Applications* 28 (1): 93–103.

Loch, C.H. and Kavadias, S. 2002. Dynamic portfolio selection of NPD programs using marginal returns. *Management Science* 48 (10): 1227–1241.

Lockett, G. and Stratford, M. 1987. Ranking of research projects: Experiments with two methods. *Omega* 15 (5): 395–400.

Mahmoodzadeh, S., Shahrabi, J., Pariazar, M., and Zaeri, M.S. 2007. Project selection by using fuzzy AHP and TOPSIS technique. *World Academy of Science, Engineering and Technology* 30: 333–338.

Mavrotas, G., Diakoulaki, D., and Caloghirou, Y. 2006. Project prioritization under policy restrictions: A combination of MCDA with 0–1 programming. *European Journal of Operational Research* 171 (1): 296–308.

McNally, R.C., Durmusoglu, S.S., Calantone, R.J., and Harmancioglu, N. 2009. Exploring new product portfolio management decisions: The role of managers' dispositional traits. *Industrial Marketing Management* 38 (1): 127–143.

Melachrinoudis, E. and Kozanidis, G. 2002. A mixed integer knapsack model for allocating funds to highway safety improvements. *Transportation Research Part A: Policy and Practice* 36 (9): 789–803.

Metaxiotis, K. and Liagkouras, K. 2012. Multi objective evolutionary algorithms for portfolio management: A comprehensive literature review. *Expert Systems with Applications* 39 (14): 11685–11698.

Mohanty, R.P., Agarwal, R., Choudhury, A.K., and Tiwari, M.K. 2005. A fuzzy ANP-based approach to R&D project selection: A case study. *International Journal of Production Research* 43 (24): 5199–5216.

Oh, J., Yang, J., and Lee, S. 2012. Managing uncertainty to improve decision-making in NPD portfolio management with a fuzzy expert system. *Expert Systems with Applications* 39 (10): 9868–9885.

Osawa, Y. and Murakami, M. 2002. Development and application of a new methodology of evaluating industrial R&D projects. *R&D Management* 32 (1): 79–85.

Ozer, M. 1999. A survey of new product evaluation models. *Journal of Product Innovation Management* 16 (1): 77–94.

Ozer, M. 2005. Factors which influence decision making in new product evaluation. *European Journal of Operational Research* 163 (3): 784–801.

Patah, L.A. and De Carvalho, M.M. 2007. Measuring the value of project management. *PICMET 2007 Proceedings*, Portland, OR, pp. 2038–2042.

Remera, D.S., Stokdykb, S.B., and Driel, M.V. 1993. Survey of project evaluation techniques currently used in industry. *International Journal of Production Economics* 32: 103–115.

Roll, Y. and Golany, B. 1993. Alternate methods of treating factor weights in DEA. *Omega* 21 (1): 99–109.

Ronkainen, I.A. 1985. Criteria changes across product development stages. *Industrial Marketing Management* 14 (3): 171–178.

Saaty, T.L. 1977. A scaling method for priorities in hierarchical structures. *Journal of Mathematical Psychology* 15(3): 234–281.

Schaffer, J.D. 1985. *Some Experiments in Machine Learning Using Vector Evaluated Genetic Algorithms*. Vanderbilt Univ., Nashville, TN.

Seiford, L.M. 1996. Data envelopment analysis: The evolution of the state of the art (1978–1995). *Journal of Productivity Analysis* 7: 99–137.

Thieme, R.J., Song, M., and Calantone, R.J. 2000. Artificial neural network decision support systems for new product development project selection. *Journal of Marketing Research* 37 (4): 499–507.

Ulrich, K.T. and Eppinger, S.D. 2004. *Product Design and Development*, 3rd edn. McGraw Hill, NY.

Wang, J. and Hwang, W.-L. 2007. A fuzzy set approach for R&D portfolio selection using a real options valuation model. *Omega* 35 (3): 247–257.

Wey, W.-M. and Wu, K.-Y. 2007. Using ANP priorities with goal programming in resource allocation in transportation. *Mathematical and Computer Modelling* 46 (7): 985–1000.

Yahaya, S.-Y. and Abu-Bakar, N. 2007. New product development management issues and decision-making approaches. *Management Decision* 45 (7): 1123–1142.

Yurdakul, M. 2003. Measuring long-term performance of a manufacturing firm using analytic network process (ANP) approach. *International Journal of Production Research* 41 (11): 2501–2529.

Index